D1085467

Environmental Physiology II

Publisher's Note

The *International Review of Physiology* remains a major force in the education of established scientists and advanced students of physiology throughout the world. It continues to present accurate, timely, and thorough reviews of key topics by distinguished authors charged with the responsibility of selecting and critically analyzing new facts and concepts important to the progress of physiology from the mass of information in their respective fields.

Following the successful format established by the earlier volumes in this series, new volumes of the *International Review of Physiology* will concentrate on current developments in neurophysiology and cardiovascular, respiratory, gastrointestinal, liver, endocrine, kidney and urinary tract, environmental, and reproductive physiology. New volumes on a given subject generally appear at two-year intervals, or according to the demand created by new developments in the field. The scope of the series is flexible, however, so that future volumes may cover areas not included earlier.

University Park Press is honored to continue publication of the *International Review of Physiology* under its sole sponsorship beginning with Volume 9. The following is a list of volumes published and currently in preparation for the series:

Volume 1: **CARDIOVASCULAR PHYSIOLOGY** (A. C. Guyton and C. E. Jones)
Volume 2: **RESPIRATORY PHYSIOLOGY** (J. G. Widdicombe)
Volume 3: **NEUROPHYSIOLOGY** (C. C. Hunt)
Volume 4: **GASTROINTESTINAL PHYSIOLOGY** (E. D. Jacobson and L. L. Shanbour)
Volume 5: **ENDOCRINE PHYSIOLOGY** (S. M. McCann)
Volume 6: **KIDNEY AND URINARY TRACT PHYSIOLOGY** (K. Thurau)
Volume 7: **ENVIRONMENTAL PHYSIOLOGY** (D. Robertshaw)
Volume 8: **REPRODUCTIVE PHYSIOLOGY** (R. O. Greep)
Volume 9: **CARDIOVASCULAR PHYSIOLOGY II** (A. C. Guyton and A. W. Cowley, Jr.)
Volume 10: **NEUROPHYSIOLOGY II** (R. Porter)
Volume 11: **KIDNEY AND URINARY TRACT PHYSIOLOGY II** (K. Thurau)
Volume 12: **GASTROINTESTINAL PHYSIOLOGY II** (R. K. Crane)
Volume 13: **REPRODUCTIVE PHYSIOLOGY II** (R. O. Greep)
Volume 14: **RESPIRATORY PHYSIOLOGY II** (J. G. Widdicombe)
Volume 15: **ENVIRONMENTAL PHYSIOLOGY II** (D. Robertshaw)
Volume 16: **ENDOCRINE PHYSIOLOGY II** (S. M. McCann)

Consultant Editor: Arthur C. Guyton, M.D., Department of Physiology and Biophysics, University of Mississippi Medical Center

INTERNATIONAL
REVIEW OF PHYSIOLOGY

Volume 15

Environmental Physiology II

Edited by

David Robertshaw, Ph.D.
Professor, Head of Physiology
Indiana University School of Medicine
Bloomington, Indiana

UNIVERSITY PARK PRESS
Baltimore • London • Tokyo

UNIVERSITY PARK PRESS
International Publishers in Science and Medicine
Chamber of Commerce Building
Baltimore, Maryland 21202

Typeset by The Composing Room of Michigan, Inc.
Manufactured in the United States of America by Universal Lithographers, Inc., and The Optic Bindery Incorporated.

Library of Congress Cataloging in Publication Data

Main entry under title:

Environmental physiology II.

(International review of physiology ; v. 15)
Edition of 1974 published under title: Environmental physiology.
Includes index.
1. Body temperature--Regulation. 2. Temperature--Physiological effect. 3. Altitude, Influence of. I. Robertshaw, D. Environmental physiology. II. Series.
QP1.P62 vol. 15 [QP135] 599'.01'08s
ISBN 0-8391-1065-0 [599'.01'91] 77-7903

Consultant Editor's Note

In 1974 the first series of the *International Review of Physiology* appeared. This new review was launched in response to unfulfilled needs in the field of physiological science, most importantly the need for an in-depth review written especially for teachers and students of physiology throughout the world. It was not without trepidation that this publishing venture was begun, but its early success seems to assure its future. Therefore, we need to repeat here the philosophy, the goals, and the concept of the *International Review of Physiology.*

The *International Review of Physiology* has the same goals as all other reviews for accuracy, timeliness, and completeness, but it also has policies that we hope and believe engender still other important qualities often missing in reviews, the qualities of critical evaluation, integration, and instructiveness. To achieve these goals, the format allows publication of approximately 2,500 pages per series, divided into eight subspecialty volumes, each organized by experts in their respective fields. This extensiveness of coverage allows consideration of each subject in great depth. And, to make the review as timely as possible, a new series of all eight volumes is published approximately every two years, giving a cycle time that will keep the articles current.

Yet, perhaps the greatest hope that this new review will achieve its goals lies in its editorial policies. A simple but firm request is made to each author that he utilize his expertise and his judgment to sift from the mass of biennial publications those new facts and concepts that are important to the progress of physiology; that he make a conscious effort not to write a review consisting of an annotated list of references; and that the important material that he does choose be presented in thoughtful and logical exposition, complete enough to convey full understanding, as well as woven into context with previously established physiological principles. Hopefully, these processes will continue to bring to the reader each two years a treatise that he will use not merely as a reference in his own personal field but also as an exercise in refreshing and modernizing his whole store of physiological knowledge.

A. C. Guyton

Contents

Preface viii

1
Physical Basis of Thermoregulation 1
D. Mitchell

2
Physiological Effects of Cold Exposure 29
G. E. Thompson

3
Temperature Regulation in Primates 71
R. Elizondo

4
Exercise and Environmental Heat Loads: Different Mechanisms for Solving Different Problems? 119
C. R. Taylor

5
Thermoregulation During Sleep and Hibernation 147
H. C. Heller and S. F. Glotzbach

6
Role of the Adrenal Medulla in Thermoregulation 189
D. Robertshaw

7
Physiological Responses and Adaptations to High Altitude 217
S. Lahiri

Index 253

Preface

With the development in recent years of an awareness of our planetary environment, the term "environmental physiology" is gradually taking on a different meaning. A few years ago, it simply referred to the study of temperature regulation and adaptations to extreme environments. The first volume of the series reflected this fact. Environmental physiology now embraces a much wider area of study and increasingly incorporates the effects of man's control of natural phenomena. The impact of pollutants, whether they be physical or chemical, on physiology systems is developing into an important part of "environmental physiology." Since this is a relatively recent definition of environmental physiology, it is premature at this stage to include it in the present volumes. This volume, therefore, retains a bias towards thermoregulation.

Three authors of the first volume accepted the challenge to update their contributions. They felt that significant advances and trends had been made in their area of expertise to warrant a new review or re-evaluation. This supports the contention of the publishers and consultant editor that progress in physiological knowledge is so rapid that a new volume every two years will be necessary to keep the articles current. Some new authors have personally evaluated progress in their field. Thus, physiological responses to cold exposure have been reviewed and two general articles on thermoregulation are also included. The adrenal medulla is known to be important in the body's response to life-threatening situations. Extremes of temperature might be considered as potentially life-threatening and the role of adrenal medullary secretions in the maintenance of body temperature under both hot and cold conditions is reviewed. Work on temperature regulation in man is limited by the very laudable regulations that now govern the use of human subjects in research. This has led to a search for primates that can be used as models for man, and a chapter that reviews progress in this field has been included. Finally, there is a chapter that discusses aspects of temperature regulation during hibernation and sleep. The physiology of sleep still continues to excite physiologists. Recent work on hibernation tends to demonstrate certain similarities between the two phenomena.

As before, the contributors are young, active, and recognized authorities in their particular areas, and their skillful presentations should serve both to stimulate others and to encourage potential research workers to consider environmental physiology as their field of endeavor.

I have enjoyed the cooperation of the various authors; their enthusiasm was certainly contagious. I have been ably assisted in the manuscript preparation by Ms. Stella McQuade, whose patience and persistence I gratefully acknowledge.

D. Robertshaw

International Review of Physiology
Environmental Physiology II, Volume 15
Edited by David Robertshaw
 University Park Press Baltimore

1
Physical Basis of Thermoregulation

D. MITCHELL

University of Witwatersrand Medical School
Johannesburg, South Africa

FIRST LAW OF THERMODYNAMICS: FUNDAMENTAL LAW 2

ENERGY EXCHANGE: PHYSICAL FULFILLMENT OF LAW 3
- Energy Budgets 3
- Metabolic Rate 6
- Physical Work Rate 7
- Sensible Heat Transfer 8
  - *Radiation* 8
  - *Convection in Air* 10
  - *Convection in Water* 11
- Evaporation 11
- Protective Covers 13
  - *Natural Covers* 13
  - *Human Clothing* 15

REGULATION: PHYSIOLOGICAL IMPLEMENTATION OF LAW 16
- Temperatures Controlled in Thermoregulation 16
- Blood Circulation and Thermoregulation 17
- Physical Models of Thermoregulation 19

The abbreviations used are: A, area (m^2); C, rate of convective heat gain per unit of total body area (W/m^2); E, rate of evaporative heat gain per unit of total body area (W/m^2); K, rate of conductive heat gain per unit of total body area (W/m^2); M, rate of metabolic energy consumption per unit of total body area (W/m^2); P, pressure (kPa); R, rate of radiant heat gain per unit of total body area (W/m^2); S, rate of heat accumulation on the body per unit of total body area (W/m^2); T, temperature (°C); W, watts generated by external forces per unit of total body area (W/m^2); h, heat transfer coefficient ($W/m^2 \cdot {}^{\circ}C$) or ($W/m^2 \cdot kPa$); $\bar{p}$, interception function for radiation (see 49) (cm^{-1}); w, wetted area; ϕ, relative humidity, as fraction; a, air ambient; b, body; c, convective; e, evaporative; r, radiant; s, skin surface; wa, water vapor, saturated at T_a; and ws, water vapor saturated at T_s.

Physics, the master, remains ineluctable. Physics, the servant, remains imperfect. Those whose work concerns the physical basis of thermoregulation must accept physics in both its roles. Their work is inevitably slow. Nevertheless, in the few years since the first reviews in this series were compiled, progress has indeed been made. This chapter points out where such progress has been made; that is the intention of this book. Some aspects which were given short shrift in the predecessor to this chapter (1) are also expanded upon. The format will be the same in principle, but not in detail, as that adopted previously.

A disproportionate share of the progress of the last few years has originated from one group of workers, namely that led by J. L. Monteith at the University of Nottingham, England. Monteith has consolidated knowledge of the basic physical principles in his book *Principles of Environmental Physics* (2); he has compiled, together with L. E. Mount, a review of the applications of these principles to thermoregulation in the conference proceedings *Heat Loss from Animals and Man* (3); and his group has undertaken an elegant analysis, supported by experiment, of heat flow in animal coats (see under "Natural Covers"). Their work is commended to the interested reader.

FIRST LAW OF THERMODYNAMICS: FUNDAMENTAL LAW

The physical law basic to all thermoregulation is the First Law of Thermodynamics, from which one can derive (1) the ubiquitous energy budget equation for the body of an animal:

$$M + W + K + R + C + E = S \tag{1}$$

Equation 1 links the rate of metabolic energy turnover (M) and the physical work rate (W) with the rates of heat exchange with the environment by conduction (K), radiation (R), convection (C), and evaporation (E). Work rate (W) has a negative sign in the usual case of an animal doing work against external forces, and the algebraic sum ($M + W$) represents the metabolic heat production. The right hand side of equation 1 represents the rate of heat storage (S) within the body of the animal and is equal to zero during thermal equilibrium (1).

Given sufficient time, any object generating heat at a fixed rate in a fixed environment will attain thermal equilibrium and, therefore, satisfy the condition $S = 0$. Often the object will adopt the temperature of the environment. In order to survive, animals must achieve thermal equilibrium without excessive deviations in body temperature. Some animals confine body temperature deviations to a smaller range than do others, but the differences between homeotherms and poikilotherms, or regulators and conformers (4), seem more and more to be differences of degree rather than of fundamental principle (5).

If over a given period the rate of heat storage is not zero, then, for an animal of fixed mass, the average temperature of the body tissues must have risen or fallen over that period. Because monotonic rises or falls in body temperature are not consistent with life, there must be time periods for every animal over which the average rate of heat storage within the body is zero. The duration of these periods remains unknown. In the case of man, the duration is certainly not of the order of minutes (6). In some circumstances, the duration may be of the order of hours (7, 8), but in general it is unlikely to be less than a day (9). The duration of periods over which equilibrium is attained depends not only upon physical variables like body mass, but also upon physiological variables determining thermoregulatory ability; it is, therefore, not possible to make general predictions governing all animals.

For an animal of fixed mass and specific heat, the rate of heat storage is directly proportional to the rate of change of mean body temperature, the average temperature of the body tissue. It would, therefore, seem possible to assess the rate of heat storage at any instant by measuring body temperature. This assumption is undoubtedly true provided temperatures are measured at sufficient sites. The question is: what constitutes a sufficiency of sites? Snellen (10) in 1966 provided direct evidence based on calorimetry that in man it is not possible to calculate mean body temperature (or heat storage) from any simple combination of the body temperatures usually measured. Nevertheless, many concerned with human thermoregulation continue to calculate average body temperature from simple combinations of rectal temperature and skin temperature. The invalidity of such calculations recently has been confirmed in several different kinds of experiments (9, 11–14). The advice to clinicians (15) that a simple nomogram based on skin temperature and core temperature may be used to calculate body heat loss or gain is, therefore, grossly misleading. To date there is no thermometric method, even approximately accurate, for calculating mean body temperature or body heat storage.

ENERGY EXCHANGE: PHYSICAL FULFILLMENT OF LAW

Energy Budgets

What is the thermal stress on seal pups born wet on the polar ice in environments with subzero temperatures and high winds (16)? Why does energy loss seem to contribute so highly to mortality in burn patients (17)? The answers to such questions reside in the assignment of numerical values to the variables in equation 1. There are two ways of assigning such numerical values; both are technically difficult. The first is to measure the variables directly and simultaneously with the use of direct calorimetry. The second is to apply the physical laws of heat exchange to each variable separately.

Progress has been made with both approaches, although, as before (1), the work has been confined mainly to the human animal. There have been at least three reports of studies carried out in the human direct calorimeters. Ryser and Jequier (18) exposed 69 newborn infants to environments with ambient temperatures between 28°C and 36°C. They demonstrated, inter alia, that zero rate of heat storage was achieved only over the very narrow range of temperatures between 30°C and 34°C and that the ambient temperature at which minimal heat loss took place was not the same as that at which metabolic rate was minimal.

The same group (14) has also determined the energy budget of men resting and exercising at ambient temperatures of 20, 25, and 30°C. Their experiments showed that thermal equilibrium was attained more rapidly during exercise (W = −30W and −90W) than during rest in the same environment. In addition, the total rate of heat loss to the environment ($R + C + E$) was the same at the three ambient temperatures at which measurements were made.

The third application of calorimetry to the measurement of human energy budgets concerned the energy exchange during acclimatization to heat (8). Four men exercised at a fixed rate for 4 hr/day on 10 consecutive days in a calorimeter set at a fixed hot environment. The average energy budgets for the

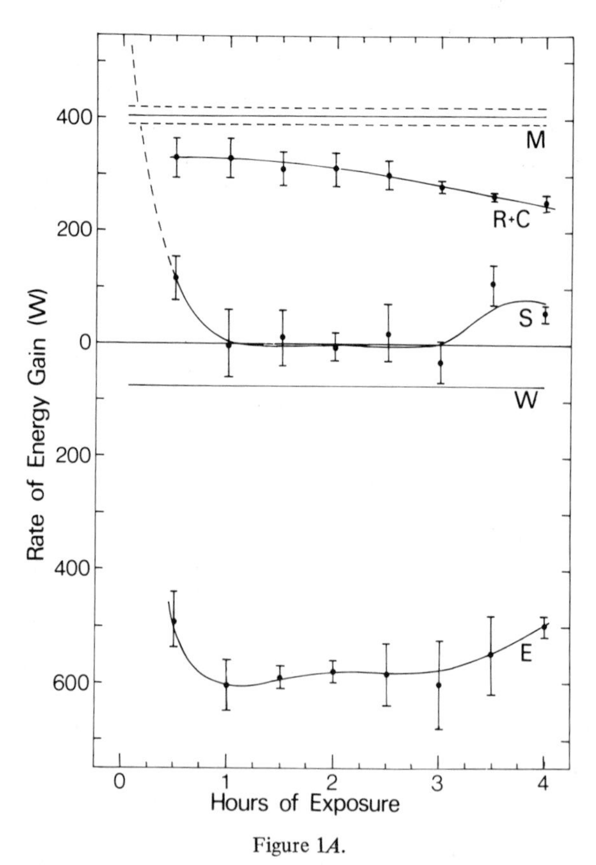

Figure 1*A*.

Figure 1. First (*A*) and tenth (*B*) day energy budgets of four men exposed for 4 hr/day to an environment of 45° C dry bulb temperature and mean radiant temperature, 4 *kPa* water vapor pressure, and 1 m/s wind speed, determined by direct calorimetry. Metabolic rate (*M*) is shown as the mean ± standard error over the 4-hr period. Rates of sensible heat transfer

men on the 1st and 10th days of exposure are shown in Figure 1. Major changes in body temperature, in sweat rate, in the cardiovascular system, and in body fluids (19, 20) took place over the 10 days of acclimatization, but as Figure 1 shows, these major changes were associated with meager changes in the energy budget. Even when unacclimatized, the men achieved thermal equilibrium during most of the exposure. Similar results were obtained in another pseudocalorimetric study of human acclimatization to heat (21).

The three experimental studies (8, 14, 18) mentioned above were carried out in complex and expensive direct calorimeters. By sacrificing some of the flexibility of such instruments, Webb and his colleagues (22) have accomplished calorimetric measurements with the use of a simple water-perfused garment. They were able to show that in men the net rate of metabolic heat gain amounted to within 1% of the net heat loss rate over 24-hr periods.

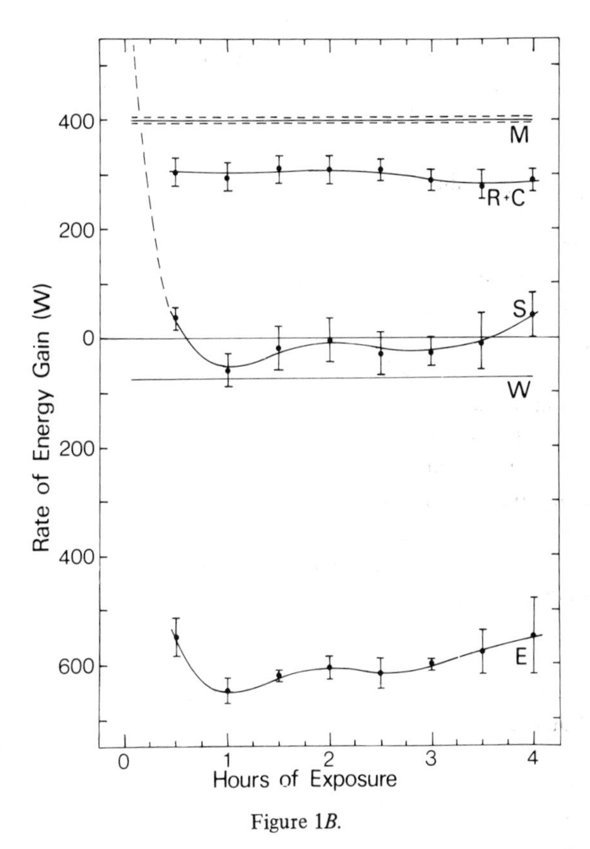

Figure 1*B*.

(*R* + *C*), evaporative heat transfer (*E*), and heat storage (*S*) are shown as the mean ± S.E. over each 30-min period. Physical work rate (*W*) was controlled at a constant value of 75 W against external forces. (Reproduced from Mitchell et al. (8), by courtesy of the American Physiological Society.)

Direct calorimetry as a means of establishing energy budgets is an impossibility in the field. The alternative is to employ physical laws of heat transfer together with estimates of metabolic rate in order to calculate all the terms of the left hand side of equation 1. Calder (23) has recently attempted such an exercise in the case of a nesting hummingbird; his attempt was severely hampered by the difficulty of obtaining corroborative measurements.

In the case of man, recourse to laboratory experiments allows such corrobative evidence. Gagge (24) has reviewed the equations necessary to calculate the energy budget of men exposed to desert environments. Varene and colleagues (25) have analyzed the budgets of man exposed to a simulated altitude of 3,800 m; among other things, they confirmed that the rate of evaporation increases with altitude (1). On a more ambitious scale, Tuller (26), using published equations for heat transfer and measured climatic variables, has attempted to set up an energy budget for a standardized man on each side of a city block. His calculations show that on sunny days solar radiation is likely to be the dominant environmental factor influencing the budget, whereas on cloudy days, wind is. Although his deductions make sense intuitively, the accuracy with which known heat transfer equations apply to the circumstances he considered is questionable.

Several advances have been made in the understanding of, and in the development of, equations for the individual terms on the left hand side of equation 1 and these are discussed below.

Metabolic Rate

The opinion was expressed in the predecessor (1) to this review that in spite of the need to measure metabolic rate accurately, there had been "little research aimed at validating the traditional measuring techniques in the last ten years, or even in the last half century." In essence, these remarks still hold true; for example, there has been no attempt to validate the measurements of the calorific value of oxygen in human combustion, especially during exercise. However, there has been an interesting controversy in the literature about the technicalities of measuring oxygen consumption, the primary quantity needed to calculate aerobic metabolic rate. Virtually all measurements of oxygen consumption in man are based on expired minute volumes and the oxygen concentration of expired air. Calculation of the rate of oxygen consumption requires the estimation of inspired volume. These estimations are usually made by means of the Haldane transformation, in which nitrogen is assumed to be metabolically inert, so that the volume of nitrogen expired is identical to the volume inspired. In 1972, the validity of the Haldane transformation was challenged by Cissik et al. (27), who claimed that the Haldane approach was in error to such an extent that a metabolic rate calculated from it could be in error by up to 30%. Their claim caused a certain amount of consternation. Were it true, virtually all measurements of metabolic rate, not only in the field of thermoregulation, but generally in all physiology, were likely to be erroneous. Fortunately, numerous subsequent checks of the Haldane transformation by a variety of independent experimental techniques have failed to lend substance to the claim of error (28–32).

Given the caveat that the calorific value of oxygen needs checking, measurement of the rate of metabolic heat production during aerobic metabolism remains a technically straightforward and relatively accurate procedure. Such complacency is unjustified in the case of anaerobic metabolism. Recent experiments have confirmed that in neither the rat (33) nor man (34) is it possible to calculate metabolic rate during a spell of anaerobic metabolism from the excess rate of oxygen consumption during recovery after the anaerobic spell. The only viable procedure seems to be to establish accurately the calorific equivalent of one of the biochemical correlates of anaerobic metabolism (35) and to measure directly the turnover rates of this metabolite during anaerobic metabolism.

It is sometimes possible to measure ventilation rate, but not oxygen consumption rate. In the case of man, metabolic rate can be estimated approximately from ventilation rate; the ratio of ventilation rate to oxygen consumption rate is approximately constant. Timbal et al. (36) have confirmed that such a constant relationship holds for conditions as widely different as shivering in cold water and exercise in air, the numerical value of the ratio being about 25.

Physical Work Rate

The physiology of exercise is reviewed elsewhere in this book (Chapter 4). In the evaluation of the energy budget (equation 1), one is interested primarily in the heat production associated with work, that is, $M + W$. The metabolic rate (M) is measurable (see under "Metabolic Rate"). The quantity W is the rate at which mechanical work is done against or by external forces and can, therefore, be calculated by analysis of the mechanics of the work being performed. Incorporation of the term $M + W$ into equation 1, in which all other factors are rates of heat flow, rests on the assumption that metabolic energy transforms either to physical work or to heat, but to no other form of energy.

There are at least two other forms of energy into which metabolic energy might transform. First, metabolic energy might be recycled into a chemical store; for instance, carbohydrate metabolism can lead to the deposition of energy in the form of fat. There is no evidence for the existence of such a process capable of proceeding at a rate comparable to the other terms in equation 1.

The second alternative fate of metabolic energy is elastic storage in muscle, tendons, or related body tissues. There is now little doubt that such storage takes place (37–40). In men performing repetitive movements of small amplitude, 60% of the energy required for each movement can be stored elastically between one cycle and the next (38). If metabolic energy is stored elastically, it does not appear immediately as heat, and equation 1 should be adjusted accordingly. As it stands, equation 1 is, therefore, not valid on a minute-to-minute basis during work involving repetitive movements, such as walking, swimming, or flying. In practice, however, the equation is never applied on a minute-to-minute basis. In the case of repetitive exercise, it is invariably applied over a large number of repeated cycles, in which circumstances there is no evidence for significant net elastic storage of energy.

Sensible Heat Transfer

The rates of conductive heat transfer (K), radiant heat transfer (R), and convective heat transfer (C) together comprise the sensible heat transfer between an animal and its environment. No progress has been made over the last few years in the physical analysis of conductive heat transfer between animals and solid surfaces with which they may be in contact. Snakes continue to know how hot the ground is even though they and the physicists are unsure of the equations governing the rate of heat gain from the ground (1). Progress has been made in the cases of radiation and convection.

Radiation Considerable progress has been made in the analysis of radiant heat transfer, particularly in the case of radiation within protective covers such as animal coats (see under "Protective Covers"). Calculation of the rate of radiant heat exchange between an animal and its environment remains relatively straightforward in the case of environments free of radiant heat having wavelengths less than about 2 μm. In environments including these shorter wavelengths, and, in particular, those in which solar radiation is present, no accurate calculations are yet possible.

The rate of longwave (>2 μm) radiant heat transfer between the surroundings and an animal can be estimated by using the following equation derived from the Stefan-Boltzmann Law (1, 2, 41):

$$R = h_r(A_r/A_b)(\overline{T}_r - \overline{T}_s) \tag{2}$$

where $\overline{T}_r$ is the mean temperature of the surrounding solid surfaces, $\overline{T}_s$ is the mean surface temperature of the animal, A_b is the total surface area of the animal, A_r the total surface area available for radiant heat exchange, and h_r is the linear radiant heat transfer coefficient. Equation 2 is only valid when animals are in enclosures much larger than themselves and in environments in which $(\overline{T}_r - \overline{T}_s) \ll (\overline{T}_r + \overline{T}_s + 546)$. The heat transfer coefficient depends slightly upon $(\overline{T}_r + \overline{T}_s)$ and is proportional to emittance of the animal surface in the wavelength region in question.

In the infrared region of wavelengths greater than about 2 μm, the emittance of all animal surfaces is thought to be high (2). Hardy (42) has recently confirmed that human skin is an almost perfect radiator (emittance approximately equal to 1), although he cautions that in environments where $\overline{T}_r$ is more than 100°C greater than $\overline{T}_s$, sufficient radiation of shorter wavelengths may be present so that emittance drops by about 5%.

Aluminum foil is often alleged to have a low emittance, and, therefore, a high reflectance, for longwave infrared radiation, for which reason it is incorporated into insulation material for buildings and has been used to make a swaddling garment ("Silver Swaddler") for neonates (43). Direct measurements

(44, 45) have shown that aluminized fabrics reflect only about 30% of longwave infrared radiation, scarcely justifying their use as radiation reflectors. "Silver Swaddlers" undoubtedly help to keep babies warm. However, they seem to do so by preventing evaporation (46), and a properly designed plastic bag would do as well.

Although no accurate calculations of radiant heat exchange in the presence of solar radiation are yet possible, techniques of assessing the optical properties of animal coats in the relevant wavelength range and of estimating the surface area exposed to such radiation have progressed to the extent that such calculations may soon become feasible. The state of the art has been reviewed thoroughly by Hutchinson et al. (47). Hutchinson and his colleagues (48) have themselves measured reflectances of the coats of heifers, water buffalo, and sheep in the presence of solar radiation. Reflectance varied with the position around the animal and with sun altitude; mean reflectance for all three species fell within the range 0.25–0.45. Mean reflectances of the excised coats of several species were measured photometrically by Cena and Monteith (49); their results are shown in Table 1.

The complex shade-seeking behavior of desert animals (50) leaves little doubt that solar radiation is a critical factor in their energy budgets. It is, therefore, surprising that such a large variety of coat colors exists in desert animals. Neither the physical consequences of the variations nor their ecological implications are yet fully understood (47, 51). On the face of it, the most surprising color for a desert animal is black. However, both theoretical calculations and experimental measurements (47, 49, 52) confirm that although black is not as thermally favorable a color as white for desert animals, it is more favorable than some intermediate colors. The reason lies in the fact that black

Table 1. Values of mean coat reflexion coefficient for various animal coats[a]

Coat	Reflexion coefficient
Sheep, Dorset Down	0.79
Sheep, Clun Forest	0.60
Sheep, Welsh Mountain	0.30
Rabbit	0.81
Badger	0.48
Calf, white	0.63
Calf, red	0.35
Toggenburg goat	0.42
Fox	0.34
Fallow deer	0.69

From Cena and Monteith (49), by permission of the Royal Society, London.

[a]Values are weighted averages for the spectrum from 0.35–0.85 μm.

coats trap thermal radiation near the outer surface; black ravens standing in the Jerusalem sun exhibit temperatures up to 84°C just below the outer feathers (53).

Just as black is a surprising color for desert animals, white would seem to be thermally unfavorable for polar animals. Their white coats are generally considered to have evolved not because they favor energy balance, but because they favor camouflage in the snow. However, what appears white to human eyes may well appear otherwise to the eyes of animals more sensitive to ultraviolet, as Lavigne and Øritsland (54) have shown by photographing polar bears in the ultraviolet region of the spectrum. Polar bears are apparently "black" to ultraviolet radiation, and absorption of the ultraviolet part of solar radiation may, therefore, contribute to their energy balance.

Convection in Air The rate of convective heat transfer between the surrounding air and an animal can be calculated (2, 55) from an equation of the form

$$C = h_c(T_a - \overline{T}_s) \tag{3}$$

where T_a is the air temperature and h_c the convective heat transfer coefficient. This coefficient depends upon wind speed and air density and, in the case of free convection (2, 55), also on surface and air temperature.

In recent years, there has been consolidation of the already substantial knowledge of the value of the convective heat transfer coefficient for naked man, but little progress for clothed man and other animals. In the case of nonhuman animals, Heller (56) has measured the convective heat transfer coefficient for the pelts of chipmunks mounted on casts of chipmunk carcasses. He found that in conditions of forced air movement the coefficient was directly proportional to wind speed, a surprising result on theoretical grounds (2, 55). Heller doubts whether his measurements would apply to live chipmunks free to adopt any body posture.

There is now some anecdotal evidence that birds employ behavioral maneuvers to increase their rate of convective heat loss in hot environments (53). One such maneuver is to fly with the legs extended, behavior observed recently in several species of passerine birds (57). One should bear in mind, as the birds undoubtedly do, that such maneuvers are beneficial only when body temperatures exceed air temperature, and some of the passerine birds were apparently flying at air temperatures of 46°C (57).

In the case of forced convection between naked man and moving air, the rate of convective heat transfer can be estimated quite accurately with the use of the engineering theory for smooth cylinders (2, 55, 58). The validity of the approach has been borne out by the correlation between theory and experiment when the air is replaced by hyperbaric helium (59). Although the pattern of free convection around man is qualitatively well described (60, 61), quantitative assessments of the rate of heat transfer are still only possible if the naphthalene sublimation technique of Nishi and Gagge (62) is applied in each case. No general predictive equation for free convection yet exists, but Rapp (63) has recently suggested that the convection from nude man at low wind speeds can be simulated by using a smooth sphere 0.75 m in diameter. His calculations show that the convective heat transfer coefficient for nude man is close to 4 $W/m^2 \cdot C$ at all low wind speeds.

Convection in Water Convective heat transfer in water takes place at a rate two or three orders of magnitude greater than that in air (1). Most aquatic animals do not possess the thermal machinery to maintain a body temperature significantly different from the surrounding water temperature. In the case of fish, the main site of metabolic heat loss to the water is not, as one might expect, the extensive body surface, but rather the gills (64). Respiratory heat exchange in fish per liter of oxygen consumed is about 100,000 times greater than that in terrestial animals (64). In fact, Stevens (65) believes that endothermy evolved through the stimulation of a thermogenic sodium pump in gill membranes, as an alternative to behavioral maneuvers, to compensate for heat loss through the gills.

Some fish, notably the skipjack tuna and mackerel shark, have specialized circulation allowing them to maintain the temperature of particular muscles and viscera somewhat about water temperature (64). Whether these fish have the ability to thermoregulate has been questioned by Neill and Stevens (66), who have suggested that what looks like thermoregulation is, in fact, a passive retardation of temperature change in the fish due to their large thermal inertia. However, the deductions of Neill and Stevens are based on an application of Newton's Law of Cooling, which, although frequently applied, is of doubtful validity in the case of animal heat transfer (67, 68).

Man is an occasionally aquatic animal, but is very poorly equipped for the purpose (68, 69). Nadel and his colleagues (70) have estimated the convective heat transfer coefficient of nude man at rest in water to be 230 $W/m^2 \cdot C$, that is about 60 times that in still air. During swimming, they estimated the coefficient to be about 580 $W/m^2 \cdot C$, irrespective of the speed of swimming. In cold water, the convective heat transfer coefficient may be increased by the movement associated with shivering (71), which occurs even during active swimming (70, 72). The argument continues as to whether the increased metabolic heat production of swimming compensates sufficiently for the increased heat transfer associated with the movement to make it thermally favorable to swim rather than remain at rest when exposed to cold water (73).

In addition to the high convective heat loss rates associated with human exposure to cold water, there is also a redistribution of heat within the body (74, 75), leading to a precipitous fall in core temperature. Survival times for men in water below 10°C are measured in minutes (73). Sloan and Keatinge (76) have shown recently that the temperature of British coastal waters is such that it is physiologically impossible for anyone to swim there, a fact known to all except the British.

Evaporation

The rate of evaporative heat exchange between the body surface of an animal and the surrounding air can be calculated (1, 2, 77) from the equation

$$E = wh_e(A_e/A_b)(\phi P_{wa} - P_{ws}) \tag{4}$$

where P_{ws} is the saturated vapor pressure at temperature $\overline{T}_s$, P_{wa} that at T_a; ϕ is the fractional humidity, A_e the surface area available for heat transfer, and h_e the evaporative heat transfer coefficient. (Evaporation normally constitutes a heat loss from the body, and E, therefore, normally has a negative sign.) The factor w is the wetted area, a number between zero and one which allows for the fact that in sweating animals the amount of water on the body surface is under physiological control; w has the value 1 when the surface is completely wet.

The evaporative coefficient h_e is directly proportional to the latent heat of evaporation of the evaporating fluid (normally sweat) and approximately proportional to the ratio of the convective heat transfer coefficient h_c to the atmospheric pressure (1). Thus, h_e varies with wind speed in the same way that h_c does.

No further experimental evidence has been provided on the question of whether the latent heat of sweat is different from that of water (1). McLean (77), reviewing the experimental evidence showing an anomalously high latent heat for human sweat in the light of the theoretical arguments against such an anomaly, concluded that the anomaly is more likely to have arisen from a systematic error in measuring metabolic rate (see under "Metabolic Rate") than it is a real peculiarity of sweat.

The alleged anomaly in the latent heat of sweat can affect the evaporative heat transfer coefficient to the extent of about 7% at most. Even in the case of nude man, the coefficient is not known to anywhere near this accuracy. In fact, there have been very few measurements of the coefficient. Belding and Kamon (78) have recently made indirect estimates for man by calculating or measuring all other terms in equation 1 for men in thermal equilibrium. They reported values for the coefficient h_e of $(184 \pm 38) \times V^{0.6}$ W/m$^2 \cdot kPa$, where V is the wind speed in meters per second for nude men and $(114 \pm 25) \times V^{0.6}$ W/m$^2 \cdot kPa$ for lightly clothed men. These values are somewhat higher than earlier values determined by a similar method (79), but they are still in reasonable agreement with values based on calculations from the convective heat transfer coefficient (1).

Even if the evaporative heat transfer coefficient was known accurately, there remains the problem of assessing wetted area w if the rate of evaporative heat transfer is to be calculated from equation 4. In all environments except those with very high water vapor pressures, the wetted area is under physiological control. In fact, control of the wetted area is the way in which sweating animals regulate their body temperatures in hot environments. Even in a fixed environment, wetted area is not entirely constant. As water evaporates from the body surface, it leaves behind solutes which depress the vapor pressure of the remaining liquid film. To maintain the same rate of evaporation, either surface temperature must rise to increase vapor pressure or wetted area must increase (see equation 4). Measurements on man show that the wetted area, in fact, does increase (80, 81).

Because of the uncertainties associated with several of its components, equation 4 is seldom used to calculate rates of evaporative heat transfer. The more usual method of calculating rates of evaporation is to measure the rate of weight loss and multiply the rate of weight loss by the latent heat of evaporation. However, even this procedure is not without error. In environments with high water vapor pressure, a significant fraction of the surface liquid may drop off without evaporating (8). Additionally, in areas covered with hair or fur, the site of evaporation may not be the skin surface, but some way removed from it, in which case it is impossible to predict how much of the heat of evaporation comes from the animal and how much from the surroundings.

One tends to consider evaporative heat loss to result only from the active secretion of water for evaporation, as in sweating or panting. In fact, all animals probably lose heat by evaporation all the time, through the continuous transepidermal water loss. Hattingh (82) recently has measured the rates of transepidermal loss in 18 species. His values range from about 0.2 $g/m^2 \cdot min$ in a snake to about 4 $g/m^2 \cdot min$ in a bat, which, if full evaporation occurs, correspond to cooling rates of about 10 W/m^2 and 150 W/m^2. Thus, transepidermal water loss can contribute significantly to the dissipation of metabolic heat. The mechanism of transepidermal water loss does not appear to be one of simple diffusion because it does not depend in any simple way upon epidermal thickness (83), and it does depend upon the state of hydration of the skin (82, 84), as well as upon blood flow (85), blood composition, and circulating hormones (86). In man, transepidermal water loss also depends upon environmental temperature and especially relative humidity, but not upon skin temperature (87); the dependence upon the relative humidity is anomalous in terms of equation 4 in that increased atmospheric humidity increases the rate of water loss.

Protective Covers

Metabolic heat for dissipation from the body surface is delivered by the circulation close to that surface, which in naked, hairless animals is also the interface between the animal and its thermal environment. In the case of haired, furred, or feathered animals or of clothed humans, the interface between the animal and the thermal environment is not so easily defined. The effective surfaces for radiant, convective, and evaporative heat transfer may be quite separate. Also, the mode of heat transfer through the protective cover may be different from the ultimate mode of transfer with the environment. The physics of heat transfer through natural animal covers is now better understood than that through human clothing, but in neither case is accurate quantitative assessment of heat exchange possible.

Natural Covers Three papers by Cena and Monteith (49, 88, 89) constitute a major advance in the knowledge of the physical basis of heat transfer in animal coats. The authors employed a heat transfer approach previously used in the study of plants, but not animals, "partly because animal physiologists seem to have made less effort than plant physiologists to understand the principles of micrometeorology." They analyzed the heat transfer by radiation, convection, and evaporation within the air microclimate between the matrix of coat fibers, made quantitative predictions of certain coat properties from the theory, and checked the theory by direct measurement on various excised animal coats.

In the case of radiation within a coat, Cena and Monteith (49) developed a theory of radiation scattering from fibers and tested the theory against measurements of the transmission and reflection of both shortwave and longwave radiation through the coats of sheep, cattle, and other animals. The optical properties of a coat appear to depend mainly upon factors such as hair length and density and not on color (see under "Radiation"). The authors defined an interception function $\bar{p}$, which depends upon the fraction of diffuse radiation intercepted per centimeter of coat depth; $\bar{p}$ ranged from about 9 cm^{-1} for sheep to 36 cm^{-1} for cattle. They confirmed the intuitive impression that the longer the coat, the less the penetration of shortwave radiation, a fact that many arid zone animals still need to learn, for they tend to have shallow coats (47). Indeed, cattle with shallow summer coats are known to be more heat tolerant in the sun than those that have failed to shed their winter coat (47).

Cena and Monteith's theory was also able to predict the mean reflectances of coats for shortwave radiation, the actual measured values of which were presented above (see under "Radiation").

In their second paper, Cena and Monteith (88) addressed themselves to the discrepancy between reported thermal conductivities of animal coats (40–150 mW/m·*K*) and the value for still air (25 mW/m·*K* at 20°C). There are at least three mechanisms by which the conductivity of a coat could increase above that of the still air trapped by the coat. The first, direct conduction through the fibers, can be dismissed instantly; Cena and Monteith calculated that such conduction is unlikely to exceed 3 mW/m·*K*. The second, radiation within the coat, could account for an increase of conductivity to values of 30–45 mW/m·*K*. Values higher than this can be explained by the third mechanism, namely free convection within the coat. In fact, to obtain agreement between theory and experiment for the conductivity of the fleece of a sheep requires an air flow within the coat of the order of only 0.01 m/s. In the course of their investigation of conductivity, Cena and Monteith showed that, for sheep and rabbits, coat conductivity was unaffected by air flow of up to 5 m/s parallel to the coat.

Finally, in their third paper (89), Cena and Monteith demonstrated that the resistance of the fleece of a sheep to water vapor diffusion is the same as that of still air for depths of up to about 30 mm. For thicker fleeces, the resistance was much lower than that of still air of equivalent thickness. Their direct measurements of the vapor permeability of fleece confirm that, provided the skin of the sheep is wet, evaporative heat flux through the natural fleece may be of the order of 200–300 W/m^2, well in excess of resting metabolic heat production for the animal.

Although the theories advanced by Cena and Monteith should apply to the coats of living animals, their confirmatory experiments were confined almost exclusively to excised coats. Obtaining confirmatory measurements on the coats of live animals is technically difficult. Cena and his colleagues devised a method involving the evaluation of coat surface temperature during heating of the coat by infrared radiation (90, 91). They applied this technique to the fleece of sheep, which they found to have a thermal resistance linearly related to coat depth and a thermal conductivity of about 190 mW/m·*K*.

Human Clothing Human clothing differs fundamentally from the natural animal covers studied by Monteith and Cena in that clothing traps layers of air roughly parallel to the body surface, whereas animal covers tend to project normal to the surface. The structure of clothing permits the relatively free movement of air within each layer, particularly during body movement.

No treatment of heat transfer in clothing with rigor approaching that achieved by Cena and Monteith has yet been published. In fact, the usual procedure is to assume that clothing offers separate resistive barriers to the sensible (that is, radiant and convective) and evaporative heat transfer between skin and environment (92). This physical model is clearly ludicrous; the complex combination of radiation, convection, conduction, and evaporation that takes place within clothing transcends immeasurably the constraints of a simple resistive barrier. It is, therefore, somewhat surprising that heat transfer calculations based on the simple model have proved quite useful in predicting human responses to the thermal environment.

If one accepts, with reservation, the simple physical model for the thermal properties of clothing, the problem remains of how to measure the resistance of the hypothetical barriers. Most measurements of this nature have been made with the use of the technique developed by Goldman et al. (93, 94), in which the garments under test are placed on a wetted copper manikin, the sensible and evaporative heat exchange rates of which can be measured directly. The copper manikin technique is simple to apply and probably is quite adequate for ranking garments in order of increasing insulative capacity to either sensible or evaporative heat. However, measurements of clothing properties based on manikins failed to predict quantitatively the thermal characteristics of the clothing when actually worn (95), particularly in the presence of appreciable body movement (96).

An alternative method of measuring the apparent resistance offered by clothing to heat transfer is to employ direct calorimetry (92). Direct calorimetry allows the required measurements to be made in situ, even in the case of exercising man. However, human calorimeters are rare instruments, so the method is not generally available.

A third method of assessing apparent resistance has been employed in the case of resistance to evaporation. The method is based on a postulated quantitative analogy between the rate of water vapor diffusion through clothing and the rate of naphthalene diffusion through the same clothing (97) and predicts a simple relationship between the resistance to evaporation and the resistance to sensible heat transfer (98). The naphthalene diffusion model for resistance to evaporation has not yet been checked accurately by direct experiment. In some circumstances, its predictions are patently invalid; for example, it fails to account for the changes in the thermal properties of clothing related to the degree of water retention in the clothing (99, 100).

In summary, accurate prediction of the thermal characteristics of clothing will not be possible until the physical basis of heat transfer within clothing is understood adequately. Until such time, published values of the thermal properties of clothing must be treated with suspicion, notwithstanding the sophisticated hardware and imposing equations upon which such values are based.

REGULATION: PHYSIOLOGICAL IMPLEMENTATION OF LAW

The major part of this chapter has been concerned with the physical basis of metabolic heat production and of heat exchange between an animal and its environment, that is, with the physical fulfillment of the First Law of Thermodynamics. All objects are subject to the First Law; living animals are subject to the additional constraint that the First Law must be satisfied without large excursions of body temperature. Thus, animals fulfill the First Law not passively, but by active regulation. The physiological mechanisms of this active regulation are discussed elsewhere in this book (Chapters 2 and 3). As in the case of its predecessor (1), it is within the province of this chapter only to comment briefly on regulations seen "through physical spectacles" (101). Three aspects will be considered—namely, what is controlled in thermoregulation, peripheral circulation and thermoregulation, and recent physical models of thermoregulation.

Temperatures Controlled in Thermoregulation

Vertebrates in most, if not all, classes and some invertebrates attempt to regulate their temperatures to some degree by various combinations of behavioral and autonomic mechanisms (5). The trend for precision of body temperature control to improve with phylogenetic rank suggests that good temperature control has some biological advantage. It is not clear what organs or tissues in higher animals benefit most from good temperature control. In the predecessor to this chapter, it was argued that the brain was a likely candidate (1). Recent physiological evidence supports this idea or at least confirms the existence of intricate physiological mechanisms which reduce the elevation of brain temperature means of which panting animals can maintain the arterial blood supply to the brain at a temperature appreciably below that of the aorta (102). Some animals, for example the dog (103) and the rhea (104), which do not have carotid retes, but do have cerebrovascular anatomy conducive to heat exchange similar to that thought to occur in the rete, also maintain the brain blood supply at a temperature below general arterial temperature. Kluger and D'Alecy (105) have claimed that the rabbit, which has neither carotid rete nor comparable structure, can employ the respiratory tract to cool arterial blood destined for the brain. However, Baker and Cronin (106) were unable to confirm this claim.

Another type of evidence which might suggest that the regulation of temperature is particularly important for brain tissue arises from experiments in which animals subjected to heat stress are given the option of cooling either the hypothalamus or the periphery by behavioral means. Corbit and his colleagues (107, 108) reported a specific preference for hypothalamic cooling in rats. Although these observations are intriguing, they must be treated with caution because there are no known neural connections between the hypothalamus and the sensory cortex by means of which animals might consciously sense the temperature of the hypothalamus.

An alternative hypothesis to account for the biological advantage of good thermoregulation concerns enzyme systems. The hypothesis is that enzymatically catalyzed biochemical reactions operate most efficiently at constant temperature and, furthermore, that reaction rates are highest at the temperatures of approximately 37–40°C found in many higher animals. This hypothesis is contradicted by recent observations made by Barnett et al. (109) on *Pachynoda sinuata* beetles. The optimal temperature for flight muscle enzyme activity in these beetles was 40–45°C, but the regulated thoracic temperature during flight was about 29°C.

Blood Circulation and Thermoregulation

One of the important functions of the circulation is the transport of body heat. In higher animals, it is the circulating blood which transfers metabolic heat from sites of generation to the potential sites of dissipation. In these animals, it is cardiovascular function and architecture which determine the distribution of temperatures within the body. Nevertheless, even the elementary relationships between blood flow, body temperature, and heat transport remain ill-defined.

Fundamental to any quantitative description of the role of the circulation in thermoregulation is a mathematical expression of the energy conservation for an element of tissue perfused with blood. It is this equation which relates blood flow rate to temperature distribution. Until recently, there was general agreement as to the form, if not the details, of the equation. Wulff (110) now believes that even the form of the equation is wrong, and if he is correct, all existing work on the thermal characteristics of perfused living tissue is invalid. However, there is reason to doubt that such a catastrophe has occurred. The equation Wulff wishes to apply is that appropriate for a porous material such as sintered metal; the equation used conventionally is that appropriate for a solid material containing pipes or channels of relatively large diameter (A. M. Patterson, personal communication). Tissue perfused with blood seems to conform more to the latter case.

A catastrophe may well have occurred in the understanding of countercurrent heat exchange in the vascular system. Countercurrent heat exchange between arterial blood destined for peripheral tissue and venous blood returning from that tissue has in the past been invoked as a mechanism for heat conservation in perfused tissues ranging from the human forearm to the porpoise fluke. Whether or not countercurrent heat exchange actually takes place in any of these situations was challenged in biophysical analysis of potential biological heat exchangers in 1968 (111). This analysis seems simply to have been ignored. Schmidt-Nielsen (112) makes no mention of it in the relevant sections of his recent authoritative book. If the analysis were incorrect, its shortcomings must be made public. Until such time, one must conclude that countercurrent heat exchange cannot occur in animals, no matter how attractive the anatomy might appear.

Whether or not countercurrent heat exchange occurs, there are vascular structures in animals, the design of which suggests they may be involved in heat exchange between different blood streams. Two such structures occur in fish such as the skipjack tuna and mackerel shark, which maintain parts of the body at temperatures above that of the sea water in which they live (64). These structures are the rete mirabile, in the vasculature leading to swimming muscles, and the hepatic sinus, in the vasculature leading to the viscera. The rete mirabile has a configuration similar to that of the carotid rete, and the heat exchange mechanism is alleged to be countercurrent heat exchange. This allegation needs proof. The hepatic sinus consists of a network of fine arteries in a pool of venous blood; here the heat exchange has not been analyzed.

Another structure with vascular organization very like that of the hepatic sinus is the mammalian placenta. The placenta may well be involved in heat exchange between mother and fetus, but no proper analysis of its thermal characteristics has been performed (113).

It is perhaps not surprising that the physical basis of heat exchange in a somewhat exotic situation such as the hepatic sinus has not been investigated. It is surprising that the thermal characteristics of the cutaneous circulation are so poorly understood. In many animals, the fine control of body temperature in moderate thermal environments entirely depends upon behavioral adjustments and on the regulation of blood flow through the skin. In sweating animals, heat tolerance is limited by the rate at which heat can be delivered to the body surface by the cutaneous circulation. The relationship between cutaneous blood flow and cutaneous heat delivery is, therefore, of fundamental importance in thermoregulation.

In mammals, the cutaneous vasculature consists of a series of parallel venous and arteriolar plexi of vessels decreasing in diameter with proximity to the surface and terminating in capillary loops which project into the epidermis. In some cutaneous regions, there are anastomoses connecting the arterial and venous supplies. The role, if any, of these arteriovenous anastomoses in thermoregulation remains unknown. Hales (114) has observed them to open in sheep exposed to hot environments; such opening may decrease cutaneous vascular resistance.

A great deal could be learned about the relationship between cutaneous blood flow and cutaneous heat exchange if the temperature profile through skin could be determined accurately in different conditions. For example, the slope of the profile at the surface determines the rate of heat exchange at the surface uniquely. In human skin, one needs a spatial resolution of about one-tenth of a millimeter; such resolution is now attainable (115). Patterson has measured profiles through the forearm skin of a resting man at various environmental temperatures (116). Typical results are shown in Figure 2 and are quite unlike the undulating profiles described by previous workers (116).

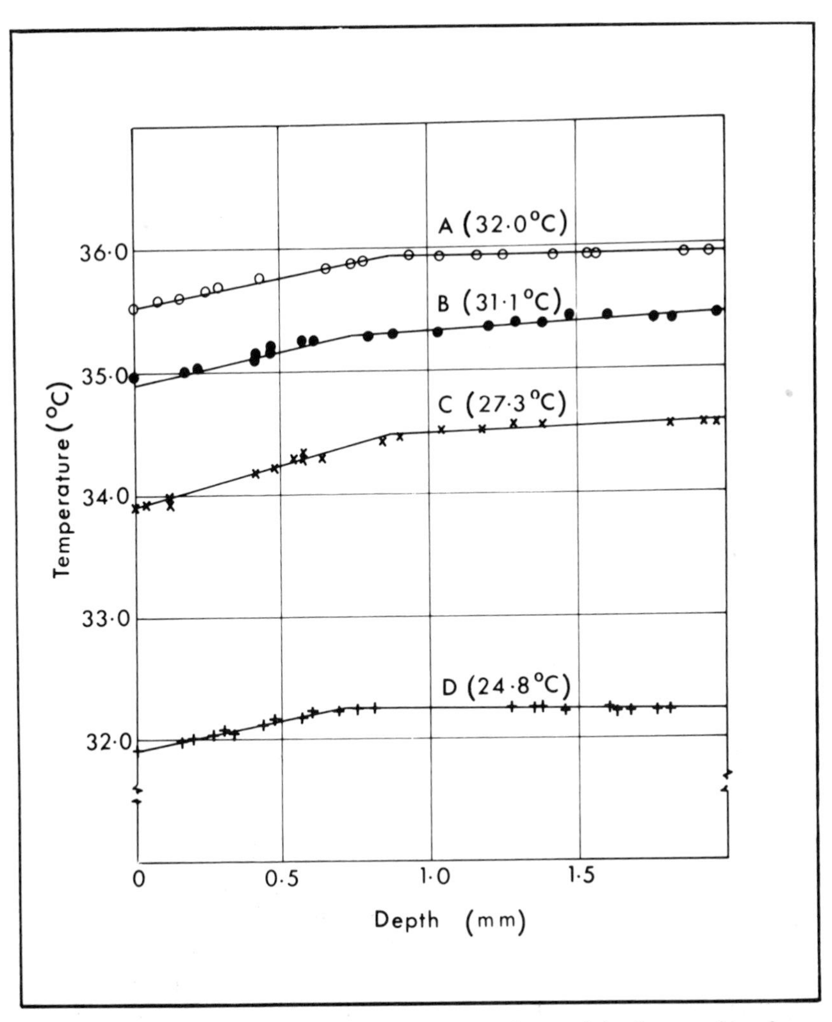

Figure 2. Temperature profiles recorded in the outer 2 mm of the forearm skin of a man after about 3-hr exposure to the ambient temperatures indicated. (Reproduced from Patterson (115), by courtesy of the South African Journal of Science.)

Physical Models of Thermoregulation

There are probably 50–100 extant models of thermoregulation (117); virtually all are concerned with human thermoregulation (118–122). Physical models of thermoregulation should predict the body temperatures and rates of heat exchange with the surroundings under all conditions of metabolism and in any environment. No model has yet attained this goal, nor does there appear to have been any attempt to consolidate into one model the best features of the various separate models.

Models of thermoregulation have continued to proliferate since 1972. With the exception of tentative models for the sea lion (123) and rat (124) based on very meager physiological and anatomical data, physical modeling has been confined to the human body. Most of the new contributions have been refinements of the now well-established negative feedback for the regulation of body temperature (125). Some of the refinements have been technical improvements to, or variations on, the thermal modeling of the controlled system (126–129). Others have attempted to incorporate new physiological observations, such as the local effect of skin temperature on sweating and circulation (125, 130) and the oscillation of cutaneous vasomotor control (131). Models which simply do not account for the known physiology continue to be published. For example, it seems futile to ignore control of shivering and control of blood flow distribution in a model for human immersion in cold water (132). Perhaps worse, models continue to be published without concomitant experimental verification (133).

Perhaps the most highly developed conventional model of human thermoregulation is the digital computer simulation of Atkins (134), which he verified by extensive comparison with physiological measurements on men exercising and resting in hot, neutral, and cold environments, varying also in humidity and wind speed. Among the important consequences of the Atkins model are, first, the confirmation of a previous claim by Brengelmann et al. (135) that vasomotor control in man is correlated with body core temperature and rate of change of skin temperature, but not with skin temperature itself. A second important consequence is that the nature of temperature regulation in man during cooling is fundamentally different from that during warming; proportional control seems to operate during warming, but not cooling.

Not all the recent models of thermoregulation have conformed to the conventional negative feedback pattern. Huckaba and Downey (118) have emphasized the possibility of feedforward, anticipatory control. A more drastic alternative is that proposed by Houdas, Guieu, and their colleagues (119, 136–138), who reject the idea that thermoregulation in man is directly concerned with the control of body temperature. Instead, they propose a servomechanism which operates to control the rate of heat flux in and out of the body; control of body temperature then follows passively. Houdas and his colleagues were able to simulate on a computer the behavior of human body temperature in response to environmental thermal stress by using their servomechanism system, but whether the human body really operates in this way is still open to question. For example, their computer model contains a feedback element which measures the rate of evaporation (not secretion) of sweat from the body surface without detecting temperature. There is no evidence for the existence of a biological device capable of this feat, nor can one easily conceive of how such a device might operate physically. Also, a model such as the one proposed by Houdas and his colleagues can never explain why body temperature is controlled at a specific operating level, e.g., 37°C (139).

In summary, physical models of thermoregulation continue to proliferate, expand, and diversify, but not always to improve.

ACKNOWLEDGMENT

I am grateful to J. M. Stewart for his help with the manuscript.

REFERENCES

1. Mitchell, D. (1974). Physical basis of thermoregulation. *In* D. Robertshaw (ed.), MTP International Review of Science, Physiology Series One, Vol. 7, Environmental Physiology, pp. 1–32. Butterworths, London.
2. Monteith, J. L. (1973). Principles of Environmental Physics. Edward Arnold, London.
3. Monteith, J. L., and Mount, L. E. (1974). Heat Loss from Animals and Man. Butterworths, London.
4. Bligh, J., and Johnson, K. G. (1973). Glossary of terms for thermal physiology. J. Appl. Physiol. 35:941.
5. Bligh, J. (1973). Temperature Regulation in Mammals and Other Vertebrates. North-Holland, Amsterdam.
6. Bailey, D., Harry, D., Johnson, R. E., and Kupprat, I. (1973). Oscillations in oxygen consumption of man at rest. J. Appl. Physiol. 34:467.
7. Snellen, J. W., Mitchell, D., and Busansky, M. (1972). Calorimetric analysis of the effect of drinking saline solution on whole body sweating. I. An attempt to measure average body temperature. Pfluegers Arch. 331:124.
8. Mitchell, D., Senay, L. C., Wyndham, C. H., van Rensburg, A. J., Rogers, G. G., and Strydom, N. B. (1976). Acclimatization in a hot humid environment: energy exchange, body temperature, and sweating. J. Appl. Physiol. 40:768.
9. Webb, P., Annis, J., and Troutman, S. J. (1972). Hourly changes in body heat storage during thermally comfortable days. Fed. Proc. 31:399 (Abstr.).
10. Snellen, J. W. (1966). Mean body temperature and the control of thermal sweating. Acta Physiol. Pharmacol. Neerl. 14:99.
11. Webb, P. (1973). Rewarming after diving in cold water. Aerosp. Med. 44:1152.
12. Horstman, D. H., and Horvath, S. M. (1972). Cardiovascular and temperature regulatory changes during progressive dehydration and euhydration. J. Appl. Physiol. 33:446.
13. Shvartz, E., and Benor, D. (1972). Heat strain in hot and humid environments. Aerosp. Med. 43:852.
14. Chappuis, P., Pittet, P., and Jequier, E. (1976). Heat storage regulation in exercise during thermal transients. J. Appl. Physiol. 40:384.
15. House, F., and Vale, R. (1972). Nomograms for calculation of heat loss. Br. Med. J. 1972(4):20.
16. Grav, H. J., Blix, A. S., and Pasche, A. (1974). How do seal pups survive birth in Arctic winter? Acta Physiol. Scand. 92:427.
17. Wilmore, D. W., Mason, A. D., Johnson, D. W., and Pruitt, B. A. (1975). Effect of ambient temperature upon heat production and heat loss in burn patients. J. Appl. Physiol. 38:593.
18. Ryser, G., and Jequier, E. (1972). Study by direct calorimetry of thermal balance on the first day of life. Eur. J. Clin. Invest. 2:176.
19. Wyndham, C. H., Rogers, G. G., Senay, L. C., and Mitchell, D. (1976). Acclimatization in a hot humid environment: cardiovascular adjustments. J. Appl. Physiol. 40:779.
20. Senay, L. C., Mitchell, D., and Wyndham, C. H. (1976). Acclimatization in

a hot humid environment: body fluid adjustments. J. Appl. Physiol. 40:786.

21. Henane, R., and Bittel, J. (1975). Changes of thermal balance induced by passive heating in resting man. J. Appl. Physiol. 38:294.
22. Webb, P., Annis, J. F., and Troutman, S. J. (1972). Human calorimetry with a water-cooled garment. J. Appl. Physiol. 32:412.
23. Calder, W. A. (1973). An estimate of the heat balance of a nesting hummingbird in a chilling climate. Comp. Biochem. Physiol. 46A:291.
24. Gagge, A. P. (1972). Partitional calorimetry in the desert. *In* M. K. Yousef, S. M. Horvath, and R. W. Bullard (eds.), Physiological Adaptations, Desert and Mountain, pp. 23–51. Academic Press, New York.
25. Varene, P., Jacquemin, C., Durand, J., and Rayneud, J. (1973). Energy balance during moderate exercise at altitude. J. Appl. Physiol. 34:633.
26. Tuller, S. E. (1975). The energy budget of man: variations with aspect in a downtown urban environment. Int. J. Biometeorol. 19:2.
27. Cissik, J. H., Johnson, R. E., and Hertig, B. A. (1972). Production of gaseous nitrogen during human steady state exercise. Aerosp. Med. 43:1245.
28. Wagner, J. A., Horvath, S. M., Dahms, T. E., and Reed, S. (1973). Validation of open-circuit method for the determination of oxygen consumption. J. Appl. Physiol. 34:859.
29. Luft, U. C., Myhre, L. G., and Loeppky, J. A. (1973). Validity of Haldane calculation for estimating respiratory gas exchange. J. Appl. Physiol. 34:864.
30. Wilmore, J. H., and Costill, D. L. (1973). Adequacy of the Haldane transformation in the computation of exercise V_{O_2} in man. J. Appl. Physiol. 35:85.
31. Herron, J. M., Saltzman, H. A., Hills, B. A., and Kylstra, J. A. (1973). Differences between inspired and expired minute volume of nitrogen in man. J. Appl. Physiol. 35:546.
32. Bachofen, H., Gurtner, H. P., and Paumgartner, G. (1973). Mixed venous-arterial difference of molecular nitrogen in man. J. Appl. Physiol. 35:791.
33. Brooks, G. A., Brauner, K. E., and Cassens, R. G. (1973). Glycogen synthesis and metabolism of lactic acid after exercise. Am. J. Physiol. 224:1162.
34. Bonde-Petersen, F., Knuttgen, H. G., and Henriksson, J. (1972). Muscle metabolism during exercise with concentric and eccentric contractions. J. Appl. Physiol. 33:792.
35. Hochachka, P. W. (1974). Regulation of heat production at the cellular level. Fed. Proc. 33:2162.
36. Timbal, J., Loncle, M., Bougues, L., and Boutelier, C. (1974). Relation entre la ventilation et la consommation entre la ventilation et la consommation d'oxygène au cours du frisson thermique chez l'Homme. C. R. Soc. Biol. 168:719.
37. Thys, H., Faraggiana, T., and Margaria, R. (1972). Utilization of muscle elasticity in exercise. J. Appl. Physiol. 32:491.
38. Thys, H., Cavagna, G. A., and Margaria, R. (1975). The role played by elasticity in an exercise involving movement of small amplitude. Pfluegers Arch. 354:281.
39. Asmussen, E., and Bonde-Petersen, F. (1974). Storage of elastic energy in skeletal muscles in man. Acta Physiol. Scand. 91:385.
40. Asmussen, E., and Bonde-Petersen, F. (1974). Apparent efficiency and

storage of elastic energy in human muscles during exercise. Acta Physiol. Scand. 92:537.

41. Cena, K. (1974). Radiative heat loss from animals and man. *In* J. L. Monteith and L. E. Mount (eds.), Heat Loss from Animals and Man, pp. 33–58. Butterworths, London.
42. Hardy, J. D. (1973). Reflectance of human skin for low temperature radiators. Arch. Sci. Physiol. 27:A21.
43. Hey, E. N. (1972). Thermal regulation in the newborn. Br. J. Hosp. Med. 8:51.
44. Frei, D. H. (1971). The influence of clothing and posture on radiant heat transfer in men. Ph.D. thesis, University of the Witwatersrand, Johannesburg, South Africa.
45. Veghte, J. H. (1973). Human exposure to high radiant environments. Aerosp. Med. 44:1147.
46. Girling, D. J., and Scopes, J. W. (1970). Value of the silver swaddler in preventing evaporative heat loss. Lancet, 1970(2):46.
47. Hutchinson, J. C. D., Brown, G. D., and Allen, T. E. (1976). Effects of solar radiation on the sensible heat exchange of mammals. *In* S. W. Tromp and H. D. Johnson (eds.), Progress in Animal Biometeorology, pp. 653–671. Swets and Zeitlinger, Amsterdam.
48. Hutchinson, J. C. D., Allen, T. E., and Spence, F. B. (1975). Measurement of the reflectances for solar radiation of the coats of live animals. Comp. Biochem. Physiol. 52A:343.
49. Cena, K., and Monteith, J. L. (1975). Transfer processes in animal coats. I. Radiative transfer. Proc. R. Soc. Lond. (Biol.) 188:377.
50. Dawson, T. J. (1972). Likely effects of standing and lying on the radiation heat load experienced by a resting kangaroo on a summer day. Aust. J. Zool. 20:17.
51. Land, M. F. (1972). The physics and biology of animal reflectors. Prog. Biophys. 24:77.
52. Kovarik, M. (1973). Radiation penetrance of protective covers. J. Appl. Physiol. 35:362.
53. Marder, J. (1973). Body temperature regulation in the brown-necked raven (*Corvus corax ruficollis*). II. Thermal changes in the plumage of ravens exposed to solar radiation. Comp. Biochem. Physiol. 45A:431.
54. Lavigne, D. M., and Øritsland, N. A. (1974). Black polar bears. Nature 251:218.
55. Mitchell, D. (1972). Convective heat transfer from man and other animals. *In* J. L. Monteith and L. E. Mount (eds.), Heat Loss from Animals and Man, pp. 59–76. Butterworths, London.
56. Heller, H. C. (1972). Measurements of convective and radiative heat transfer in small mammals. J. Mammol. 53:289.
57. Frost, P. G. H., and Siegfried, W. R. (1975). Use of legs as dissipators of heat in flying passerines. Zool. Afr. 10:101.
58. Missenard, F. A. (1973). Coefficient d'échange de chaleur du corps humain par convection, en fonction de la position, de l'activité du sujet et de l'environnement. Arch. Sci. Physiol. 27:A45.
59. Timbal, J., Vieillefond, H., Guenard, H., and Varene, P. (1974). Metabolism and heat losses of resting man in hyperbaric helium atmosphere. J. Appl. Physiol. 36:444.
60. Cox, R. N., and Clark, R. P. (1973). The natural convection flow about the human body. Rev. Gen. Thermique 133:11.
61. Clark, R. P., and Cox, R. N. (1974). The application of aeronautical

techniques to physiology. I. The human microenvironment and convective heat transfer. Med. Biol. Eng. 12:270.
62. Nishi, Y., and Gagge, A. P. (1970). Direct evaluation of convective heat transfer coefficient by naphthalene sublimation. J. Appl. Physiol. 29:830.
63. Rapp, G. M. (1973). Convective heat transfer and convective coefficients of nude man, cylinders and spheres at low air velocities. ASHRAE Trans. 79(I):75.
64. Carey, F. G. (1973). Fishes with warm bodies. Sci. Am. 228:36.
65. Stevens, E. D. (1973). The evolution of endothermy. J. Theor. Biol. 38:597.
66. Neill, W. H., and Stevens, E. D. (1974). Thermal inertia versus thermoregulation in "warm" turtles and tunas. Science 184:1008.
67. Kleiber, M. (1972). A new Newton's law of cooling. Science 178:1283.
68. Calder, W. A., and King, J. R. (1972). Body weight and the energetics of temperature regulation: a re-examination. J. Exp. Biol. 56:775.
69. Holmer, I. (1974). Physiology of swimming man. Acta Physiol. Scand. (Suppl.) 407:9.
70. Nadel, E. R., Holmer, I., Berge, U., Astrand, P.-O., and Stolwijk, J. A. J. (1974). Energy exchanges of swimming man. J. Appl. Physiol. 36:465.
71. Gee, G. K., and Goldman, R. F. (1973). Heat loss of man in total water immersion. Physiologist 16:318.
72. Nielsen, B. (1973). Metabolic reactions to cold during swimming at different speeds. Arch. Sci. Physiol. 27:A207.
73. Hayward, J. S., Eckerson, J. D., and Collis, M. L. (1975). Thermal balance and survival time prediction of man in cold water. Can. J. Physiol. Pharmacol. 53:21.
74. Craig, A. B., and Dvorak, M. (1973). Effects of exercise on heat balance during head out immersion in water. Physiologist 16:291.
75. Hayward, J. S., Collis, M., and Eckerson, J. D. (1973). Thermographic evaluation of relative heat loss areas of man during cold water immersion. Aerosp. Med. 44:708.
76. Sloan, R. E. G., and Keatinge, W. R. (1973). Cooling rates of young people swimming in cold water. J. Appl. Physiol. 35:371.
77. McLean, J. A. (1974). Loss of heat by evaporation. *In* J. L. Monteith and L. E. Mount (eds.), Heat Loss from Animals and Man, pp. 19–31. Butterworths, London.
78. Belding, H. S., and Kamon, E. (1973). Evaporative coefficients for prediction of safe limits in prolonged exposures to work under hot conditions. Fed. Proc. 32:1598.
79. Belding, H. S., and Kamon, E. (1971). Prediction of safe limits for prolonged exposure to heat. Fed. Proc. 30:209 (Abstr.).
80. Berglund, L. G., Gallagher, R. R., and McNall, P. E. (1973). Simulation of the thermal effects of dissolved materials in human sweat. Comput. Biomed. Res. 6:127.
81. Berglund, L. G., and McNall, P. E. (1973). Human sweat film area and composition during prolonged sweating. J. Appl. Physiol. 35:714.
82. Hattingh, J. (1972). A comparative study of transepidermal water loss through the skin of various animals. Comp. Biochem. Physiol. 43A:715.
83. Hattingh, J. (1972). The correlation between transepidermal water loss and the thickness of epidermal components. Comp. Biochem. Physiol. 43A:719.
84. Hattingh, J. (1973). The relationship between skin structure and transepidermal water loss. Comp. Biochem. Physiol. 45A:685.

85. Hattingh, J. (1972). The influence of blood flow on transepidermal water loss. Acta Derm. Venereol. 52:365.
86. Hattingh, J. (1975). The influence of hormones and blood flow on transepidermal water loss. Comp. Biochem. Physiol. 50A:439.
87. Hattingh, J. (1972). The influence of skin temperature, environmental temperature and relative humidity on transepidermal water loss. Acta Derm. Venereol. 52:438.
88. Cena, K., and Monteith, J. L. (1975). Transfer processes in animal coats. II. Conduction and convection. Proc. R. Soc. Lond. (Biol.) 188:395.
89. Cena, K., and Monteith, J. L. (1975). Transfer processes in animal coats. III. Water vapour diffusion. Proc. R. Soc. Lond. (Biol.) 188:413.
90. Cena, K., and Clark, J. A. (1974). Heat balance and thermal resistances of sheep's fleece. Phys. Med. Biol. 19:51.
91. Clark, J. A., Cena, K., and Monteith, J. L. (1973). Measurements of the local heat balance of animal coats and human clothing. J. Appl. Physiol. 35:751.
92. Mitchell, D., and van Rensburg, A. J. (1973). Assessment of clothing insulation: the problem and its solution using direct calorimetry on exercising men. Arch. Sci. Physiol. 27:A149.
93. Goldman, R. F. (1973). Clothing, its physiological effects, adequacy in extreme thermal environments, and possibility of future improvements. Arch. Sci. Physiol. 27:A137.
94. Goldman, R. F. (1974). Clothing design for comfort and work performance in extreme thermal environments. Trans. N. Y. Acad. Sci. 36:531.
95. Martin, H. deV., and Goldman, R. F. (1972). Comparison of physical, biophysical and physiological methods of evaluating the thermal stress associated with wearing protective clothing. Ergonomics 15:337.
96. Haisman, M. F., and Goldman, R. F. (1974). Physiological evaluations of armoured vests in hot-wet and hot-dry climates. Ergonomics 17:1.
97. Nishi, Y., and Gagge, A. P. (1970). Moisture permeation of clothing–a factor governing thermal equilibrium and comfort. ASHRAE Trans. 76(I):137.
98. Nishi, Y. (1973). Vapour permeation efficiency of clothing by naphthalene sublimation. Arch. Sci. Physiol. 27:A163.
99. Craig, F. N. (1972). Evaporative cooling of men in wet clothing. J. Appl. Physiol. 33:331.
100. Craig, F. N., and Moffitt, J. T. (1974). Efficiency of evaporative cooling from wet clothing. J. Appl. Physiol. 36:313.
101. Hill, A. V. (1956). Why biophysics? Science 124:1233.
102. Baker, M. A. (1972). Influence of the carotid rete on brain temperature in cats exposed to hot environments. J. Physiol. Lond. 220:711.
103. Baker, M. A., Chapman, L. W., and Nathanson, M. (1974). Control of brain temperature in dogs: effects of tracheostomy. Respir. Physiol. 22:325.
104. Kilgore, D. L., Bernstein, M. H., and Schmidt-Nielsen, K. (1973). Brain temperature in a large bird, the rhea. Am. J. Physiol. 225:739.
105. Kluger, M. J., and D'Alecy, L. G. (1975). Brain temperature during reversible upper respiratory bypass. J. Appl. Physiol. 38:268.
106. Baker, M. A., and Cronin, M. J. (1976). Can panting mammals without a carotid rete cool their brains below deep body temperature? Fed. Proc. 35:481.
107. Corbit, J. D. (1973). Voluntary control of hypothalamic temperature. J. Comp. Physiol. Psychol. 83:394.
108. Corbit, J. D., and Ernits, T. (1974). Specific preference for hypothalamic cooling. J. Comp. Physiol. Psychol. 86:24.

109. Barnett, P. S., Heffron, J. J. A., and Hepburn, H. R. (1975). Some thermal characteristics of insect flight: enzyme optima versus intra-thoracic temperature. S. Afr. J. Sci. 71:373.
110. Wulff, W. (1974). The energy conservation equation for living tissue. IEEE Trans. Biomed. Eng. 21:494.
111. Mitchell, J. W., and Myers, G. E. (1968). An analytical model of the counter-current heat exchange phenomena. Biophys. J. 8:897.
112. Schmidt-Nielsen, K. (1975). Animal Physiology. Cambridge University Press, Cambridge.
113. Sinclair, J. C. (1972). Thermal control in premature infants. Annu. Rev. Med. 23:129.
114. Hales, J. R. S. (1973). Effects of exposure to hot environments on the regional distribution of blood flow and on cardiorespiratory function in sheep. Pfluegers Arch. 344:133.
115. Patterson, A. M. (1976). Measurements of temperature profiles in human skin. S. Afr. J. Sci. 72:78.
116. Patterson, A. M. (1976). The behaviour of skin temperature profiles in the forearm of a nude resting subject at air temperatures from 24°C to 34°C. Report No. 69, School of Mechanical Engineering, University of the Witwatersrand, Johannesburg, South Africa. (ISBN O 85494 396 X).
117. Hardy, J. D. (1972). Models of temperature regulation–a review. *In* J. Bligh and R. E. Moore (eds.), Essays on Temperature Regulation, pp. 163–186. North-Holland, Amsterdam.
118. Huckaba, C. E., and Downey, J. E. (1973). Overview of human thermoregulation. *In* A. S. Iberall and A. C. Guyton (eds.), Regulation and Control in Physiological Systems, pp. 212–216. I.S.A., Pittsburgh.
119. Houdas, Y., and Guieu, J. D. (1975). Physical models of human thermoregulation. *In* P. Lomax, E. Schönbaum, and J. Jacob (eds.), Temperature regulation and Drug Action, pp. 12–21. Karger, Basle.
120. Fan, L.-P., Hsu, F.-T., and Hwang, C.-L. (1971). A review on mathematical models of the human thermal system. IEEE Trans. Biomed. Eng. 18:218.
121. Iberall, A. S. (1972). Comments on "A review on mathematical models of the human thermal system." IEEE Trans. Biomed. Eng. 19:67.
122. Shitzer, A. (1973). Addendum to "A review on mathematical models of the human thermal system." IEEE Trans. Biomed. Eng. 20:65.
123. Luecke, R. H., Natarajan, V., and South, F. E. (1975). A mathematical biothermal model of the California sea lion. J. Therm. Biol. 1:35.
124. Fuller, C. A., Horowitz, J. M., and Horwitz, B. A. (1975). Sorting of signals from thermosensitive areas. Comput. Programs Biomed. 4:264.
125. Mitchell, D., Atkins, A. R., and Wyndham, C. H. (1972). Mathematical and physical models of thermoregulation. *In* J. Bligh and R. E. Moore (eds.), Essays on Temperature Regulation, pp. 37–54. North-Holland, Amsterdam.
126. Chan, A. K., Sigelmann, R. A., Guy, A. W., and Lehmann, J. F. (1973). Calculation by the method of finite differences of the temperature distribution in layered tissues. IEEE Trans. Biomed. Eng. 20:86.
127. Gordon, R. G., and Roemer, R. B. (1975). The effect of radial node spacing on finite difference calculations of temperatures in living tissues. IEEE Trans. Biomed. Eng. 22:77.
128. Gordon, R. G., and Roemer, R. B. (1975). The effect of radial node spacing on the performance of a mathematical model of the human temperature regulatory system. IEEE Trans. Biomed. Eng. 22:80.
129. Atkins, A. R. (1973). Application of the "Hopscotch" algorithm for

solving the heat flow equation for the human body. Comput. Biol. Med. 3:397.
130. Stolwijk, J. A. J., Nadel, E. R., Wenger, C. B., and Roberts, M. F. (1973). Development and application of a mathematical model of human thermoregulation. Arch. Sci. Physiol. 27:A303.
131. Kitney, R. I. (1974). The analysis and simulation of the human thermoregulatory control system. Med. Biol. Eng. 12:57.
132. Miller, N. C., and Seagrave, R. C. (1974). A model of human thermoregulation during water immersion. Comput. Biol. Med. 4:165.
133. Horowitz, J. M., and Erskine, L. K. (1973). Central regulation of temperature in cold environments. I. A. dynamic model with two temperature inputs. Comput. Biomed. Res. 6:57.
134. Atkins, A. R. (1975). A study of temperature control in the human body by experiment and computer simulation. Ph.D. thesis, University of the Witwatersrand, Johannesburg, South Africa.
135. Brengelmann, G. L., Wyss, C., and Rowell, L. B. (1973). Control of the skin forearm blood flow during periods of steadily increasing skin temperature. J. Appl. Physiol. 35:77.
136. Houdas, Y., and Guieu, J. D. (1973). Le Système thermo-régulateur de l'Homme: système régulé ou système asservi. Arch. Sci. Physiol. 27:A311.
137. Houdas, Y., Sauvage, A., Bonaventure, M., Ledru, C., and Guieu, J. D. (1973). Thermal control in man: regulation of central temperature of adjustments of heat exchanges by servomechanism. J. Dyn. Syst. Meas. Contr. 95:331.
138. Houdas, Y., Sauvage, A., Bonaventure, M., and Guieu, J. D. (1973). Modèle de la réponse évaporatoire à l'augmentation de la charge thermique. J. Physiol. Paris 66:137.
139. Gray, B. F., Gray, P., and Kirwan, N. A. (1972). Applications of combustion theory to biological systems. Combust. Flame 18:439.

CORRIGENDUM

In the predecessor to this chapter (1), in the section on physical work rate, the words "concentric" and "eccentric" should be interchanged throughout. I am grateful to J. W. Snellen for pointing out this error.

International Review of Physiology
Environmental Physiology II, Volume 15
Edited by David Robertshaw
 University Park Press Baltimore

2
Physiological Effects of Cold Exposure

G. E. THOMPSON
Hannah Research Institute
Ayr, Scotland

AFFERENT NERVES 30
NERVE CENTERS 32

EFFERENT NERVES 33
Somatic Nerves 33
Sympathetic Nerves 35
Skin Vessels 35
Heart 36
Hair 37
Adipose Tissue 37
Splanchnic Nerves and Adrenal Medulla 38

EFFERENT HORMONES 40
Thyroid 41
Adrenal Cortex 42
Insulin 42
Growth Hormone 42
Prolactin 42
Antidiuretic Hormone 42

EFFECTOR ORGANS 43
Skin and Hair 43
Muscle 45
Brown Adipose Tissue 47
White Adipose Tissue 50
Gut 51
Liver 52
Lipid Metabolism 52
Carbohydrate Metabolism 54
Protein Metabolism 55

Heart and Circulation 56
Kidney and Water Balance 57

This review describes some of the most important physiological events that occur in animals when they are exposed to a cold environment for a short time or a long time. In taking a broad view it is hoped that the reader's attention is drawn to the best areas for further work on this topic. Newborn or hibernating animals or hypothermia or cold injury have not been considered. Some related facts about anatomy, physics, biochemistry, and norepinephrine are included, but references are given only to experiments that involve cold exposure.

AFFERENT NERVES

When an animal's surroundings become cold, the skin also cools, and this is signaled to the brain by receptors in the skin. Heat-conserving and heat-producing reflexes in the body are thus stimulated.

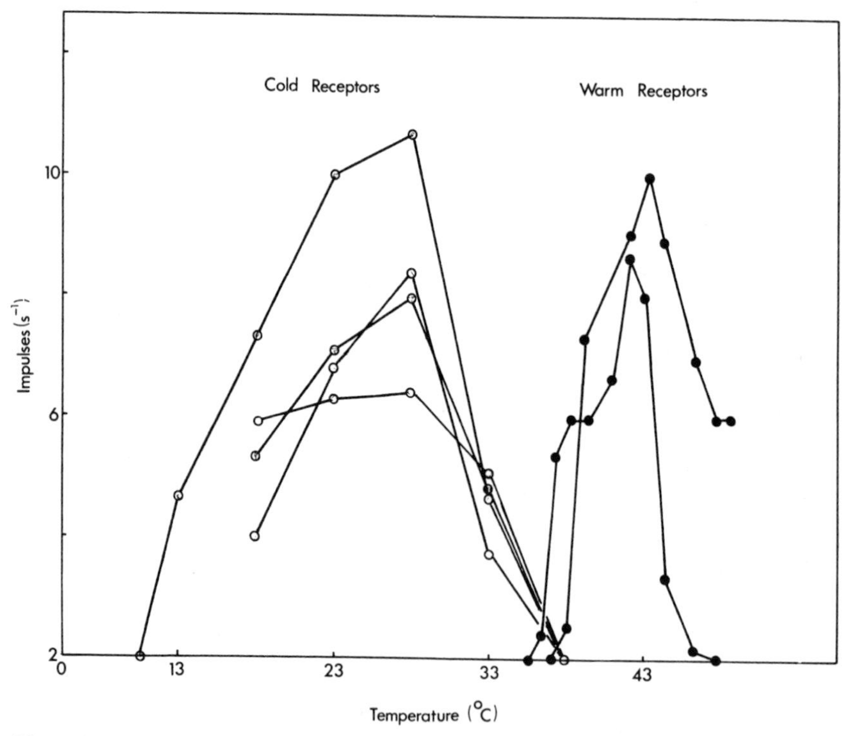

Figure 1. Static sensitivity curves for four cold receptors and two warm receptor preparations in rat skin. (Reproduced from Iggo (3), by courtesy of J. Physiol. (Lond.).)

A cold receptor appears as a dermal papilla containing a single, small myelinated nerve fiber which divides into a number of unmyelinated terminals which penetrate into the cytoplasm of the basal epithelial cells (1). The cold sensitivity of the cold receptor appears to be based on the properties of its transmembrane sodium pump (2). Its signals increase in frequency as skin temperature falls below a "normal" value, whereas "warm" receptors reduce their activity (Figure 1). Cold receptors typically produce a transient burst of impulses with a frequency that is related to the rate of fall in temperature, followed by a steady discharge of impulses, their frequency being related to the absolute temperature (3). Single cold fibers usually innervate one cold receptor in the skin (1, 3); the primary cold fiber enters the spinal cord through a dorsal nerve root and a dorsal horn. Here, fibers from a surprisingly large area of skin converge on (4), and synapse with, secondary neurons which cross to the opposite side of the cord and ascend, in the lateral spinothalamic tract, to the thalamus, hypothalamus, and somatosensory cortex (Figure 2). The signal in the primary cold fiber is already modified in the relay neurons of the dorsal horn (5) and further modified in the thalamus (6).

Cold receptors are also present in the hypothalamus (7), spinal cord (8), and elsewhere in the body, but these are unlikely to be active at the beginning of

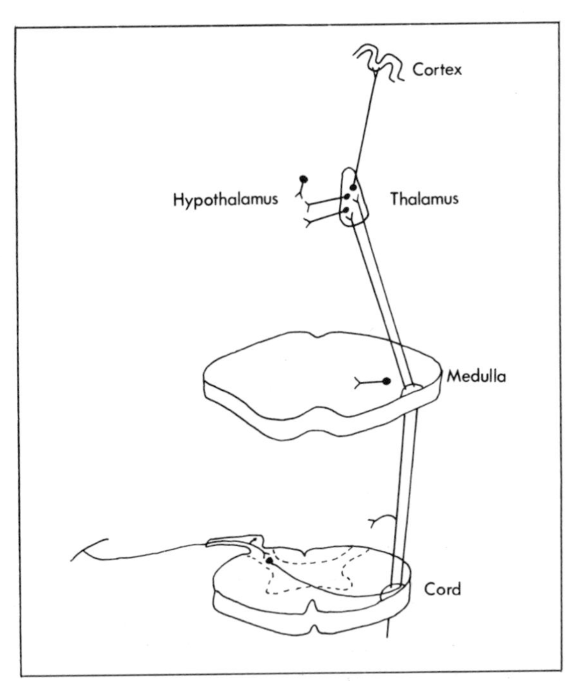

Figure 2. Some (simplified) afferent cold receptor pathways from receptors in the skin, medulla, and hypothalamus. An ascending fiber of a neuron sensitive to spinal cooling is also illustrated, but its cold receptor is not included. Convergence of pathways is not illustrated. Spinal and medullary reflex arcs are included, although the evidence for their existence is incomplete. Descending neurons that might process signals in the dorsal horn are not included.

cold exposure when the animal's hypothalamic and core temperatures normally increase (9). As cold exposure continues and especially in situations such as swimming (10), the skin and core may both be cold and the cold receptors at all sites may be active. The brain does not receive separate information from central and peripheral cold receptors in the body; convergence of cold signals from skin and spine (11) and from skin and hypothalamus (12) is commonly found. This could have the effect of integrating the cold "signal" before it reaches interneurons in the brain. The interaction between peripherally and centrally originating cold stimuli has been studied by artifically cooling or warming the hypothalamus and skin and by measuring the size of the response. It was found that a high brain temperature reduces, but does not eliminate, the vasoconstriction and shivering brought about by cooling the skin (13). When the brain is cooled as well as the skin, the two factors are said by some to add (14) and by others to multiply (13) the effects on heat production and heat conservation. Cooling the hypothalamus and spinal cord together produces an additive effect (15).

NERVE CENTERS

Afferent cold fibers enter the anterior hypothalamus, synapse with interneurons, and these in turn synapse with efferent sympathetic and somatic nerve fibers leaving the hypothalamus. Many of these fibers enter or leave the hypothalamus laterally because cutting through the nerves there is most effective in disrupting the normal reflex responses to a cold environment (16). The interneurons have not been properly characterized yet, but it has been suggested that they can be identified by the signals that they carry (7, 17). Interneurons that take part in cold-reflex activity are found in the posterior hypothalamus, which is considered by some people to be the "cold center" in the brain, but almost as many occur in the anterior hypothalamus (18). The likely functions of these interneurons are to transmit impulses from cold receptors to cold effectors, inhibit heat loss effector nerves during cold stimulation and, perhaps, to provide a reference signal for comparison with cold receptor signals.

Experiments are in progress to find out which neurotransmitters act at the synapses between the different neurons that mediate the reflex responses to cold. Injections of various neurotransmitters into the hypothalamus of the conscious animal can stimulate or inhibit shivering. Norepinephrine is one of the neurotransmitters that has an effect (19), and an increased turnover of norepinephrine has been detected in the hypothalamus during cold exposure (20). When neurotransmitters are applied to specific neurons in the reflex arc, impulses are sometimes generated, but the results have not been consistent enough to suggest which transmitters stimulate or inhibit specific synapses (21–23).

The hypothalamus is the major "center" for cold-stimulated reflexes, but there is evidence for less important centers lower and higher in the central nervous system. Shivering can be induced by cooling the medulla oblongata of animals that have had the hypothalamus destroyed (24) and by cooling the spinal cord of animals that have had the spinal cord disconnected from the brain (25). However, it has not been shown electrophysiologically that reflex arcs circuit through these lower centers. The cold sensitivity of the spinal cord has been studied extensively, and there is no evidence for thermoreceptors at the surface of the spinal cord with afferent fibers in the dorsal roots (26). The cold receptors in the spinal cord (see the above section) may feed into a spinal reflex or, alternatively, the cold sensitivity of the cord may be a result of local cooling of the spinal motoneurons, which increases their excitability (27).

The frontal areas of the brain are involved in the habituation changes that are brought about by repeated cold exposure (28), and perhaps in the behavioral responses to cold exposure.

Describing the reflex pathways that are stimulated by cold exposure can be a relatively precise science. Defining the function of individual pathways is more difficult, but an attempt to analyze all of the likely inputs and outputs that are necessary to regulate body temperature in the cold has been made (29). It has been suggested by Brück and co-workers (30) that the nervous pathways that control shivering in muscle are functionally separate from those that control nonshivering thermogenesis in brown adipose tissue in animals such as the young guinea pig that produce extra heat from these two sources when they are exposed to a cold environment. This is their evidence: recordings in the cold of the temperature of brown adipose tissue and of the electrical activity of muscle showed that heat was produced in the adipose tissue first and the electrical activity of the muscle increased later (30). In the initial stage, when there was no shivering, the temperature in the vertebral canal increased, but the temperature in the hypothalamus fell (31). In the guinea pig, brown adipose tissue is positioned around the cervical and upper thoracic part of the spinal cord where there are warm receptors (32). Artificial heating of the brown adipose tissue abolishes shivering during cold exposure. A small area of white matter in the medioventral area of the spinal cord at the level of the fifth cervical vertebra must be intact for this inhibitory response to take place; it presumably contains the afferent fibers that connect the spinal warm receptors to the hypothalamus (33). From the evidence, the following sequence of events has been proposed: when the guinea pig is exposed to a cold environment, its skin and hypothalamic cold receptors initiate nonshivering thermogenesis, but do not initiate shivering, which is relatively insensitive to a fall in hypothalamic temperature. Because of the position of brown adipose tissue in the body, its heat stimulates warm receptors in the spinal cord which inhibit shivering. Only when the environment is cold enough to reduce the temperature of the cervical spinal cord does shivering occur.

EFFERENT NERVES

Somatic Nerves

Experimentally made lesions placed at different levels in the brain show that those somatic nerves that are necessary for shivering run caudally from the posterior hypothalamus, through the midbrain adjacent to the red nucleus, then sweep out laterally to occupy the most lateral parts of the reticular formation in the pons and medulla, down the spinal cord in the lateral columns, to synapse with motoneurons which pass out of the cord in the ventral nerve roots and innervate the voluntary muscles. If the anterior hypothalamus is electrically stimulated, this inhibits shivering in the cold (34), but if the caudal hypothalamus, midbrain, or pons is electrically stimulated, it causes shivering in a neutral environment (35). This suggests that the nerve fibers descending from the brain control shivering. However, if an animal is exposed to cold and the electrical signal is recorded from fibers in the "shivering path" in the caudal hypothalamus or midbrain, it is found that the rhythm of their electrical activity bears no relation to the rhythm of shivering (36). This suggests that the rhythm of shivering is generated at the spinal level and must arise either as an intrinsic activity of the spinal cord or through some spinal reflex. To separate these two possibilities, the dorsal roots of the spine were cut on one side of an animal to interrupt any spinal reflexes, and it was exposed to cold. The rhythm of shivering was lost on the operated side and substituted by totally random movements. From this result, it was deduced that muscle contractions, produced initially by descending fibers acting on motoneurons, are synchronized by proprioceptors in the muscle (37) acting through reflex pathways that do not involve the cerebellum (37, 38). Later, when the role of Renshaw cells in the normal muscle spindle reflex became known, it was suggested that the rhythmic activity of the motoneurons during shivering is mostly controlled by Renshaw cells, and a less important role was ascribed to proprioceptors in the muscle (39). More recently, however, it has been found that the beginning of shivering in muscle is preceded by spontaneous activity in muscle spindle endings, suggesting that γ motoneurons may, through the proprioceptors, generate shivering at the spinal level as well as control its rhythm and amplitude (40). The activity of the γ efferent system was investigated indirectly by making electrical recordings from primary and secondary afferent spindle endings and Golgi tendon organs in the dorsal nerve roots of anesthetized cats. Static discharge patterns were recorded, as were dynamic responses to muscle stretch before, during, and after shivering that was brought about by surrounding the animal with ice. When the animal was rewarmed, the changes in static and dynamic discharge brought about by shivering were imitated by artificially stimulating γ motoneurons in the ventral roots. It was concluded that nerves descending from the hypothalamus may produce shivering at the spinal level by stimulating dynamic γ motoneurons and inhibiting static γ motoneurons which would allow an instability in the loop of neurons between muscle fibers, muscle spindles, and the spinal cord and tremor in the muscle fibers (Figure 3). The discharge from Golgi tendon organs showed small changes during shivering which could be explained by changes in the viscous properties of the muscle fibers (41).

When shivering is produced by locally cooling the spinal cord, it may be mediated through a reflex arc (see under "Afferent Nerves" and "Nerve Centers") or it may be generated by a local effect of cold which increases the

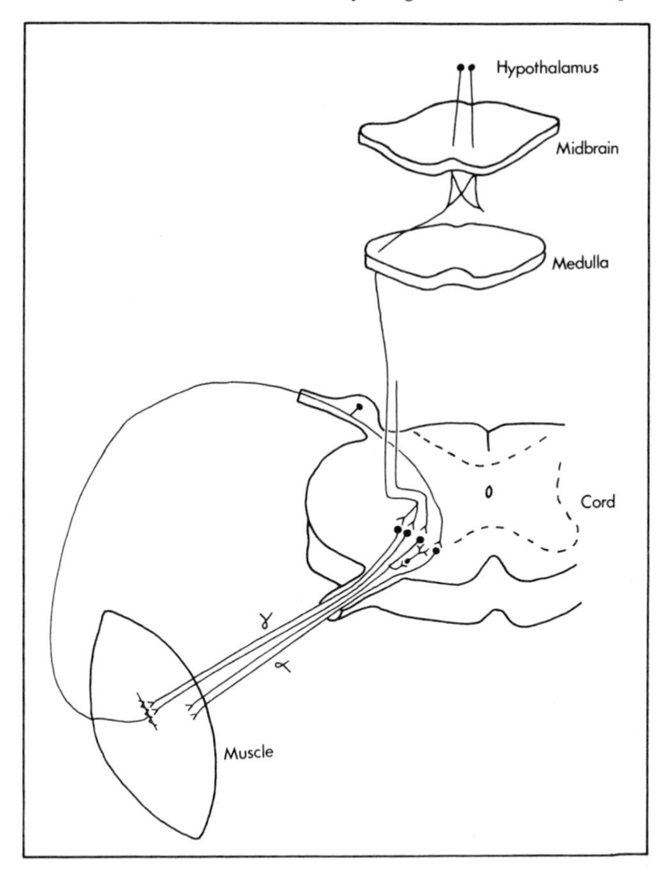

Figure 3. A scheme of efferent somatic nerve pathways that might control shivering. Spinal and medullary inputs, for which the evidence is incomplete, have not been included.

excitability of spinal motoneurons (27). Locally cooling the spinal cord also inhibits Renshaw cells (42), which could allow a more diffuse α motoneuron activity and tremor in muscle fibers.

In the intact animal, shivering, produced by cold exposure, can be facilitated or inhibited by other physiological influences. Baroreceptor stimulation, caused by raising the blood pressure in the carotid sinus, augments shivering (43), whereas chemoreceptor stimulation, caused by perfusion of the carotid body and sinus with an hypoxic solution, diminishes shivering (44). Stretching a muscle facilitates shivering (45). All of these stimuli are likely to have their effect at the level of the α or γ motoneurons.

Sympathetic Nerves

Skin Vessels The cutaneous blood vessels normally constrict when animals are exposed to a cold environment, but this reaction is disrupted if nervous tissue in the posterior hypothalamus is destroyed (46), suggesting that it is mediated by a nervous reflex that passes through the hypothalamus. The efferent nerves that produce this vasoconstriction are sympathetic nerves (47). Many sympathetic efferent fibers descend from the posterior hypothalamus to the spinal cord, but their paths are varied; some are direct fibers and others are multisynaptic with projections into tegmental nuclei and the medulla. They also mediate other peripheral vasoconstrictor reflexes such as the baroreceptor reflex and the defense reaction, and it is not known which of these pathways specifically mediates the cold reflex. In the spinal cord, the efferent sympathetic (preganglionic) nerves are located in the intermediolateral grey column. They are normally myelinated, leave the thoracolumbar spinal cord through the ventral roots, and pass down the ramus communicans to the sympathetic chain, where each preganglionic neuron synapses with a number of postganglionic neurons. The postganglionic fibers, usually nonmyelinated, return up the ramus communicans, join with cutaneous nerve bundles, and innervate the smooth muscle of blood vessels in the skin. The neurotransmitter at the synapse between preganglionic and postganglionic neurons is acetylcholine, but at the postganglionic neuron-smooth muscle receptor junction it is norepinephrine. If the central nervous system is artificially cooled, this produces an increase in the electrical activity in postganglionic sympathetic fibers supplying arteries of the skin (48). Sympathetic nervous activity also produces constriction of veins in the skin during cold exposure (49) and, probably, constriction of arteriovenous anastomoses. The vasoconstrictor receptors in skin arteries, arteriovenous anastomoses, and veins are all α receptors. Locally cooling blood vessels can modify their response to sympathetic nervous activity. For example, in cutaneous arteries (50) and veins (51), a given frequency of impulses sent down the sympathetic nerves that supply them produces more constriction if the vessels are cool (17–27°C) than if they are normal (37°C). If cooling continues (10°C), however, arteries do not constrict in response to sympathetic nerve stimulation (49) or to exogenous catecholamines (52). This offers a simple explanation for the phenomenon of "cold vasodilation," which occurs in brief episodes in the skin of the extremities when they approach freezing point. The brevity of the dilation can be explained by assuming that the paralysis of the artery is reversed when warm blood is delivered to it, which would restore the normal sympathetic constriction (52).

As well as an increased release of norepinephrine from sympathetic nerves onto the outside of cutaneous blood vessels in the cold, there is also more norepinephrine and epinephrine from the adrenal medulla in the blood stream during cold exposure (53).

Chronic exposure to a cold environment has been reported to reduce the sensitivity of animals' ear skin vessels to norepinephrine (54), but other similar experiments have not shown this effect (55).

Heart It is not clear how cold stimulation affects the activity of nerves that supply the heart, and different conclusions have been reached from experiments on different animal preparations. When the spinal cord is artificially cooled in the anesthetized dog, immobilized with succinylcholine, no increase in arterial blood pressure occurs even though cutaneous vasoconstriction occurs. This has been attributed to a nervous reflex, involving the carotid sinus and vagus nerves, which slows the heart and thus regulates blood pressure (56). In agreement with this, a reduced electrical activity has been recorded in the cardiac sympathetic nerves (48). In contrast, when unanesthetized rats are put into a cold (5°C) environment, the turnover of tritiated norepinephrine in the heart increases (57, 58), which indicates an increased activity of cardiac sympathetic nerves. The difference between these experiments can be reconciled if one assumes that the artifically cooled animals were only mildly stimulated by cold and that the animals exposed to a cold environment were more strongly stimulated (see under "Heart and Circulation").

In animals that have been chronically exposed to a cold environment, there is evidence of an increased sympathetic tone to the heart. In the first few days in the cold, increased production of tyrosine hydroxylase, a rate-limiting enzyme for the synthesis of norepinephrine, is induced in the superior cervical ganglion (59), and after a few weeks of cold exposure the turnover of norepinephrine in the heart is increased (58). It is not certain whether or not the heart of the cold-acclimated animal has a reduced sensitivity to norepinephrine (60, 61). There may be, also, an increased vagal tone to the heart, because in cold-acclimated animals at room temperature vagotomy produces a marked increase in blood pressure in contrast to its lack of effect in control, warm-acclimated animals (62). Perhaps as a result of a chronically increased vagal tone, the heart of the cold-acclimated animal loses some of its sensitivity to acetylcholine (63).

Hair Erection of the hair coat normally takes place when mammals are exposed to a cold environment; this response can be reproduced by electrically stimulating the hypothalamus (64). Sympathetic nerves innervate the piloerector muscles of hair, and cutting these nerves prevents piloerection in a cold environment (65).

Adipose Tissue Both brown and white adipose tissues are rapidly stimulated by sympathetic nerves during cold exposure. The stimulus probably originates from the hypothalamus, because locally cooling the anterior hypothalamus appears to stimulate brown adipose tissue in the young animal (66) and white adipose tissue in the adult (67), and centrally administered neurotransmitters stimulate (68) or inhibit (69) fat mobilization at the same time that they stimulate or inhibit shivering. The efferent nerve pathways from the hypothalamus to adipose tissue have not been followed, and the literature describing the distribution of sympathetic nerves in the two types of adipose tissue contains many contradictions. There is evidence that, in brown adipose tissue, the sympathetic nerves supplying the blood vessels in the adipose tissue have a separate origin from those supplying the adipose cells, although the nerves around the blood vessels are often seen to branch into the tissue. It has been reported that the nerves supplying the brown adipose tissue have their preganglionic-postganglionic synapse in the adipose tissue itself, whereas the nerves supplying the vessels are known to have their ganglia in the sympathetic chain ventral to the spinal cord. However, subsequent descriptions of the histochemistry of brown adipose tissue have not confirmed the presence of ganglia. It has also been reported that in white adipose tissue there is only a sympathetic nerve supply to the blood vessels, but, more recently, fine sympathetic nerves supplying the adipose cells have also been described. The receptors that mediate lipolysis in adipose tissue cells and the receptors that dilate the blood vessels in adipose tissue are both β receptors.

Sensory nerves in brown and white adipose tissue have been described. When animals are exposed to a cold environment, the turnover of norepinephrine in the brown adipose tissue increases, reflecting an increased sympathetic nervous activity (58). Cutting the sympathetic nerves to brown adipose tissue prevents the normal depletion of fat that occurs in this tissue during cold exposure (70). When animals that have white adipose tissue are exposed to cold, free fatty acids, which probably originate from adipose tissue triglycerides, are mobilized into the blood stream; this mobilization can be prevented by pretreating the animals with autonomic ganglion-blocking drugs (71). Probably, the increased rate of blood flow through adipose tissue in the cold is also directly or indirectly due to increased activity of sympathetic nerves supplying this tissue.

When rats are chronically exposed to cold, the norepinephrine content of, and norepinephrine turnover in, the brown adipose tissue increases, which reflects development of its sympathetic innervation (58).

Splanchnic Nerves and Adrenal Medulla Some preganglionic sympathetic efferent fibers, arising from the spinal cord as high as the fourth thoracic segment and as low as the first lumbar segment, pass through the sympathetic chain without synapsing and form the splanchnic nerves. Some of these fibers synapse with postganglionic neurons in the celiac ganglion, and the postganglionic fibers innervate several different organs, perhaps modifying their blood flow or metabolism or both during cold exposure. However, the evidence is contradictory.

Recordings of the electrical activity in preganglionic sympathetic efferent fibers in the splanchnic nerves of anesthetized, immobilized animals during spinal cooling have shown a reduced nervous activity (48). Their postganglionic fibers innervate the arteries and veins of the gut, and a reduction in their activity is consistent with the diversion of blood from skin vessels to gut vessels, which is known to occur during cold stimulation (see under "Heart and Circulation") and thought to aid conservation of heat in the body. There is a similar response in the splanchnic nerve when the hypothalamus is artificially cooled. It has been suggested that differentiation of sympathetic activity, increasing to the skin and decreasing to the gut vessels, is a special thermoregulatory response of the vasomotor system (72). Recordings from postganglionic fibers of a branch of the splanchnic nerve, the splenic nerve, have also shown a reduced nervous activity during spinal cooling (48), suggesting that the same response is transmitted through the various branches of the splanchnic nerve. However, studies of the effect of cold exposure on the turnover of norepinephrine in various organs supplied by the splanchnic nerve conflict with these conclusions. The turnover of norepinephrine in the intestine does not change in rats after 1 day of cold exposure (58), suggesting that the increased blood flow through the gut is a passive consequence of cutaneous vasoconstriction. In the same situation, the turnover of norepinephrine in the spleen and kidney is increased by cold exposure (57), which indicates an increased activity in their nerve supply and shows that the branches of the splanchnic nerve supplying these organs have a different activity from those branches that supply the intestine. In addition, there are many physiological consequences of cold exposure which are difficult to explain unless one assumes that other branches of the splanchnic nerve also have an increased activity. For example, postganglionic fibers from the celiac ganglion innervate the islets of Langerhans in the pancreas and, if stimulated, inhibit their secretion of insulin. There is normally an inhibition of insulin secretion during acute cold exposure (see under "Insulin"), although there is at the moment no direct evidence that this is brought about by the sympathetic nervous system. In addition, postganglionic sympathetic fibers from the celiac ganglion innervate the liver, and their stimulation produces glycogenolysis in the

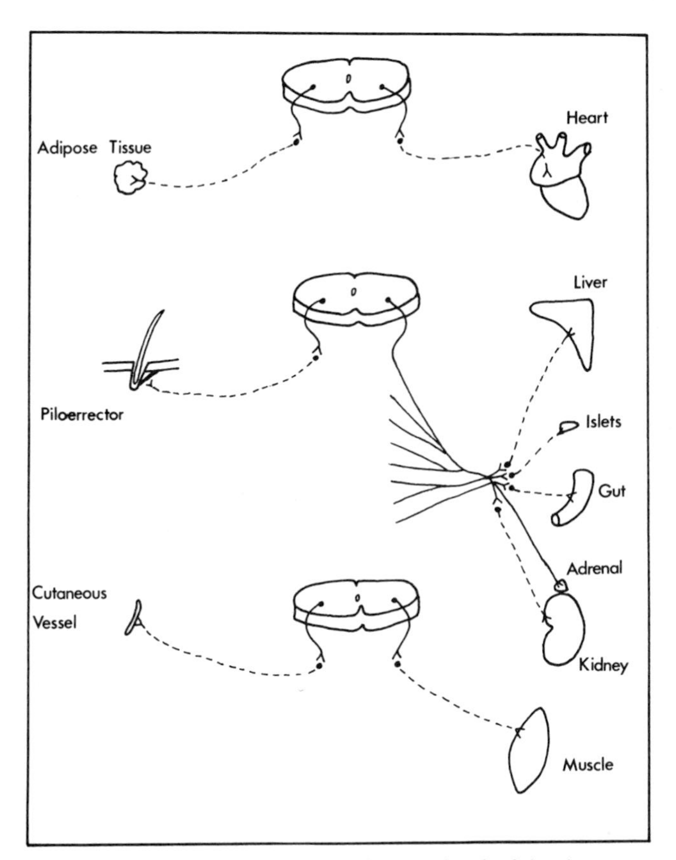

Figure 4. Spinal efferent sympathetic nerves that are involved in the response to cold exposure, showing the peripheral distribution of preganglionic (——) and postganglionic (– – –) neurons.

liver and release of glucose into the circulation. Normally, there is a release of glucose from the liver early in cold exposure (see under "Carbohydrate Metabolism"), although there is, again, no evidence at the moment that this is brought about by these nerves. Finally, some preganglionic efferent sympathetic fibers in the splanchnic nerves pass through the celiac ganglion without synapsing and innervate the adrenal medulla (Figure 4). These nerves are known to control the secretion of catecholamines from the adrenal medulla. It is especially difficult to believe that their activity is reduced during cold exposure in view of the evidence that catecholamines are released from the adrenal medulla into the blood stream during cold exposure (53, 73). The normal increase in blood glucose concentration during acute cold exposure is prevented by adrenal demedullation (71, 74) which suggests that, in the cold, the splanchnic nerves stimulate the adrenal medulla to release catecholamines into the circulation and, in this way, stimulate glycogenolysis in the body. These same circulating catecholamines are not, however, essential for the mobilization of adipose tissue fatty acids into the blood during cold exposure, because this mobilization is not prevented in animals that have had the splanchnic nerves cut (75) or the adrenal medulla removed (71).

These various facts raise two questions. a) Does splanchnic nerve activity increase, or decrease, or are there variations between its branches during acute cold exposure? b) If splanchnic nerve activity increases, why does hepatic glycogenolysis depend upon the adrenal medulla when the liver has its own branch of the splanchnic nerve?

When cold exposure continues for several days, catecholamines are depleted from the adrenal medulla (53) and tyrosine hydroxylase enzyme activity is induced (59). This lasts for a few weeks, but then catecholamine concentrations and enzyme activity return to normal (76).

EFFERENT HORMONES

When an animal is exposed to a cold environment, there are changes in the concentrations of catecholamines, insulin, thyroid hormones, adrenal corticoids, and prolactin in the blood. In the short term, these changes are brought about by nervous reflexes which are initiated by cold receptors and their afferent impulses and centered on the hypothalamus. However, some of the hormonal responses develop slowly and, therefore, develop fully only in response to chronic cold exposure. The efferent pathway most likely to mediate the release of catecholamines and insulin is via the posterior hypothalamus, spinal sympathetic efferent nerves, and splanchnic nerves (see under "Splanchnic Nerves and Adrenal Medulla"). The efferent pathway most likely to control the release of thyroid hormones, adrenal corticoids, and prolactin is via the posterior hypothalamus, the anterior pituitary, and, in the case of the thyroid hormones and corticosteroids, tropic hormones in the peripheral circulation which are taken up by their respective endocrine glands and influence the secretion of hormones.

Thyroid

The evidence that thyroid activity is stimulated by exposure to a cold environment is contradictory, controversial, and voluminous. The following is a brief selection from the literature to illustrate the confusion. Environmental cold has been reported to reduce (77), not affect (78) and increase (79), the circulating level or rate of secretion of thyrotropic hormone. Plasma thyroxine concentration has been reported to decrease (79) and increase (80) in the cold. Plasma triiodothyronine is reported to increase in the cold (80), and it has been suggested that there may be a preferential secretion of triiodothyronine in the cold (81).

Analyzing the effect of cold exposure on thyroid function is made more difficult by the fact that animals increase their food intake during prolonged cold exposure and may, incidentally, increase their intake of iodine (82) and/or produce more fecal bulk and lose more thyroxine in their feces than control animals (83). These factors, by themselves, alter thyroid function. Attempts to control the iodine intake and bulk of the diet have not, so far, clarified the problem. In some controlled experiments, cold exposure stimulated thyroid activity (84), but in others it had no effect (85).

Prolonged exposure to a cold environment may also affect the distribution of thyroid hormones in the body and their rate of breakdown. Triiodothyronine increases more than thyroxine in the extrathyroid tissues during cold exposure which could be due to a preferential secretion of triiodothyronine or conversion of thyroxine to triiodothyronine in the tissues (81). The binding of L-triiodothyronine to plasma proteins in cold-exposed rats has been reported to be greater than normal (86) and less than normal (87). Its binding to heart (88), liver, kidney, and skeletal muscle is increased by cold exposure; in kidney tissue, the increased binding is to microsomes which are believed to be a major site of deiodination of thyroxine and triiodothyronine (89). For this reason, there may be an increased degradation of thyroxine and triiodothyronine in the peripheral tissues during cold exposure (86), although this is not a consistent finding (90). An increased binding of thyroxine by the gut or liver and increased output through the bile might explain the greater peripheral uptake of thyroid hormone and the loss of thyroxine in the feces during cold exposure (90–92). The urinary excretion of iodide has been reported to increase (93), as well as to decrease (90), in a cold environment.

Although it is not certain that thyroid activity is increased during cold exposure, studies continue into the likely role of such an increase. These are mostly directed toward an examination of why thyroid hormones increase the metabolism and heat production of a number of tissues, such as skeletal muscle, liver, and kidney, but not brain tissue. The available evidence suggests that induction of ribonucleic acid and protein synthesis in the tissues is an essential part of this reaction, and an increase in the number and size of mitochondria in skeletal muscle has been observed after the administration of thyroid hormones. The metabolic process in the tissues that is increased by the hormones is probably the transmembrane pumping of sodium ions, because inhibition of sodium transport by ouabain drastically reduces the calorigenic action of triiodothyronine and thyroxine. Exactly how these hormones stimulate sodium pumping is not known, but they may activate, or increase the concentration of, sodium potassium-activated adenosine triphosphatase enzyme in the cell membrane. Other likely roles of the thyroid hormones during cold exposure are to potentiate some of the effects of sympathetic stimulation and to stimulate the development of a winter coat.

Adrenal Cortex

In the first few hours of cold exposure, the blood levels of adrenal corticoids increase, but after several days, and perhaps because food intake has increased in response to the cold, the levels return to normal (94). Recently shorn sheep that die in severely cold, wet, and windy conditions have lipid-laden and hemorrhagic adrenals, suggesting that they have been excessively stimulated by adrenocorticotropic hormone. The survival of sheep in these conditions can be improved by pretreating them with cortisone (95). Plasma cortisol concentration and rate of secretion into the circulation increase greatly if cold stress is severe and hypothermia is imminent (96), and uptake of cortisol by the liver and kidneys is increased (97).

The increase in plasma corticosterone in rats in a cold environment is mediated through the hypothalamus and can be prevented by cutting the lateral and posterior nerve connections of the hypothalamus (98).

Injections of exogenous adrenal corticoids stimulate the total heat production of animals, but it is more often assumed that their value in the early stage of cold exposure is to enhance the mobilization of fatty acids that is initiated by the sympathetic nervous system and to stimulate gluconeogenesis in the liver.

Insulin

The concentration of insulin in the blood falls during acute (99) and chronic (100) cold exposure. This also favors mobilization of fatty acids from white adipose tissue and gluconeogenesis in the liver. It does not prevent the shivering leg from increasing its uptake of glucose during cold exposure (see under "Muscle").

Growth Hormone

In a number of experiments, cold exposure had no effect on the concentration of growth hormone in blood plasma (101–104).

Prolactin

Cold exposure reduces the concentration of prolactin in the blood of heifers (105) and male rats (103).

Antidiuretic Hormone

Exposure to a cold environment for a few hours reduces the concentration of antidiuretic hormone in the blood of rats (106) and humans (107). Increased hormone concentrations were found in dogs that were partially restrained in a cold environment (108), but not when they were given more freedom to move (109). The reduced antidiuretic hormone levels might be brought about by a reflex arc involving cold receptors and the hypothalamic nuclei that control the secretion of antidiuretic hormone. More likely, a shift of blood from the skin to the heart in the cold might stimulate stretch receptors in the left atrium and reflexly reduce antidiuretic hormone release. Alternatively, the increased concentration of glucocorticoid hormone in the blood could inhibit release of antidiuretic hormone from the posterior pituitary.

EFFECTOR ORGANS

Skin and Hair

Heat energy from the tissues is mainly convected by the circulatory system to the skin surface. In cool, thermoneutral environmental temperatures, vasoconstriction in the skin regulates this flow of heat. The skin of the human hand constricts very effectively, and the ratio of maximum to minimum hand blood flow can be as high as 60:1 (110), whereas the average for the entire human shell is approximately 6:1 (111). Some areas of skin, such as the human face, show no detectable ability to reduce heat loss in a cold environment (112). Cutaneous vasoconstriction mostly involves constriction of smaller vessels (113), the arteries before the veins (114), and closure of arteriovenous anastomoses, which diverts blood from the capillaries in the skin of the extremities (115). Gradations of vasoconstriction, as the environmental temperature falls, are probably achieved by extending the time spent in vasoconstriction (116) and the extent of vasoconstriction, from distal to proximal parts of the body. The effectiveness of blood in convecting heat depends upon the path taken by the blood vessels; blood flowing along isothermal lines will not transfer heat, whereas blood flowing through a steep temperature gradient convects most effectively. In many parts of the body, arteries and veins run side by side, and heat is exchanged between them. This is especially important in the blood vessels supplying the limbs, which are parts of the body that have a relatively high surface to mass ratio and for this reason contribute disproportionately to total heat loss. In the limbs, heat contained in the arterial blood transfers to returning venous blood; the limb vessels, therefore, maintain a steep temperature gradient along their length and less heat is lost from the limb surface to the environment (117). The change from maximal peripheral vasodilation in a warm environment to maximal vasoconstriction in a cold environment, in cattle for example (118), produces an increase in mean tissue insulation from 0.04–0.14 m^2 kW^{-1}. Cold vasodilation does not occur extensively or for long in the body, and this author is not aware of any measurements showing that its occurrence, at environmental temperatures just above freezing, appreciably reduces the mean tissue insulation of the whole body. It may have some value in reducing cold injury to peripheral tissues.

Heat energy is transferred from the skin surface through the hair coat to the environment, where it is lost by convection and radiation. The insulation provided by a variety of coat types has been measured in vitro. The feature of a coat that correlates best with its insulative value is the orientation of the hairs; hairs that lie flat across the direction of heat flow insulate better than hairs that stand erect, parallel to the direction of heat flow (119). The thermal insulation between the skin of the animal and the surface of the coat is changed by piloerection. In young cattle, for example, the insulation of the hair coat is 0.11 $m^2 kW^{-1}$ when the hair is not erect and 0.16 $m^2 kW^{-1}$ when erect (120). Evaporative heat transfer through the coat in cold environmental temperatures is probably governed by the minimal rate at which water diffuses out of the skin. Measurements of the evaporative heat loss in the cold have shown very little variation with environmental temperature (118). Free convective heat transfer through the coat increases with increasing temperature difference between skin and air (121) and, therefore, can be a major avenue of heat loss through the coat at low environmental temperatures. Radiative transfer through the coat depends upon a number of factors, those of physiological importance being the color of the skin, the color of the hair, and the extent to which the hair intercepts radiation. A cold, sunny climate favors a white coat growing from a dark skin (122). Conductive heat transfer through the hair coat is probably small and is estimated at less than 10% of the total flow (121). The insulation of the boundary layer at the coat surface does not vary with environmental temperature or the physiological state of the animal, and in cattle, for example (118), is about 0.12 $m^2 kW^{-1}$. In animal species that have very little hair, such as the human or the pig, the insulation of the boundary layer is a larger part of the total body insulation. Relatively low wind velocities (up to 4 $m s^{-1}$) destroy the boundary layer and, therefore, chill these animals if the environment is cold. In animal species that have a thick layer of wool or hair, the insulation of the coat remains unaffected by wind velocities up to about 11 $m s^{-1}$; faster wind speeds penetrate the air in the coat and force convection in it (123).

Applying what is known about the insulative properties of parts of the animal to the whole animal is not easy. Changing the body's posture affects the rate of heat loss from it (124), and postural changes which reduce the effective area of heat loss normally occur in cold environments. Animals in groups lose less heat than animals alone (125), and "huddling" is also a normal behavioral response to cold exposure in some animals. The way in which an animal chooses to orient itself with respect to the sun and wind is also important, but the consequences are difficult to analyze. Shivering may increase the body's heat loss, by increasing convection (126) inside or outside the hair coat or both.

Chronic exposure to a cold environment can reduce the insulation of the tissues of the extremities (127) and increase their vascularity (see under "Heart and Circulation"), but measurements of the mean tissue insulation of cold-acclimated animals have not detected any significant change from normal values (128). The insulative value of the hair coat increases greatly in the winter. The winter coat is produced in response to a seasonal change in day length, as well as to a fall in environmental temperature, and develops by growth of new hairs with a different configuration (129) and by reduced shedding of old hairs (130). A thick coat is advantageous in reducing the wasteful metabolic response to cold exposure, but disadvantageous in reducing the animal's sensitivity to peripheral cold.

Muscle

When the environmental temperature falls, the animal's first responses are peripheral vasoconstriction and piloerection, but if the temperature continues to fall below a "critical" value, the rate of heat loss from the animal to the environment exceeds the rate of heat production from the animal's resting metabolism. To maintain a "normal" central body temperature, extra heat must be produced, the amount being dictated by the difference between body and air temperatures and the maximal thermal insulation of the body. The most important mechanism for producing extra heat in most adult mammals is shivering in voluntary muscles.

Muscular contraction is initiated by an action potential at the motor nerve end plate. The action potential propagates through the muscle fiber, calcium ions are liberated in the fiber bringing the contractile proteins into an active state, and they shorten the length of the muscle fiber. Before shivering is visible in muscle, a "preshivering tone" is detectable as intermittent and uncoordinated electrical impulses in single muscle fibers. As cold exposure continues, the fibers increase their frequency of electrical activity and more fibers become active until, during shivering, there is coordinated activity of groups of muscle fibers (45). There is no general rule governing which muscles shiver first or most; some reports favor the masseter muscles, some the pectoral muscles, and others the proximal and extensor muscles of the limbs.

Muscle contraction is coupled to the conversion of adenosine triphosphate to adenosine diphosphate and inorganic phosphate. The rate of supply of adenosine diphosphate to the tissue is thought to govern the rate at which this tissue uses energy substrate, consumes oxygen, and produces heat. In cattle, when the total oxygen consumption of the animal is increased by 40–50% during cold exposure, the oxygen consumption of the shivering hind leg increases 400% (131) (Table 1). Increases in the consumption of oxygen by the leg are met first by increases in the extraction of oxygen from the blood stream and then, at higher rates of oxygen consumption, by increases in the rate of blood flow (132). Sheep can increase their metabolic heat production 7-fold (133), almost all by shivering as far as is known. The type of substrate that is used by muscle for oxidative metabolism probably depends upon the type of substrate that is available inside the muscle cells. Triglycerides (134) and glycogen (135), normally present in muscle cells, are partially depleted early in cold exposure. In addition, shivering muscle takes up extra nutrients from the circulation. In cattle, extra free fatty acids (136), glucose (137), and volatile fatty acids (138), which are products of digestion, are taken up by the shivering leg. The relative importance of these nutrients differs in fed and fasted animals. Measurements of the uptake of blood plasma triglycerides by shivering muscle have not been

Table 1. Effect of acute cold exposure on the rate of oxygen consumption by the shivering hind leg in cattle

	Total oxygen consumption (ml min^{-1})	Leg oxygen consumption (ml min^{-1})	Leg blood flow (ml min^{-1})	Leg oxygen extraction (ml 100 ml^{-1})
Neutral environment	719	18.8	582	3.2
Cold (1 °C + wind)	1,056	74.9	1,172	6.4

From Bell, Clarke, and Thompson (131), by courtesy of Q. J. Exp. Physiol.

made, but it is unlikely that they contribute substantially to oxidative metabolism because there is no change in lipoprotein lipase activity in skeletal muscle during cold exposure (139). Muscle can act as a store of amino acids in the body which, in acute cold exposure, may be released from muscle, converted to glucose by the liver, and become available again to muscle and other peripheral tissues. Measurements of the free amino acid content of muscle after a few hours of cold exposure have not shown any change (140). Measurements of blood flow and arteriovenous difference in blood amino acid content during a few hours of cold exposure also have not shown any significant effect on amino acid output from the shivering leg (137). However, it may be deduced from analyses of the carcasses of rats exposed to cold for a few days that there is catabolism and net loss of muscle protein (141, 142). Increased blood corticosteroid and reduced blood insulin levels would favor gluconeogenesis from amino acids.

It has been known for some time that infusing catecholamines into an animal will facilitate its shivering in a cold environment, but there is a variety of ways in which the catecholamines could have this effect, ranging from altering the discharge from muscle spindles to depolarizing the muscle cells. It has not been possible to decide whether these effects imitate a normal response to cold exposure. Fresh interest in the interaction between catecholamines and shivering may result from the finding of an increased turnover of norepinephrine in skeletal muscle during cold exposure (57) and an increase in epinephrine-stimulated adenylate cyclase activity in skeletal muscle of rats during the first few days of cold exposure, when shivering is at its maximal. After a few weeks in the cold, this adenylate cyclase activity disappears (143). Adenylate cyclase is a catecholamine-receptor molecule and mediates the metabolic effects of catecholamines in tissues.

If rabbits are continuously exposed to cold, their shivering decreases with time and, after 5 weeks, almost disappears (144). Studies of the metabolism of muscle have been made in animals that have been chronically exposed to a cold environment to find out whether the skeletal muscles are still producing extra heat even though shivering has declined. When anesthetized cold-acclimated rats, curarized to prevent shivering, were exposed to an environment of 10°C, the oxygen consumption of the hind leg and the rest of the body doubled. This effect was reproduced by infusing norepinephrine into the animals (145). Similarly, the oxygen consumption of the autoperfused hind limb of the dog was found to increase by about 90% when the anesthetized animal was exposed to an environment of −10°C. This increase in metabolism was abolished by denervating the leg muscles of warm-acclimated dogs, but was sustained after denervation of the muscles in cold-acclimated dogs, suggesting endocrine stimulation (146). These experiments indicate that there may be a qualitatively different kind of heat-generating mechanism in the muscles of cold-acclimated animals, but neither experiment demonstrates a metabolic response comparable in size to that in shivering muscle, and both are especially inadequate when one considers the increased metabolic capability of cold-acclimated animals in the cold (128).

The metabolic pathways in skeletal muscle that might be specifically stimulated by chronic exposure to a cold environment remain unknown. The resting potential in skeletal muscle cells is slightly increased after chronic cold exposure (147). The average cytochrome *c* concentration of muscles is higher (148), and the turnover of some mitochondrial proteins is increased (149). However, the rate of incorporation of labeled amino acid into muscle is not altered (150), nor is the intramuscular concentration of carnitine (151), which is important for free fatty acid oxidation.

The vascularity of the red fibers, but not the white fibers, in gastrocnemius and soleus muscles is increased during chronic cold exposure (see under "Heart and Circulation"). Therefore, it might be wise to separate the two fiber types in any study of the effect of cold acclimation on muscle metabolism.

Brown Adipose Tissue

Brown adipose tissue can, within a few minutes, vary its heat production. This tissue occurs in the newborn of most animal species, but the newborn is not considered in this review. The adults of some of these species, usually the smaller ones, retain brown adipose tissue. In these adults, its development is stimulated by chronic exposure to a cold environment. The location of brown adipose tissue in the body varies; it may be close to the spinal cord, heart, aorta, or kidneys.

Measurements of the flow of heat from the interscapular brown fat pad in anesthetized, cold-acclimated rats have been used to calculate that the brown adipose tissue of the body supplies only about 8% of the total body heat in a cold (5°C) environment (152). This is a small contribution compared to the contribution of the same tissue in some newborn animals, but the cold-acclimated rat contains relatively little brown adipose tissue in its body, viz. 1–2% of the body weight.

Assuming that brown adipose tissue is the only tissue that produces extra heat when the body is stimulated by the sympathetic nervous system or injection of norepinephrine, then the surgical removal of some of this tissue should reduce the maximal metabolic response of the animal to norepinephrine in proportion to the fraction of brown adipose tissue that is removed. This is a dubious basis for an experiment, because it is known that, in the rat, for example, the metabolic response to catecholamines is too big to be produced by such a small amount of tissue. Nevertheless, a number of people have tried this approach, and the results have produced an interesting idea. In some experiments, removal of the interscapular brown adipose tissue, which makes up 30–40% of the total in cold-acclimated rats, reduced the maximal metabolic response to norepinephrine appreciably and permanently; the biggest reduction was 60%. In other experiments, removal of the same tissue did not have any effect on this response or reduced it very little. In yet another similar experiment, it was found that the removal of the tissue had no immediate effect, but the response diminished over a period of a few days. In subsequent experiments, this gradual decline was partially prevented by replacing pieces of brown adipose tissue in the peritoneal cavity. It was concluded that the brown adipose tissue is not itself a major site of oxygen consumption in the body when norepinephrine is administered, but it is an endocrine gland producing secretions that increase the capability of some other tissue or tissues to respond to norepinephrine. This idea is, unfortunately, difficult to test; animals that have brown adipose tissue are small, and very little information is available on the heat production of organs such as muscle and liver in these animals (see under "Muscle" and "Liver").

The metabolic events that occur in brown adipose tissue when it is stimulated by cold exposure have been studied in detail. Chronically exposing animals to a cold environment stimulates an increase in the mass of brown adipose tissue and in the number of adipose cells (153), which probably originate from reticuloendothelial cells. The activity of some cytoplasmic enzymes increases (154); there is also a 4- to 5-fold increase in mitochondrial mass (155) and an increased amount of inner mitochondrial membrane material (156). There is an increase in the turnover of some mitochondrial proteins, but their function is not known (149). More specifically, the total activity of a number of mitochondrial respiratory enzymes increases (153, 157), and the whole tissue develops a higher rate of oxidative metabolism measured in vitro (158). All of these changes take time to develop, and they may be produced as a result of chronic stimulation by the sympathetic nervous system or, more likely, by hormonal stimulation.

During acute cold exposure, the extra output of heat from brown adipose tissue is a result of sympathetic nervous stimulation. There are basically two mechanisms, currently being studied, by which norepinephrine and cold exposure could stimulate heat production in brown adipose tissue cells. The first is based on the fact that norepinephrine depolarizes brown adipocytes by reducing the electrical resistance of their cell membranes, and at the same time it activates a sodium potassium-activated adenosine triphosphatase enzyme which uses the energy in adenosine triphosphate to transport ions across the cell membrane and restore the resting potential. The net effect is to produce adenosine diphosphate and thus stimulate cellular metabolism. The second mechanism results from the fact that norepinephrine also activates adenylate cyclase at the cell membrane and increases the intracellular concentration of cyclic adenosine monophosphate

(159), which in turn activates hormone-sensitive lipase. This breaks down triglycerides to diglycerides and free fatty acids, the rate-limiting step in triglyceride breakdown. This provides free fatty acid substrate for oxidative metabolism. However, free fatty acids may also reduce the efficiency of oxidative phosphorylation (i.e., they act as uncouplers), so that the oxidative metabolism of the tissue is not limited by the supply of adenosine diphosphate. Mitochondria from the brown adipose tissue of guinea pigs that have been exposed to a cold environment for a few days show evidence of a loosening of the coupling between oxidation and phosphorylation, which tightens again when the animals are returned to a warm environment (155). The mechanism by which long chain, free fatty acids increase the dissipation of energy in the brown adipose tissue mitochondria during cold exposure is not clear, but there is evidence that it is related to the ability of free fatty acids to conduct hydrogen ions through the inner mitochondrial membrane into the mitochondrial matrix (160). According to the chemiosmotic theory of oxidative phosphorylation, the movement of hydrogen ions out of the mitochondrial matrix is part of the normal process by which the energy from electron transfer is used to synthesize adenosine triphosphate.

The extra substrate that is oxidized by brown adipose tissue in the cold is probably free fatty acids from the locules of triglyceride that are stored in the adipose tissue cell and mobilized by norepinephrine. It is difficult to demonstrate this because the adipose tissue triglycerides cannot be selectively labeled with radioactive isotopes. If the animal's energy intake is adjusted to meet its increased energy output in a cold environment, then increased lipogenesis should match increased lipolysis, and the stored triglycerides will not be depleted. Lipogenesis in the adipose tissue of the rat is mainly from circulating glucose, and the incorporation of injected [^{14}C]glucose into brown fat lipid is increased by chronic cold exposure. However, because it is not agreed whether the label appears mostly in the fatty acid (161) or the glycerol (162) moiety of the triglyceride, it is not possible to decide whether fatty acid synthesis or triglyceride re-esterification is stimulated or both. In the brown fat of cold-adapted hamsters, the incorporation of [1-^{14}C] acetate into lipid is increased in vitro, suggesting an accelerated fatty acid synthesis (163). If, during cold-acclimation of the rat, there is a net depletion of lipid from the brown adipose tissue, there is much more depletion of neutral lipid (mainly triglyceride) than phospholipid (161).

Cold exposure, for 1 day or for 28 days, produces a very large increase in the total lipoprotein lipase activity of brown adipose tissue, indicating that this tissue may increase its uptake of triglycerides from plasma in the cold. The increase in lipoprotein lipase activity is partially reversed by insulin if it is injected into the animal (164); therefore, it may arise in part from the fall in plasma insulin in the cold (see under "Insulin") and in part from sympathetic stimulation of brown adipose, which is known to increase its lipoprotein lipase activity.

White Adipose Tissue

White adipose tissue synthesizes and stores triglycerides, which are mobilized during starvation or cold exposure. In the first hours or days of cold exposure, the immediate increase in the animal's energy loss is not met by its gradual increase in energy intake and adipose tissue triglycerides are depleted (165, 166). The mitochondria in the cells of white adipose tissue are fewer in number and have a different structure from those in brown adipose tissue. Enzyme activity for fatty acid synthesis is slightly higher in white adipose tissue than in brown adipose tissue, but the enzyme activities for glycolysis and fatty acid oxidation are much lower in white adipose tissue. White adipose cells store more triglyceride, which is mobilized during cold exposure through the action of catecholamines on adenylate cyclase. Thus, hormone-sensitive lipase is activated in the cold (167), and the released free fatty acids appear in the blood stream and are taken up by peripheral tissues (136). The turnover of free fatty acids in the blood is increased, as is the rate of oxidation of blood free fatty acids by the body (168), producing a fall in respiratory quotient. Mobilization of free fatty acids from white adipose tissue into the circulation by catecholamines can increase fatty acid oxidation in the body, but the extra oxidation of fatty acids, presumably substitutes for other fuels because the total oxidation rate of the animal is not greatly increased. In cold exposure, the increased total metabolism, brought about by stimulation of shivering muscle or brown adipose tissue or both, and the increased fatty acid oxidation, brought about by mobilization of free fatty acids into the circulation, are coincidental.

The lipoprotein lipase activity of white adipose tissue is decreased during cold exposure, which probably diverts blood plasma triglycerides away from white adipose tissue to brown adipose tissue and other tissues. This effect of cold exposure on white adipose tissue lipoprotein lipase activity is reversed by administering insulin to the animal (164).

In some animals, the relative proportions of the different individual fatty acids is not identical in adipose triglycerides and in blood plasma free fatty acids, so when the adipose tissue fatty acids are mobilized into the blood stream they may change the relative proportions of the individual free fatty acids in the blood. Also, the different free fatty acids do not leave the circulation at the same rate; those with a shorter chain length and those with double bonds pass through membranes fastest.

It has been known for some time that animals that have been continuously exposed to a cold environment have relatively more unsaturated and less saturated fatty acids in adipose tissue triglycerides than normal. In rats, for example, after 2 or 3 days of cold exposure the percentage of oleic acid (18 carbons to 1 double bond) increases and the percentage of palmitic acid (16 carbons to 0 double bonds) falls in the subcutaneous fat. This is attributed to preferential mobilization of saturated fatty acids (166). An alternative suggestion, that the local temperature of the tissue determines the saturation or unsaturation of the fatty acids that are deposited in it, received support when it was found that in the backfat of pigs there is a steep fall in temperature with distance from the inside to the outside of the tissue and a corresponding increase in unsaturation of the fatty acids (169). However, in a similar experiment it was found that rearing pigs in a cold environment also increased the unsaturation of the fatty acids in an internal, perirenal, fat pad which is unlikely to be directly cooled by the environment. When the effect on all the fat depots was added and compared with the amount of unsaturated fat taken in the diet, it was concluded that a cold environment must favor de novo synthesis of unsaturated fatty acids in the whole body (170).

When young rats are chronically exposed to a cold environment, extra white adipose tissue cells appear; however, there is no corresponding accumulation of triglyceride, so the tissue is composed of many small cells. These smaller cells are more sensitive to the lipolytic effect of norepinephrine than are larger white fat cells from rats maintained in a neutral environment (171).

When the animal that is chronically exposed to a cold environment has increased its intake of food enough to meet its increased expenditure of energy, the mass of adipose tissue triglyceride should remain constant. Then, the rate at which white adipose tissue incorporates [1-^{14}C]palmitate into triglyceride in vitro is the same as the rate in control animals in a neutral environment (172). The rate of fatty acid biosynthesis by white adipose tissue was found to be accelerated in in vitro studies (172), but not in in vivo studies (173) on cold-acclimated rats.

Gut

Animals usually (174), but not invariably (175), increase their intake of food during prolonged exposure to a cold environment. If the environment is cold enough, growth is inhibited in spite of an increased food intake (174). Mice may choose to take their extra food as carbohydrate (176), but rats may choose extra fat (177). Different experiments have shown high fat diets to be advantageous (177) and disadvantageous (178) for the survival of rats at low environmental temperatures. If a diet has an amino acid imbalance, increasing its intake in a cold environment can result in an adequate amount of the limiting amino acid being consumed and the imbalance being overcome (179).

Gastric motility in the dog is affected by cold exposure, and gastric emptying is faster (180).

Experiments with pigs in a calorimeter have shown that the extra heat produced in the body after eating can, at least partly, substitute for the extra heat that is produced to maintain body temperature in the cold. In the same experiments, it was found that the digestibility of food energy in the gut is reduced by cold exposure (181). However, when fed sheep are exposed to a cold environment for a few hours, this increases the flow of volatile fatty acids from the gut into the portal vein (182, 183). The gut tissues normally take up volatile fatty acids for their own metabolism, but, perhaps in a cold environment, when there are more circulating free fatty acids, these may partially substitute for volatile fatty acids and allow more volatile fatty acids through into the portal vein for utilization by other tissues.

The viscera consume little or no extra oxygen when the animal is exposed to a cold environment (182, 184), but in animals that have been chronically exposed to cold the gastrointestinal tract makes up a larger proportion of the total body weight than normal (185). For this reason, its metabolism may be a greater fraction of the animal's total metabolism.

Liver

In the normal animal, the liver has a high rate of oxygen consumption for its small size; in sheep, for example, the liver is 1–2% of the body weight, but it consumes about 14% of the body's total oxygen uptake. The liver increases its oxygen consumption when the animal is acutely exposed to a cold environment, but this increase accounts for only 5% of the total body increase (182) (Table 2). The importance of the liver, therefore, is not in producing extra heat, but in converting extra nutrients to a form that is suitable for utilization by peripheral heat-producing tissues. It is not known whether the liver is important for producing extra heat in the cold-acclimated animal. When a rat acclimated to 20°C is put into an environment of 30°C, the liver's heat production falls (186); it would be interesting if this experiment were repeated with rats acclimated to a lower environmental temperature, and quantitative results were obtained for liver and total body heat production. Evidence for "uncoupling" of oxidation and phosphorylation in the liver mitochondria of cold-acclimated animals (187) has not been confirmed (188).

Lipid Metabolism During acute exposure to cold, when the concentration of free fatty acids in the blood is high, the liver takes up substantially more free fatty acids from the circulation than normal, especially if the animal is also fasted (182). Free fatty acids accumulate in the liver tissue to some extent, but triglycerides do not (134) unless the animal is fasted (189). This occurs at the time when liver glycogen is depleted (see under "Carbohydrate Metabolism"), and experiments with liver tissue in vitro show more conversion of [1-^{14}C] palmitate to $^{14}CO_2$, less conversion of [U-^{14}C]glucose to $^{14}CO_2$ (190), and a marked fall in biosynthesis of fatty acids from [1-^{14}C]acetate (191). These

Table 2. Effect of acute cold exposure on the rate of oxygen consumption by the liver in sheep

	Total oxygen consumption (ml min^{-1})	Liver oxygen consumption (ml min^{-1})	Liver blood flow (ml min^{-1})	Liver oxygen extraction (ml 100 ml^{-1})
Neutral environment	231	32.2	1,284	2.51
Cold (1° C + wind)	460	43.7	1,576	2.77

From Thompson, Gardner, and Bell (182), by courtesy of Q. J. Exp. Physiol.

changes are consistent with the fact that animals are prone to ketosis if they are fasted at the beginning of cold exposure (192).

In the liver tissue of cold-acclimated animals, fatty acids continue to be oxidized at a high rate, and glucose oxidation is partially restored to normal (190), as is fatty acid biosynthesis (191). The liver of the cold-acclimated rat contains increased amounts of carnitine and acylcarnitine (151), which suggests an increased capability to deliver free fatty acids into the inside of the mitochondria for oxidation. Cold-acclimated animals are more resistant to fasting ketosis than acutely cold-exposed animals (192).

Not all fatty acids are oxidized in the liver; some are esterified to triglycerides which are released into the blood, in association with small amounts of protein, phospholipid, and other lipids, as very low density lipoproteins. The value of converting blood free fatty acids to blood triglycerides during cold exposure is that the uptake of blood triglycerides by peripheral tissues is preferential to tissues with the most lipoprotein lipase activity, whereas the uptake of free fatty acids is not, as far as one can tell, so selective. The circulating level of very low density lipoproteins falls immediately and rapidly when rats are exposed to cold (193), showing that their utilization exceeds production. The absolute changes in utilization and production, however, have not been measured. The total utilization of very low density lipoproteins will depend upon the balance between tissues, such as brown adipose tissue and cardiac muscle (194), that have an increased lipoprotein lipase activity in the cold and tissues such as white adipose tissue that have a reduced lipoprotein lipase activity. On balance, there is probably more peripheral uptake of triglycerides, and this may be the reason why the circulating level falls. The liver's own lipoprotein lipase activity is unchanged (164).

During chronic exposure to a cold environment, the circulating level of very low density lipoproteins continues to be low (193). The rate at which they are produced in the body from [1-^{14}C]palmitate and the rate of turnover of very low density ^{14}C-lipoprotein in the body are both reported to be reduced (195), but these measurements were made on cold-acclimated rats that were anesthetized and removed from the cold for the experiment, which could have affected the results.

In the normal animal, dietary choline is necessary for phospholipid synthesis, and phospholipids are, in turn, necessary for the liver's production of very low density lipoproteins. When animals are fed a diet that contains very little choline or its precursors, fat accumulates in the liver. However, if choline-deficient animals are exposed to a cold environment, this, at least partially, prevents or cures the accumulation of liver lipid (196). There is no simple explanation for this occurrence; it may be due to the natural stimulation of fatty acid oxidation in the liver by cold exposure, which may be further enhanced by increased intake of magnesium due to the naturally increased intake of food in a cold environment (197). Alternatively, it may be due to choline precursors becoming available in the body because growth rate is reduced in the cold, but there is conflicting evidence for this idea. Some experiments have shown that the prevention of liver lipid accumulation by cold exposure does not take place if the diet is low in methionine, a precursor of choline, as well as choline (198, 199), but other experiments have shown that cold exposure does have its effect in rats that are fed a low protein (5% casein) diet (200).

Cold exposure also prevents liver lipid accumulation in rats that are fed a high fat-cholesterol diet (201), which is probably due to the increased fatty acid oxidation and reduced fatty acid synthesis in the liver in this environment (202).

Liver slices from rats that have been acutely or chronically exposed to a cold environment have a normal capacity to synthesize phospholipids. The incorporation of [1-^{14}C]acetate into phospholipid is depressed due to reduced de novo synthesis of fatty acid, but the incorporation of [1-^{14}C]glycerol, [2-^{14}C]glycine, or inorganic ^{32}P is normal (203).

Carbohydrate Metabolism The peripheral tissues metabolize extra glucose when animals are exposed to a cold environment, and the liver is important in containing a store of glycogen which can be mobilized as glucose into the blood and in converting a variety of different substrates to glucose. Gluconeogenesis during cold exposure is likely to be particularly important in ruminants and in fasted nonruminants, which derive little or no glucose from the gut, and in pregnant or lactating animals, which have additional peripheral demands for glucose.

During acute cold exposure, the liver increases its output of glucose into the circulation (204), and the concentration and turnover of glucose in the circulatory system usually increase (205). The rate at which glucose is oxidized in the body also increases, more or less in proportion to the increase in total body metabolism (205, 206). The shivering leg takes up extra glucose (see under "Muscle"); other tissues, such as brain, may not, but this has not been measured.

The glucose coming from the liver in the first minutes and hours of cold exposure probably originates from stored glycogen. Key enzymes which are believed to be rate-limiting for gluconeogenesis in the liver begin to increase their activity 24–48 hr after the beginning of cold exposure (141, 207, 208). Estimates of the rate of gluconeogenesis in the rat, by measuring the conversion of [2,3-^{14}C]succinate into [^{14}C]glucose, have shown a large increase as early as days 1–7 of cold exposure (209). The substrates for gluconeogenesis have not been clearly identified. Extra glycerol and extra propionate are taken up by the liver in the first few hours of cold exposure in the recently fed ruminant (182), and these could be converted to glucose.

In adult animals acutely exposed to a cold environment, there is no direct evidence for an accelerated Cori cycle, i.e., circulation of muscle lactate to the liver for glucose synthesis. In cattle that were exposed for a few hours to an environment cold enough to double their resting oxygen consumption (137) or for a few days to a milder environment (131), there was no consistent increase in output of lactate from the shivering leg in the normally-fed animal. In sheep that were, for a few hours, cold enough to double their oxygen consumption, there was an insignificantly small increase in hepatic uptake of lactate (204). When dogs were acutely exposed to cold, sufficient to increase their metabolism

5-fold, there was a large increase in [^{14}C]lactate turnover and conversion to $^{14}CO_2$ in the body, but no increase in circulating level of lactate, suggesting that this increased lactate production and utilization were confined to the muscles (210). It is not known whether there is an alanine-glucose exchange between muscle and liver at some stage in cold exposure. There is evidence that amino acids are substrate for gluconeogenesis early in cold exposure, but this is not simply a direct result of cold stimulation. The gradually increased intake of food at the beginning of cold exposure does not often meet the immediately increased caloric expenditure; there is also catabolism of body proteins (see under "Muscle") and deamination of their amino acids in the liver to yield glucose and keto acids (see under "Protein Metabolism").

In the chronically cold-exposed animal, the liver enzymes that control glycogenesis and glycolysis do not change their activity (211), but some of the key enzymes controlling gluconeogenesis maintain the high activity found in the acutely cold-exposed animal (207, 209, 211), even though isotope tracer measurements of the rate of gluconeogenesis show much slower rates in cold-acclimated rats than in acutely cold-exposed rats (209).

Protein Metabolism When considering the effect of acute cold exposure on hepatic enzyme activities, there are three factors to take into account: 1) Animals eat more food when they are in a cold environment and, therefore, more amino acids circulate to the liver. 2) Even though the intake of food increases, the animal's intake of calories may not match its output. 3) The concentration of insulin in the blood falls, the concentration of glucocorticoids increases, and there may be other hormonal changes also (see under "Efferent Hormones"). The first factor has been studied by comparing cold-exposed rats with rats made hyperphagic by hypothalamic lesions or with rats on a high protein diet. The second factor has been studied by comparing cold-exposed rats with rats fed a restricted caloric intake. The results of such experiments do not agree in detail, but they suggest that increases in transaminase activity, urea cycle activity, and oxidation of amino acids during cold exposure can almost all be attributed to an increased supply of amino acids and to caloric deficiency in the body, and not to an effect of cold exposure per se (208, 212–214).

During the first few days of cold exposure, when there is gluconeogenesis from protein in the body, protein is probably depleted from muscle (see under "Muscle") more than liver (141, 142). The turnover of body protein has been estimated in acutely and chronically cold-exposed mice by injecting them with [^{75}Se]selenomethionine, an amino acid analog which rapidly incorporates into protein, and then measuring the rate of decline of total body γ-radiation. An 18% increase in turnover rate was found during acute cold exposure and a 29% increase during chronic cold exposure; the latter was not prevented by restricting the intake of food (215). Subsequently, it was found that most of the [^{75}Se] selenomethionine incorporation is into liver and kidney protein. In these tissues, and not in muscle, incorporation is much faster in cold-acclimated rats. On this basis, it is suggested that amino acid-protein-amino acid recycling is accelerated in the liver and kidney and may contribute to extra heat production in the cold-acclimated animal (150). Mitochondrial proteins do not appear to take part in this (149).

Heart and Circulation

In cool, neutral environmental temperatures (i.e., when cutaneous vasoconstriction occurs, but not shivering), the total peripheral resistance to blood flow increases. At the same time, cardiac output falls and blood pressure remains constant (216), suggesting that the constriction of cutaneous blood vessels by sympathetic nerves does not increase blood pressure, because a nervous reflex involving the carotid sinus and vagus nerves secondarily produces a predominantly parasympathetic influence on the heart (see under "Heart") and reduces cardiac output. There is, probably, an increase in hepatic portal blood flow in this situation.

At environmental temperatures significantly below the "critical" environmental temperature, and when there is a substantial metabolic response, the resistance to flow through muscle (217), and probably some other tissues, falls and so the total peripheral resistance falls (216). It is not known why blood vessels dilate in shivering muscle, but a major factor is likely the production of metabolites in muscle. The extra consumption of oxygen by the body is met by increases in the extraction of oxygen from the blood and in cardiac output (217–220). The way in which the heart increases its output seems to vary between species. In man it is achieved almost entirely by an increase in stroke volume (218, 219), with very little increase in heart rate (216, 219). In cattle (217) and sheep (182) it is achieved by an increase in heart rate, and stroke volume may even fall slightly. In all species, there is an increase in blood pressure (182, 217–219) and probably a predominantly sympathetic influence on the heart (see under "Heart"). Most of the studies of the effect that exposure to a cold environment has on the flow of blood through various organs have been carried out on animals that are in this physiological state (viz., an increased cardiac output and increased blood pressure). Organs that, in the acutely cold-exposed animal, receive more blood include the muscular organs such as skeletal muscle (217, 220) (some more than others (221)), heart and diaphragm (220), white adipose tissue (221), brown adipose tissue, kidneys, adrenal glands (220), viscera, and liver (182). If blood flow is expressed as a fraction of the cardiac output, the muscular organs receive a greater fraction of the cardiac output (217, 220), and the internal organs (182, 220) and skin (220) a smaller fraction (Figure 5).

When cold-acclimated and warm-acclimated rats are compared in a neutral environment, the cold-acclimated animals have a higher cardiac output but the same (arterial-mixed venous) difference in blood oxygen content, a greater blood flow to the internal organs but less to the carcass. When cold-acclimated rats are exposed to the cold, cardiac output further increases, but the organs of the body receive practically the same fraction of the cardiac output that they receive in a neutral environment, except for the skin which receives less (220). The density of blood capillaries (mm^{-2}) in a variety of tissues is not affected by acute

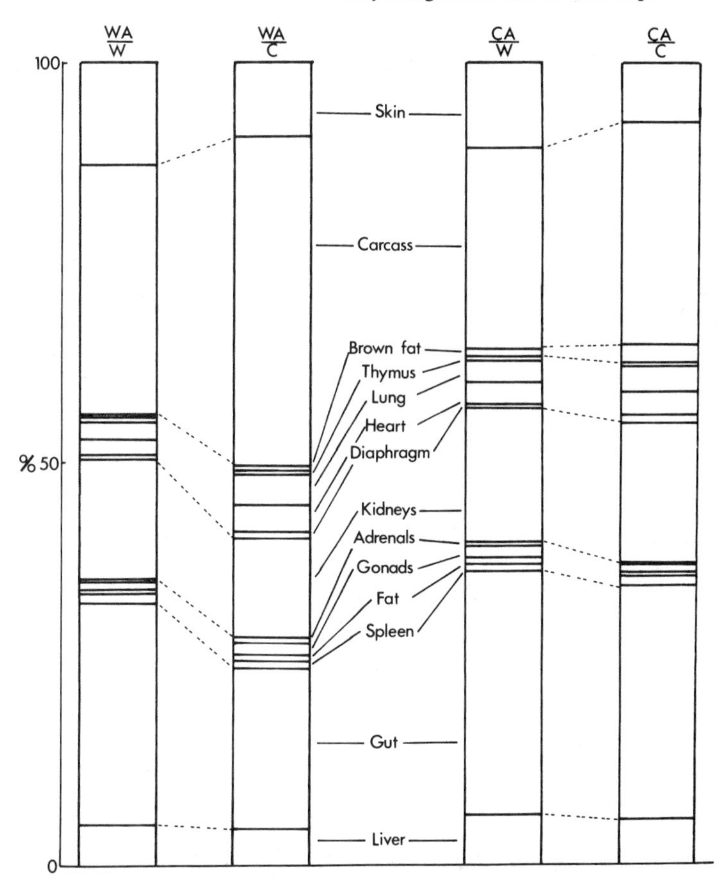

Figure 5. The fractional distribution of cardiac output in warm-acclimated (WA) and cold-acclimated (CA) rats exposed to warm (W) and cold (C) environments. (Reproduced from Jansky and Hart (220), by permission of the National Research Council of Canada from the Canadian Journal of Physiology and Pharmacology.)

exposure to a cold environment, as would be expected. Chronic cold exposure, however, increases the vascularity of ear skin and the red fiber regions of the gastrocnemius and soleus muscles but not the white fiber regions of these muscles, or the heart or liver (222).

Kidney and Water Balance

When animals are exposed to a cold environment, they become dehydrated. Their intake of water does not increase in the cold, even though their intake of food increases (223). Although the loss of water by evaporation from the body surface is usually minimal, there is a greater output of water in the urine during the first few hours (224) or days (223) of cold exposure, even if the subject is already partially dehydrated (225). Measurements of the concentration of antidiuretic hormone in blood have found reduced levels in a cold environment (see under "Antidiuretic Hormone"); in addition, the renal tubules are less sensitive to the antidiuretic action of injected hormone (106, 226), perhaps because this action is antagonized by circulating glucocorticoids. Together with the greater output of water in the urine, there is a greater loss of sodium (223–225), potassium (223, 224), chloride, calcium, phosphate (224), and nitrogen (141) in the urine and a greater output of sodium and potassium in the feces (223). The net effect of the changes of intake and output of water is a more noticeable fall in blood plasma volume that in extracellular fluid volume (227) and increases in plasma protein concentration and hematocrit (228). The net effect of the changes in intake (bearing in mind the increased intake of food) and output of electrolytes and the reduction in the body water is to leave the blood with an increased osmolality (229). The rat, for some unknown reason, does not satisfy its thirst until it is removed from the cold (229).

REFERENCES

1. Hensel, H., Andres, K. H., and Düring, M. (1974). Structure and function of cold receptors. Pfluegers Arch. 352:1.
2. Pierau, F.-K., Torrey, P., and Carpenter, D. O. (1974). Mammalian cold receptor afferents: rôle of an electrogenic sodium pump in sensory transduction. Brain Res. 73:156.
3. Iggo, A. (1969). Cutaneous thermoreceptors in primates and sub-primates. J. Physiol. (Lond.) 200:403.
4. Hellon, R. F., and Mitchell, D. (1974). Convergence in a thermal afferent pathway. J. Physiol. (Lond.) 239:61P.
5. Hellon, R. F., and Misra, N. K. (1973). Neurones in the dorsal horn of the rat responding to scrotal skin temperature change. J. Physiol. (Lond.) 232:375.
6. Hellon, R. F., and Misra, N. K. (1973). Neurones in the ventrobasal complex of the rat thalamus responding to scrotal skin temperature changes. J. Physiol. (Lond.) 232:389.
7. Eisenman, J. S., and Jackson, D. C. (1967). Thermal response patterns of septal and preoptic neurones in cats. Exp. Neurol. 19:33.
8. Simon, E., and Iriki, M. (1970). Ascending neurones of the spinal cord activated by cold. Experientia 26:620.
9. Lomax, P., Malveaux, E., and Smith, R. E. (1964). Brain temperature in rat during exposure to low environmental temperatures. Am. J. Physiol. 207:736.
10. Keatinge, W. R., and Sloan, R. E. G. (1972). Effect of swimming in cold water on body temperature of children. J. Physiol. (Lond.) 226:55P.
11. Simon, E. (1972). Temperature signals from skin and spinal cord converging on spinothalamic neurones. Pfluegers Arch. 337:323.
12. Nutik, S. (1973). Convergence of cutaneous and preoptic region thermal afferent on posterior hypothalamic neurones. J. Neurophysiol. 36:250.
13. Jacobson, F. H., and Squires, R. D. (1970). Thermoregulatory responses of the cat to preoptic and environmental temperatures. Am. J. Physiol. 218:1575.
14. Hellstrøm, B., and Hammel, H. T. (1967). Some characteristics of temperature regulation in the unanaesthetised dog. Am. J. Physiol. 213:547.

15. Jessen, C., and Ludwig, O. (1971). Spinal cord and hypothalamus as core sensors of temperature in the conscious dog. II. Addition of signals. Pfluegers Arch. 324:205.
16. Lipton, J. M., Dwyer, P. E., and Fossler, D. E. (1974). Effects of brainstem lesions on temperature regulation in hot and cold environments. Am. J. Physiol. 226:1356.
17. Guieu, J. D., and Hardy, J. D. (1971). Integrative activity of preoptic units. I. Response to local and peripheral temperature changes. J. Physiol. (Paris) 63:253.
18. Edinger, H. M., and Eisenman, J. S. (1970). Thermosensitive neurones in tuberal and posterior hypothalamus of cats. Am. J. Physiol. 219:1098.
19. Myers, R. D., and Yaksh, T. L. (1969). Control of body temperature in the unanaesthetised monkey by cholinergic and aminergic systems in the hypothalamus. J. Physiol. (Lond.) 202:483.
20. Simmonds, M. A. (1969). Effect of environmental temperature on the turnover of noradrenaline in hypothalamus and other areas of rat brain. J. Physiol. (Lond.) 203:199.
21. Beckman, A. L., and Eisenman, J. S. (1970). Microelectrophoresis of biogenic amines on hypothalamic thermosensitive cells. Science 170:334.
22. Hori, T., and Nakayama, T. (1973). Effects of biogenic amines on central thermoresponsive neurones in the rabbit. J. Physiol. (Lond.) 232:71.
23. Murakami, N. (1973). Effects of iontophoretic application of 5-hydroxytryptamine, noradrenaline and acetylcholine upon hypothalamic temperature-sensitive neurones in rats. Jpn. J. Physiol. 23:435.
24. Lin, M. T., and Chai, C. Y. (1974). Independence of spinal cord and medulla oblongata on thermal activity. Am. J. Physiol. 226:1066.
25. Kosaka, M., and Simon, E. (1968). The central nervous spinal mechanism of cold shivering. Pfluegers Arch. 302:357.
26. Klussman, F. W. (1969). Influence of temperature upon the afferent and efferent motor innervation of the spinal cord. Pfluegers Arch. 305:295.
27. Pierau, F.-K., Klee, M. R., and Klussman, F. W. (1969). Effects of local hypo- and hyperthermia on mammalian spinal motoneurones. Fed. Proc. 28:1006.
28. Glaser, E. M., and Griffin, J. P. (1962). Influence of the cerebral cortex on habituation. J. Physiol. (Lond.) 160:429.
29. Fuller, C. A., Horowitz, J. M., and Horwitz, B. (1974). Central regulation of temperature in cold environments. II. A dynamic model with three temperature inputs. Comp. Biomed. Res. 7:164.
30. Brück, K., and Wünnenberg, B. (1965). Blockade of chemical thermogenesis and the elicitation of muscle shivering in newborn guinea pigs. Pfluegers Arch. 282:376.
31. Brück, K., and Wünnenberg, W. (1966). Relationship between thermogenesis in brown fat tissue, temperature in the cervical portion of the vertebral sinus and shivering. Pfluegers Arch. 290:167.
32. Wünnenberg, W., and Brück, K. (1968). Single unit activity evoked by thermal stimulation of the cervical spinal cord in the guinea-pig. Nature (Lond.) 218:1268.
33. Wünnenberg, W., and Brück, K. (1968). Function of thermoreceptive structures in the cervical spinal cord of the guinea-pig. Pfluegers Arch. 299:1.
34. Hemingway, A., Forgrave, P., and Birzis, L. (1954). Shivering suppression by hypothalamic stimulation. J. Neurophysiol. 17:375.

35. Birzis, L., and Hemingway, A. (1957). Shivering as a result of brain stem stimulation. J. Neurophysiol. 20:91.
36. Birzis, L., and Hemingway, A. (1957). Efferent brain discharge during shivering. J. Neurophysiol. 20:156.
37. Perkins, J. F. (1945). The rôle of the proprioceptors in shivering. Am. J. Physiol. 145:264.
38. Dworkin, S. (1930). Observations on the central nervous control of shivering and heat regulation in the rabbit. Am. J. Physiol. 93:227.
39. Stuart, D. G., Oh, K., Ishikawa, K., and Eldred, E. (1966). The rhythm of shivering. I. General sensory contribution. Am. J. Phys. Med. 45:61.
40. Schafer, S. S., and Schafer, S. (1973). The rôle of the primary afference in the generation of a cold shivering tremor. Exp. Brain Res. 17:381.
41. Schafer, S. S., and Schafer, S. (1973). The behaviour of the proprioceptor of the muscle and the innervation of the fusimotor system during cold shivering. Exp. Brain Res. 17:364.
42. Pierau, K.-F., and Spaan, G. (1970). Renshaw inhibition during local spinal cord cooling and warming. Experientia 26:978.
43. Ishii, K., and Ishii, K. (1960). Carotid sinus reflex acting upon shivering. Tohoku J. Exp. Med. 72:229.
44. Mott, J. C. (1963). The effects of baroreceptor and chemoreceptor stimulation on shivering. J. Physiol. (Lond.) 166:563.
45. Burton, A. C., and Bronk, D. (1937). The motor mechanism of shivering and thermal muscular tone. Am. J. Physiol. 119:284.
46. Keller, A. D., and Hare, W. K. (1932). The hypothalamus and heat regulation. Proc. Soc. Exp. Biol. Med. 29:1069.
47. Hemingway, A., and Lillehei, C. W. (1950). Thermal cutaneous vasomotor response in dogs. Am. J. Physiol. 162:301.
48. Walther, O.-E., Iriki, M., and Simon, E. (1970). Antagonistic changes of blood flow and sympathetic activity in different vascular beds following central thermal stimulation. II. Cutaneous and visceral sympathetic activity during spinal cord heating and cooling in anaesthetised rabbits and cats. Pfluegers Arch. 319:162.
49. Webb-Peploe, M. M., and Shepherd, J. T. (1968). Response of the superficial limb veins of the dog to changes in temperature. Circ. Res. 22:737.
50. Martin, R. P., and Wallace, W. F. M. (1970). The effect of temperature on the response of the perfused rabbit ear artery to electrical stimulation. J. Physiol. (Lond.) 207:80P.
51. Webb-Peploe, M. M., and Shepherd, J. T. (1968). Peripheral mechanism involved in response of dogs cutaneous veins to local temperature change. Circ. Res. 23:701.
52. Keatinge, W. R. (1958). The effect of low temperatures on the responses of arteries to constrictor drugs. J. Physiol. (Lond.) 142:395.
53. Leduc, J. (1961). Catecholamine production and release in exposure and acclimation to cold. Acta Physiol. Scand., Suppl. 53:183.
54. Nagasaka, T., and Carlson, L. D. (1971). Effects of blood temperature and perfused norepinephrine on vasomotor responses of rabbit ear. Am. J. Physiol. 220:289.
55. Harada, E., and Kanno, T. (1975). Rabbit's ear in cold acclimation: studies on the change in ear temperature. J. Appl. Physiol. 38:389.
56. Simon, E. (1969). Circulatory effects of spinal cord hypothermia. J. Neurovisc. Relat. 31:223.
57. Bralet, J., Beley, A., and Lallemant, A. M. (1972). Alterations in nor-

epinephrine turnover in various peripheral organs of the rat during exposure and acclimation to cold. Pfluegers Arch. 335:186.

58. Cottle, W. H., Nash, C. W., Veress, A. T., and Ferguson, B. A. (1967). Release of noradrenaline from brown fat of cold-acclimated rats. Life Sci. 6:2267.
59. Thoenen, H. (1970). Induction of tyrosine hydroxylase in peripheral and central adrenergic neurones by cold exposure of rats. Nature 228:861.
60. Tirri, R., Harri, M. N. E., and Laitinen, L. (1974). Lowered chronotropic sensitivity of rat and frog hearts to sympathomimetic amines following cold acclimation. Acta Physiol. Scand. 90:260.
61. Himms-Hagen, J., and Mazurkiewicz-Kwilecki, I. M. (1970). Unchanged sensitivity to noradrenaline in isolated tissues from cold-acclimated rats. Can. J. Physiol. Pharmacol. 48:657.
62. Le Blanc, J., and Côté, J. (1967). Increased vagal activity in cold-adapted animals. Can. J. Physiol. Pharmacol. 45:745.
63. Harri, M. N. E., and Tirri, R. (1974). Lowered sensitivity to acetylcholine in hearts from cold-acclimated rats and frogs. Acta Physiol. Scand. 90:509.
64. Walker, E. A. (1940). The hypothalamus and pilomotor regulation. Proc. Assoc. Res. Nerv. Ment. Dis. 20:400.
65. Thompson, G. E., Robertshaw, D., and Findlay, J. D. (1969). Noradrenergic innervation of the arrectores pilorum muscles of the ox. Can. J. Physiol. Pharmacol. 47:310.
66. Brück, K., and Schwennicke, H. P. (1971). Interaction of superficial and hypothalamic thermosensitive structures in the control of non-shivering thermogenesis. Int. J. Biometeorol. 15:156.
67. Calvert, D. T., Clough, D. P., Findlay, J. D., and Thompson, G. E. (1972). Hypothalamic cooling, heat production and plasma lipids in the ox. Life Sci. 11:223.
68. Darling, K. F., Findlay, J. D., and Thompson, G. E. (1974). Effect of intraventricular acetylcholine and eserine on the metabolism of sheep. Pfluegers Arch. 349:235.
69. Thompson, G. E., and Clough, D. P. (1971). The effect of intraventricular noradrenaline on plasma unesterified fatty acid concentration in the ox. Life Sci. 10:415.
70. Slavin, B. G., and Bernick, S. (1974). Morphological studies on denervated brown adipose tissue. Anat. Rec. 179:497.
71. Gilgen, A., Maickel, R. P., Nikodi-Jevik, O., and Brodie, B. B. (1962). Essential rôle of catecholamines in the mobilisation of free fatty acids and glucose after exposure to cold. Life Sci. 1:709.
72. Iriki, M., Riedel, W., and Simon, E. (1971). Regional differentiation of sympathetic activity during hypothalamic heating and cooling in anaesthetised rabbits. Pfluegers Arch. 328:320.
73. Guidotti, A., Zivkovic, B., Pfeiffer, R., and Costa, E. (1973). Involvement of 3′5-cyclic adenosine monophosphate in the increase of tyrosine hydroxylase activity elicited by cold exposure. Naunyn Schmiedebergs Arch. Exp. Pathol. Pharmacol. 278:195.
74. Jarratt, A. M., and Nowell, N. W. (1969). The effect of adrenal demedullation on the blood sugar level of rats subjected to long-term cold stress. Can. J. Physiol. Pharmacol. 47:1.
75. Bost, J., and Dorleac, E. (1965). Effect of cold on plasma free fatty acids in sheep. C. R. Soc. Biol. 159:2209.

76. Kvetnanský, R., Gewirtz, G. P., Weise, V. K., and Kopin, I. J. (1971). Catecholamine-synthesizing enzymes in the rat adrenal gland during exposure to cold. Am. J. Physiol. 220:928.
77. Sellers, E. A., Flattery, K. V., and Steiner, G. (1974). Cold acclimation of hypothyroid rats. Am. J. Physiol. 226:290.
78. Bakke, J. L., and Lawrence, N. L. (1971). Effects of cold-adaptation, rewarming and heat exposure on thyrotropin (TSH) secretion in rats. Endocrinology 89:204.
79. Koch, B., Jobin, M., and Fortier, C. (1966). TSH and thyroxine secretion rates following prolonged exposure to cold in the rat. Fed. Proc. 25:516.
80. Eastman, C. J., Ekins, R. P., Leith, I. M., and Williams, E. S. (1974). Thyroid hormone response to prolonged cold exposure in man. J. Physiol. (Lond.) 241:175.
81. Bernal, J., and Del Rey, F. (1975). T_3/T_4 ratios and α-glycerophosphate activity in intact rats exposed to a cold environment. Horm. Metab. Res. 7:222.
82. Magwood, S. G. A., and Héroux, O. (1968). Fecal excretion of thyroxine in warm- and cold-acclimated rats. Can. J. Physiol. Pharmacol. 46:601.
83. Intoccia, A., and Van Middlesworth, L. (1959). Thyroxine excretion increase by cold exposure. Endocrinology 64:462.
84. Straw, J. (1969). Effects of fecal weight on thyroid function in cold-exposed rats. J. Appl. Physiol. 27:630.
85. Héroux, O., and Petrovic, V. M. (1969). Effect of high- and low-bulk diets on the thyroxine turnover rate in rats with acute and chronic exposure to different temperatures. Can. J. Physiol. Pharmacol. 47:963.
86. Balsam, A. (1974). Augmentation of the peripheral metabolism of L-triiodothyronine and L-thyroxine after acclimation to cold: multifocal stimulation of the binding of iodothyronines by tissues. J. Clin. Invest. 53:980.
87. Cottle, W. H., and Veress, A. T. (1966). Serum binding and biliary clearance of triiodothyronine in cold-acclimated rats. Can. J. Physiol. Pharmacol. 44:571.
88. Harland, W. A., Orr, J. S., Dunnigan, M. G., and Fyfe, T. (1971). Cardiac effects of cold-induced hyperthyroidism and the seasonal variation in incidence of myocardial infarction. Br. J. Exp. Pathol. 52:147.
89. Balsam, A., and Leppo, L. E. (1974). Enhanced binding of 3,5,3′-L-triiodothyronine by rat kidney microsomes after acclimation to cold. Proc. Soc. Exp. Biol. Med. 145:1401.
90. Galton, V. A., and Nisula, B. C. (1969). Thyroxine metabolism and thyroid function in the cold-adapted rat. Endocrinology 85:79.
91. Cottle, W. H. (1964). Biliary and fecal clearance of endogenous thyroid hormone in cold-acclimated rats. Am. J. Physiol. 207:1063.
92. Galton, V. A., and Nisula, B. C. (1972). The enterohepatic circulation of thyroxine. J. Endocrinol. 54:187.
93. Cadot, M., Julien, M.-F., and Chevillard, L. (1969). Estimation of thyroid function in rats exposed or adapted to environments at 5 or 30°C. Fed. Proc. 28:1228.
94. Boulouard, R. (1963). Effects of cold and starvation on adrenocortical activity of rats. Fed. Proc. 22:750.
95. Panaretto, B. A., and Ferguson, K. A. (1969). Pituitary adrenal interactions in shorn sheep exposed to cold wet conditions. Aust. J. Agric. Res. 20:99.

96. Panaretto, B. A., and Vickery, M. R. (1970). The rates of plasma cortisol entry and clearance in sheep before and during their exposure to a cold wet environment. J. Endocrinol. 47:273.
97. Panaretto, B. A. (1974). Relationship of visceral blood flow to cortisol metabolism in cold-stressed sheep. J. Endocrinol. 60:235.
98. Chowers, I., Conforti, N., and Feldman, S. (1972). Body temperature and adrenal function in cold-exposed hypothalamic-disconnected rats. Am. J. Physiol. 223:341.
99. Bassett, J. M., personal communication.
100. Beck, L. V., Zaharko, D. S., Kalser, S. C., Miller, R., Van Gemert, M., and Kunig, R. (1967). Variation in serum insulin and glucose of rats with chronic cold exposure. Life Sci. 6:1501.
101. Okada, Y., Miyai, K., Iwatsubo, H., and Kumahara, Y. (1970). Human growth hormone secretion in normal adult subjects during and after exposure to cold. J. Clin. Endocrinol. Metab. 30:393.
102. Golstein-Golaire, J., Vanhaelst, L., Bruno, O. D., Leclercq, R., and Copinschi, G. (1970). Acute effects of cold on blood levels of growth hormone, cortisol and thyrotropin in man. J. Appl. Physiol. 29:622.
103. Mueller, G. P., Chen, H. T., Dibbet, J. A., Chen, H. J., and Meites, J. (1974). Effects of warm and cold temperatures on release of TSH, GH and prolactin in rats. Proc. Soc. Exp. Biol. Med. 147:698.
104. McIntyre, H. B., and Odell, W. D. (1974). Physiological control of growth hormone in the rabbit. Neuroendocrinology 16:8.
105. Wettemann, R. P., and Tucker, H. A. (1974). Relationship of ambient temperature to serum prolactin in heifers. Proc. Soc. Exp. Biol. Med. 146:908.
106. Itoh, S. (1954). The release of antidiuretic hormone from the posterior pituitary body on exposure to heat. Jpn. J. Physiol. 4:185.
107. Segar, W. E., and Moore, W. W. (1968). The regulation of antidiuretic hormone release in man. I. Effects of change in position and ambient temperature on blood ADH levels. J. Clin. Invest. 47:2143.
108. Sadowski, J., Nazar, K., and Szczepańska-Sadowska, E. (1972). Reduced urine concentration in dogs exposed to cold: relation to plasma ADH and 17-OHCS. Am. J. Physiol. 222:607.
109. Sadowski, J., Kurkus, J., and Chwalbińska-Moneta, J. (1975). Plasma hormone and renal function changes in unrestrained dogs exposed to cold. Am. J. Physiol. 228:376.
110. Forster, R. E., Ferris, B. G., and Day, R. (1946). The relationship between total heat exchange and blood flow in the hand at various ambient temperatures. Am. J. Physiol. 146:600.
111. Burton, A. C., and Bazett, H. C. (1936). A study of the average temperature of the tissues, of the exchanges of heat and vasomotor responses in man by means of a bath calorimeter. Am. J. Physiol. 117:36.
112. Froese, G., and Burton, A. C. (1957). Heat losses from the human head. J. Appl. Physiol. 10:235.
113. Haddy, F. J., Fleishman, M., and Scott, J. B. (1957). Effect of change in air temperature upon systemic small and large vessel resistance. Circ. Res. 5:58.
114. Wood, J. E., and Eckstein, J. W. (1958). A tandem forearm plethysmograph for study of acute responses of the peripheral veins of man: the effect of environmental and local temperature change and the effect of pooling blood in the extremities. J. Clin. Invest. 37:41.

115. Zanick, D. C., and Delaney, J. P. (1973). Temperature influences on arteriovenous anastomoses. Proc. Soc. Exp. Biol. Med. 144:616.
116. Burton, A. C., and Taylor, R. M. (1940). A study of the adjustment of peripheral vascular tone to the requirements of the regulation of body temperature. Am. J. Physiol. 129:565.
117. Bazett, H. C., Love, L., Newton, M., Eisenberg, L., Day, R., and Forster, R. (1948–9). Temperature changes in blood flowing in arteries and veins in man. J. Appl. Physiol. 1:3.
118. Blaxter, K. L., and Wainman, F. W. (1961). Environmental temperature and the energy metabolism and heat emission of steers. J. Agric. Sci. Camb. 56:81.
119. Berry, I. L., and Shanklin, M. D. (1961). Physical factors affecting thermal insulation of livestock hair coats. Research Bulletin of the Missouri Agricultural Experiment Station, No. 802.
120. Gonzalez-Jimenez, E., and Blaxter, K. L. (1962). The metabolism and thermal regulation of calves in the first month of life. Br. J. Nutr. 16:199.
121. Cena, K., and Monteith, J. L. (1975). Transfer processes in animal coats. II. Conduction and convection. Proc. Roy. Soc. Lond. (Biol.) 188:395.
122. Cena, K., and Monteith, J. L. (1975). Transfer processes in animal coats. I. Radiative transfer. Proc. Roy. Soc. Lond. (Biol.) 188:377.
123. Ames, D. R., and Insley, L. W. (1975). Wind-chill effect for cattle and sheep. J. Anim. Sic. 40:161.
124. Benzinger, T. H., and Kitzinger, C. (1963). Gradient layer calorimetry and human calorimetry. *In* Temperature: Its Measurement and Control in Science and Industry, Vol. 3, Part 3, pp. 87–109. J. D. Hardy (ed.), Reinhold Publishing Corporation, New York.
125. Mount, L. E. (1960). The influence of huddling and body size on the metabolic rate of the young pig. J. Agric. Sci. Camb. 55:101.
126. Mount, L. E. (1964). Radiant and convective heat loss from the new born pig. J. Physiol. (Lond.) 173:96.
127. Le Blanc, J., Hildes, J. A., and Héroux, O. (1960). Tolerance of Gaspé fisherman to cold water. J. Appl. Physiol. 15:1031.
128. Webster, A. J. F., Hicks, A. M., and Hays, F.L. (1969). Cold climate and cold temperature induced changes in the heat production and thermal insulation of sheep. Can. J. Physiol. Pharmacol. 47:553.
129. Dowling, D. F., and Nay, T. (1960). Cyclic changes in the follicles and hair coat in cattle. Aust. J. Agric. Res. 11:1064.
130. Webster, A. J. F., Chlumecky, J., and Young, B. A. (1970). Effects of cold environments on the energy exchanges of young beef cattle. Can. J. Anim. Sci. 50:89.
131. Bell, A. W., Clarke, P. L., and Thompson, G. E. (1975). Changes in the metabolism of the shivering hind leg of the young ox during several days of continuous cold exposure. Q. J. Exp. Physiol. 60:267.
132. Bell, A. W., Thompson, G. E., and Findlay, J. D. (1974). The contribution of the shivering hind leg to the metabolic response to cold of the young ox. Pfluegers Arch. 346:341.
133. Bennett, J. W. (1973). The maximum metabolic response of sheep to cold: effects of rectal temperature, shearing, feed consumption, body posture and body weights. Aust. J. Agric. Res. 23:1045.
134. Therriault, D. G., and Poe, R. H. (1965). The effects of acute and chronic cold exposure on tissue lipids in the rat. Can. J. Physiol. Pharmacol. 43:1427.

135. Klain, G. J., and Hannon, J. P. (1969). Gluconeogenesis in cold-exposed rats. Fed. Proc. 28:965.
136. Bell, A. W., and Thompson, G. E. (1974). Effects of cold exposure and feeding on net exchange of plasma free fatty acids and glycerol across the hind leg of the young ox. Res. Vet. Sci. 17:265.
137. Bell, A. W., Gardner, J. W., Manson, W., and Thompson, G. E. (1975). Acute cold exposure and the metabolism of blood glucose, lactate and pyruvate and plasma amino acids in the hind leg of the fed and fasted young ox. Br. J. Nutr. 33:207.
138. Bell, A. W., Gardner, J. W., and Thompson, G. E. (1974). The effects of acute cold exposure and feeding on volatile fatty acid metabolism in the hind leg of the young ox. Br. J. Nutr. 32:471.
139. Abe, K., and Yoshimura, K. (1972). Changes in lipoprotein lipase activity in tissues of rat exposed to cold of various durations. J. Physiol. Soc. Jpn. 34:81.
140. Williams, J. N., Schurr, P. E., and Elvehjan, C. A. (1950). The influence of chilling and exercise on free amino acid concentrations in rat tissues. J. Biol. Chem. 182:55.
141. Beaton, J. (1963). Nitrogen metabolism in cold-exposed rats. Can. J. Biochem. Physiol. 41:1169.
142. Beaton, J. (1963). The relation of dietary protein level to liver enzyme activities in cold-exposed rats. Can. J. Biochem. Physiol. 41:1865.
143. Muirhead, M., and Himms-Hagen, J. (1974). Changes in the adenyl cyclase system of skeletal muscle of cold-acclimated rats. Can. J. Biochem. 52:176.
144. Héroux, O. (1967). Metabolic adjustments to low temperatures in New Zealand white rabbits. Can. J. Physiol. Pharmacol. 45:451.
145. Janský, L., and Hart, J. S. (1963). Participation of skeletal muscle and kidney during non-shivering thermogenesis in cold-acclimated rats. Can. J. Biochem. Physiol. 41:953.
146. Davis, T. R. A. (1967). Contribution of skeletal muscle to non-shivering thermogenesis in the dog. Am. J. Physiol. 213:1423.
147. Sherebrin, M. H., and Burton, A. C. (1966). Changes in the resting potential of skeletal muscle in rats with cold acclimation. Can. J. Physiol. Pharmacol. 44:791.
148. Depocas, F. (1966). Concentration and turnover of cytochrome c in skeletal muscles of warm- and cold-acclimated rats. Can. J. Physiol. Pharmacol. 44:875.
149. Bukowiecki, L., and Himms-Hagen, J. (1971). Decreased half life of some mitochondrial proteins in skeletal muscle and brown adipose tissue of cold-acclimated rats. Can. J. Physiol. Pharmacol. 49:1015.
150. Yousef, M. K., and Chaffee, R. R. J. (1970). Studies on protein-turnover rates in cold-acclimated rats. Proc. Soc. Exp. Biol. Med. 133:801.
151. Delisle, G., and Radomski, M. W. (1968). The rôle of carnitine in the animal exposed to cold. Can. J. Physiol. Pharmacol. 46:71.
152. Imai, Y., Horwitz, B. A., and Smith, R. E. (1968). Calorigenesis of brown adipose tissue in cold-exposed rats. Proc. Soc. Exp. Biol. Med. 127:717.
153. Thomson, J. F., Habeck, D. A., Nance, S. L., and Beetham, K. L. (1969). Ultrastructural and biochemical changes in brown fat in cold-exposed rats. J. Cell Biol. 41:312.
154. Hahn, P., and Skala, J. (1972). Changes in interscapular brown adipose tissue of the rat during perinatal and early postnatal development and after cold acclimation. Comp. Biochem. Physiol. 41:147.

155. Andersen, H. T., Christiansen, E. N., Grav, H. J., and Pedersen, J. I. (1970). Thermogenesis of brown adipose tissue effected in mitochondrial respiratory control. Acta Physiol. Scand. 80:1.
156. Skala, J., Barnard, T., and Lindberg, O. (1970). Changes in interscapular brown adipose tissue of the rat during perinatal and early post natal development and after cold acclimation. II. Mitochondrial changes. Comp. Biochem. Physiol. 33:509.
157. Barnard, T., Skala, J., and Lindberg, O. (1970). Changes in interscapular brown adipose tissue of the rat during perinatal and early post natal development and after cold acclimation. I. Activities of some respiratory enzymes. Comp. Biochem. Physiol. 33:499.
158. Smith, R. E., and Roberts, J. C. (1964). Thermogenesis of brown adipose tissue in cold-acclimated rats. Am. J. Physiol. 206:143.
159. Skala, J., Novak, E., Hahn, P., and Drummond, G. I. (1972). Changes in interscapular brown adipose tissue of the rat during perinatal and early postnatal development and after cold acclimation. V. Adenyl cyclase, cyclic AMP, protein kinase, phosphorylase, phosphorylase kinase and glycogen. Int. J. Biochem. 3:229.
160. Flatmark, T., and Pedersen, J. I. (1975). Brown adipose tissue mitochondria. Biochim. Biophys. Acta 416:53.
161. Steiner, G., Schönbaum, E., Johnson, G. E., and Sellers, E. A. (1968). Lipid metabolism: effects of immunosympathectomy and acclimation to cold. Can. J. Physiol. Pharmacol. 46:453.
162. Himms-Hagen, J. (1965). Lipid metabolism in warm-acclimated and cold-acclimated rats exposed to cold. Can. J. Physiol. Pharmacol. 43:379.
163. Baumber, J., and Denys, A. (1964). Acetate-I-^{14}C utilisation by brown fat from hamsters in cold exposure and hibernation. Can. J. Biochem. 42:1397.
164. Radomski, M. W., and Orme, T. (1971). Response of lipoprotein lipase in various tissues to cold exposure. Am. J. Physiol. 220:1852.
165. Masoro, E. J., Asuncion, C. L., Brown, R. K., and Rapport, D. (1957). Lipogenesis from carbohydrate in the negative caloric balance state induced by exposure to cold. Am. J. Physiol. 190:177.
166. Kodama, A. M., and Pace, N. (1964). Effect of environmental temperature on hamster body fat composition. J. Appl. Physiol. 19:863.
167. Maickel, R. P., Sussman, H., Yamada, K., and Brodie, B. B. (1963). Control of adipose tissue lipase activity by the sympathetic nervous system. Life Sci. 2:210.
168. Masironi, R., and Depocas, F. (1961). Effect of cold exposure on respiratory $^{14}CO_2$ production during infusion of albumin-bound palmitate-I-^{14}C in white rats. Can. J. Biochem. 39:219.
169. McGrath, W. S., van der Noot, G. W., Gilbreath, R. L., and Fisher, H. (1968). Influence of environmental temperature and dietary fat on backfat composition of swine. J. Nutr. 96:461.
170. Fuller, M. F., Duncan, W. R. H., and Boyne, A. W. (1974). Effect of environmental temperature on the degree of unsaturation of depot fats of pigs given different amounts of food. J. Sci. Food Agric. 25:205.
171. Therriault, D. G., and Mellin, D. B. (1971). Cellularity of adipose tissue in cold-exposed rats and the calorigenic effect of norepinephrine. Lipids 6:486.
172. Patkin, J. K., and Masoro, E. J. (1961). Effects of cold acclimation on lipid metabolism in adipose tissue. Am. J. Physiol. 200:847.
173. Patkin, J. K., and Masoro, E. J. (1964). Fatty acid synthesis in normal and cold-acclimated rats. Can. J. Physiol. Pharmacol. 42:101.

174. Sugahara, M., Baker, D. H., Harmon, B. G., and Jensen, A. H. (1969). Effect of ambient temperature and dietary amino acids on carcass fat deposition in rats. J. Nutr. 98:344.
175. Milligan, J. D., and Christison, G. I. (1974). Effects of severe winter conditions on performance of feedlot steers. Can. J. Anim. Sci. 54:605.
176. Donhoffer, S. Z., and Vonotzky, J. (1947). The effect of environmental temperature on food selection. Am. J. Physiol. 150:329.
177. Dugal, L. P., Leblond, C. P., and Thérien, M. (1945). Resistance to extreme temperatures in connection with different diets. Can. J. Res. E. 23:244.
178. Lang, K., and Grab, W. (1946). Resistance to cold and nutrition. III. Effect of feeding fat on the resistance to cold. Klin. Wochenochr. 24:37.
179. Anderson, H. L., Benevenga, N. J., and Harper, A. E. (1969). Effect of cold exposure on the response of rats to a dietary amino acid imbalance. J. Nutr. 99:184.
180. Sleeth, C. K., and Van Liere, E. J. (1937). The effect of environmental temperature on the emptying time of the stomach. Am. J. Physiol. 118:272.
181. Fuller, M. F., and Boyne, A. W. (1972). The effects of environmental temperature on the growth and metabolism of pigs given different amounts of food. Br. J. Nutr. 28:373.
182. Thompson, G. E., Gardner, J. W., and Bell, A. W. (1975). The oxygen consumption, fatty acid and glycerol uptake of the liver in fed and fasted sheep during cold exposure. Q. J. Exp. Physiol. 60:107.
183. Thompson, G. E., Bassett, J. M., and Bell, A. W., unpublished observations.
184. Janský, L., Dvořak, R., and Zeisberger, E. (1969). Metabolic significance of splanchnic organs for non-shivering thermogenesis. Physiol. Bohemoslov. 18:242.
185. Barnett, S. A., and Widdowson, E. M. (1965). Organ-weights and body-composition in mice bred for many generations at −3°C. Proc. Roy. Soc. Lond. (Biol.) 162:502.
186. Stoner, H. B. (1973). The rôle of the liver in non-shivering thermogenesis in the rat. J. Physiol. (Lond.) 232:285.
187. Panagos, S., Beyer, R. E., and Masoro, E. J. (1958). Oxidative phosphorylation in liver mitochondria prepared from cold-exposed rats. Biochim. Biophys. Acta 29:204.
188. Chaffee, R. R. J., Kaufman, W. C., Kratochvil, C. H., Sorenson, M. W., Conaway, C. H., and Middleton, C. C. (1969). Comparative chemical thermoregulation in cold- and heat-acclimated rodents, insectivores, proto-primates and primates. Fed. Proc. 28:1029.
189. Masoro, E. J. (1960). Alterations in hepatic lipid metabolism induced by acclimation to low environmental temperatures. Fed. Proc., Suppl. 5, 19:115.
190. Baumber, J., and Denyes, A. (1965). Oxidation of glucose-U-^{14}C and palmitate-I-^{14}C by liver kidney and diaphragm from hamsters in cold exposure and hibernation. Can. J. Biochem. 43:747.
191. Masoro, E. J., Cohen, I. A., and Panagos, S. S. (1954). Effect of exposure to cold on some aspects of hepatic acetate utilisation. Am. J. Physiol. 179:451.
192. Scott, S. L., and Engel, F. L. (1953). The influence of the adrenal cortex and cold stress on fasting ketosis in the rat. Endocrinology 53:410.
193. Radomski, M. W. (1966). Short and long term effect of cold exposure on serum lipids and lipoproteins in the rat. Can. J. Physiol. Pharmacol. 44:711.

194. Rogers, M. P., and Robinson, D. S. (1974). Effects of cold exposure on heart clearing factor lipase and triglyceride utilization in the rat. J. Lipid Res. 15:263.
195. McBurney, L. J., and Radomski, M. W. (1969). Metabolism of serum free fatty acid and low-density lipoproteins in the cold acclimated rat. Am. J. Physiol. 217:19.
196. Sellers, E. A., and You, R. W. (1952). Effects of cold environment on deposition of fat in liver in choline deficiency. Biochem. J. 51:573.
197. Platner, W. S., and Shields, J. L. (1969). Pseudolipotropic effect of cold exposure and dietary magnesium on choline-deficient rats. Fed. Proc. 28:978.
198. Radomski, M. W., and Wood, J. D. (1964). The lipotropic action of cold. I. The influence of cold and choline deficiency on liver lipids of rats at different intakes of dietary methionine. Can. J. Physiol. Pharmacol. 42:769.
199. Chahl, J. S., and Kratzing, C. C. (1966). Environmental temperature and choline requirement in rats. II. Choline and methionine requirements for lipotropic activity. J. Lipid Res. 7:22.
200. Treadwell, C. R., Flick, D. F., and Vahouny, G. V. (1958). Nutrition studies in the cold. II. Curative effect of cold on fat fatty livers. Proc. Soc. Exp. Biol. Med. 97:434.
201. Treadwell, C. R., Flick, D. F., and Vahouny, G. V. (1957). Nutrition studies in the cold. I. Influence of diet and low environmental temperature on growth and on the lipid content of livers in the rat. J. Nutr. 63:611.
202. Bobek, P., and Ginter, E. (1966). Lipotropic effect of cold on rat liver. Nature 210:204.
203. Kline, D., McPherson, C., Pritchard, E. T., and Rossiter, R. J. (1956). Effect of exposure to a cold environment on labeling of phospholipids in rat liver slices. Proc. Soc. Exp. Biol. Med. 92:756.
204. Thompson, G. E., Manson, W., Clarke, P. L., and Bell, A. W., unpublished observations.
205. Depocas, F., and Masironi, R. (1960). Body glucose as fuel for thermogenesis in the white rat exposed to cold. Am. J. Physiol. 199:1051.
206. Minaire, Y., Vincent-Falquet, J.-C., Pernod, A., and Chatonnet, J. (1973). Energy supply in acute cold-exposed dogs. J. Appl. Physiol. 35:51.
207. Klain, G. J., and Hannon, J. P. (1969). Gluconeogenesis in cold-exposed rats. Fed. Proc. 28:965.
208. Whitten, B. K., Burlington, R. F., Posiviata, M. A., Sidel, C. M., and Beecher, G. R. (1970). Amino acid catabolism in environmental extremes: effect of temperature and calories. Am. J. Physiol. 219:1046.
209. Penner, P. E., and Himms-Hagen, J. (1968). Gluconeogenesis in rats during cold acclimation. Can. J. Biochem. 46:1205.
210. Minaire, Y., Pernod, A., Jomain, M. J., and Mottaz, M. (1971). Lactate turnover and oxidation in normal and adrenal demedullated dogs during cold exposure. Can. J. Physiol. Pharmacol. 49:1063.
211. Morrison, G. R., Brock, F. E., Sobral, D. T., and Shank, R. E. (1966). Cold-acclimatization and intermediary metabolism of carbohydrates. Arch. Biochem. Biophys. 114:494.
212. Klain, G. J., and Vaughan, D. A. (1963). Alterations of protein metabolism during cold acclimation. Fed. Proc. 22:862.
213. Klain, G. J., Vaughan, D. A., and Vaughan, L. N. (1969). Effect of protein intake and cold exposure on selected liver enzymes associated with amino acid metabolism. J. Nutr. 80:107.

214. Beaton, J. R., and Feleki, V. (1964). Tissue enzyme response to cold and to hyperphagia in the rat. Can. J. Physiol. Pharmacol. 42:787.
215. Yousef, M. K., and Luick, J. R. (1969). ^{75}Se-selenomethionine turnover as an index of protein metabolism during cold exposure of white mice. Can. J. Physiol. Pharmacol. 47:273.
216. Wezler, K., and Thauer, R. (1943). Der Kreislauf im Dienste der Wärmeregulation. Z. Gesamte Exp. Med. 112:354.
217. Bell, A. W., and Thompson, G. E. (1974). The effect of acute cold exposure and feeding on the circulation of the young ox (*Bos taurus*) with special reference to the hind leg. Res. Vet. Sci. 17:384.
218. Raven, P. B., Niki, T., Dahms, T. E., and Horvath, S. M. (1970). Compensatory cardiovascular responses during environmental cold stress 5°C. J. Appl. Physiol. 29:417.
219. Raven, P. B., Wilkerson, J. E., Horvath, S. M., and Boldman, N. W. (1975). Thermal metabolic and cardiovascular responses to various degrees of cold stress. Can. J. Physiol. Pharmacol. 53:293.
220. Janský, L., and Hart, J. S. (1968). Cardiac output and organ blood flow in warm- and cold-adapted rats exposed to cold. Can. J. Physiol. Pharmacol. 46:653.
221. Bell, A. W., Hilditch, T. E., Horton, P. W., and Thompson, G. E. (1976). The distribution of blood flow between individual muscles in the hind limb of the young ox: values at thermoneutrality and during exposure to cold. J. Physiol. (Lond.), 257:229.
222. Héroux, O., and St. Pierre, J. (1957). Effect of cold acclimation on vascularization of ears, heart, liver and muscle of white rats. Am. J. Physiol. 188:163.
223. Fregly, M. J. (1968). Water and electrolyte exchange in rats exposed to cold. Can. J. Physiol. Pharmacol. 46:873.
224. Lennquist, S., Granberg, P. O., and Wedin, B. (1974). Fluid balance and physical work capacity in humans exposed to cold. Arch. Environ. Health 29:241.
225. Wallenberg, L. R. (1975). Effect of cold exposure on free water reabsorption in hydropenic man. Acta Physiol. Scand. 95:21.
226. Fregly, M. J., and Tyler, P. E. (1972). Renal response of cold-exposed rats to Pitressin and dehydration. Am. J. Physiol. 222:1065.
227. Conley, C. L., and Nickerson, J. L. (1945). Effects of temperature change on the water balance in man. Am. J. Physiol. 143:373.
228. Spealman, C. R., Newton, M., and Post, R. L. (1947). Influence of environmental temperature and posture on volume and composition of blood. Am. J. Physiol. 150:628.
229. Fregly, M. J., and Waters, I. W. (1966). Water intake of rats immediately after exposure to a cold environment. Can. J. Physiol. Pharmacol. 44:651.

International Review of Physiology
Environmental Physiology II, Volume 15
Edited by David Robertshaw

3
Temperature Regulation in Primates

R. ELIZONDO

Indiana University School of Medicine,
Bloomington, Indiana

TEMPERATURE SENSORS 73
Skin Sensors 74
Central Nervous System Temperature Sensors 75
Deep Body Temperature Sensors Outside the Central Nervous System 77

THERMOREGULATORY EFFECTOR RESPONSES 81
Homeothermy in Higher Primates 82
Cardiovascular Responses 84
Metabolic Responses 92
Sweating 96

The deep body temperature of many mammalian species is controlled within a narrow range of so-called "normal" body temperature between 36–39°C; these species are classified as homeotherms. Man, for example, can survive core temperatures of 36–38°C. Higher temperatures of 40–41°C, such as associated with strenuous exercise and febrile states, can be tolerated for only short periods of time, whereas core temperatures of 44–46°C are associated with rapid denaturation of vital cellular proteins and death (1). Likewise, man can tolerate hypothermic states associated with body temperatures of 30°C for only a few days before death occurs, presumably due to failure of the cardiovascular system (2). The temperature conditions of the environments inhabited by mammals, on the other hand, vary widely and may range from approximately 50°C on a hot day in the desert to −60°C on the Antarctic waste lands. It is clear, therefore, that the regulation of body temperature by homeotherms provides them considerable independence from the thermal environment.

The control of body temperature in homeotherms has received and continues to receive a great deal of interest. The interest of the clinical scientist stems, to a large extent, from the accepted clinical practice of using core temperature as a general diagnostic index for disease states. For the physiologist interested in homeostatic functions, homeothermy has some advantages as a system for study in that core temperature can be readily monitored. Furthermore, many of the effector responses, such as sweating, panting, metabolic rate, and vasomotor changes induced by disturbances in the internal and/or external thermal environment of the organism, can also be directly monitored and quantified.

Interest in the control mechanisms associated with temperature regulation is reflected by the numerous reviews and symposia (3–12) concerning specific thermoregulatory responses to heat and cold stress of a wide variety of species. Some of this information was most recently reviewed at an international symposium held in 1974 (13). The present review emphasizes a discussion of major recent advances in the understanding of the control of autonomic temperature regulatory responses of mammals. Because of restricted space and due to the author's personal bias, emphasis is placed on a discussion of the autonomic responses of man and infrahuman primates to acute thermal stress. The reader should consult the suggested recent reviews for more detailed information on any specific subject.

Any discussion of such a complex system as that which regulates body temperature in homeotherms and involves integrated responses from numerous subsystems can be unified through the use of a conceptual model. In recent years, it has become fashionable to describe the thermoregulatory system of homeotherms in terms of control systems theory.

As indicated above, internal body temperature in homeotherms is regulated with a certain degree of accuracy. This means that internal body temperature is compared in some way with a reference point, and when differences occur between the reference and internal body temperature, an appropriate effector response is initiated which tends to reduce the difference. In other words, the overall characteristics of the thermoregulatory system of higher vertebrates can be described in the terminology of control theory as a negative feedback control system. Figure 1 illustrates a generalized diagram of a feedback system that is commonly used as a model for thermoregulation. It consists of a controlled system which is the body itself and can be subjected to a disturbance, usually in the form of environmental heat, cold, or internal heat, as in exercise and fever. The disturbance causes a change in the controlled variable. The nature of the controlled variable, however, remains to be defined. Although hypothalamic temperature has often been identified as the controlled variable, it now appears that a combination of central, peripheral, and deep body temperatures may be involved. The disturbance causes a change in the controlled variable which is detected by temperature sensors, and a corresponding feedback signal is produced. The deviation of the controlled variable from the reference value constitutes an error load which activates a central comparator, which in turn elicits a

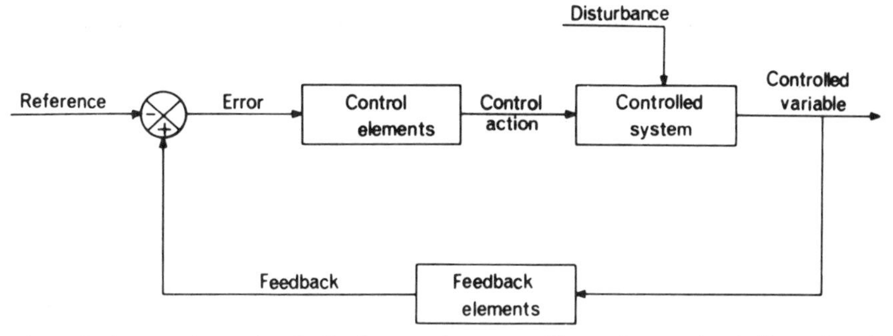

Figure 1. A general negative feedback control system with a reference signal. (Reproduced from Mitchell et al.'s chapter in *Essays on Temperature Regulation* (Bligh and Moore, eds.) (3) by courtesy of North-Holland.)

control effector action which counteracts the disturbance, tending to return the controlled variable toward its reference point. It is not yet clear whether the comparator which establishes the reference point for the system uses an internal reference (set point) or operates by integrating two feedback loops, one from cold-sensitive elements in the system and one from warm-sensitive elements.

The simple model described above, like all models of thermoregulation, is only an approximation of the real system. Such models, on the one hand, have value in that they provide a means of clearly conceptualizing the major characteristics of thermoregulation in homeotherms. On the other hand, such analogies can also be misleading because the basic principles used in a given model are only part of the many theoretical possibilities of control systems. Finally, because control systems models are purely analog descriptions of the physiological thermoregulator, they cannot replace the task of physiological research designed to elucidate and quantify empirically the circuitry involved in the temperature-regulating system of homeotherms.

TEMPERATURE SENSORS

The control of body temperature within narrow limits depends upon the coordinated activities of many subsystems within the body. Current understanding of the control of coordinated thermoregulatory responses in homeotherms is based upon evidence that the integrity of specific areas of the nervous system is essential for the coordinated control of normal body temperature and, in addition, upon the observation that these neural regions themselves exhibit the characteristic of thermal sensitivity. Because all biological processes are temperature-dependent, it is imperative that several functional characteristics, such as the magnitude of the response to local cooling or warming and the degree of specificity of the response, be considered when characterizing a certain neural tissue as a thermoreceptor. Finally, it should be noted that thermosensitivity in neural tissue does not, a priori, indicate a role in thermoregulation. Such a role is ultimately dependent upon the demonstration of a clear correlation between neural activity and a specific thermoregulatory effector response. In homeotherms, temperature sensors have been identified in the skin, the central nervous system, and the deep tissues of the body. Their direct link to thermoregulatory effector mechanisms, however, is less clear and often based on indirect evidence.

Skin Sensors

The importance of skin temperature sensors in thermoregulation is well established (14, 15). Cutaneous temperature sensors generally are characterized as illustrated in Figure 2 by a) a static firing rate at a constant temperature; b) a dynamic response to the rate of change of temperature, with so-called "warm receptors" having a positive temperature coefficient and "cold receptors" having a negative coefficient; and c) essentially no response to mechanical stimuli (16). There are also significant differences in the static response curves for the warm and cold cutaneous temperature receptors, the maximal discharge frequency being much lower for cold receptors than for warm receptors. Hensel and Iggo (17) and Iggo (18) have recently electrophysiologically examined the skin receptors in rhesus monkeys. Single warm and cold receptors on the hairy skin having spotlike receptive fields were identified. The receptors were highly specific to thermal stimuli and did not respond to mechanical deformation of the skin. The cutaneous warm receptors had a range of static activity between 30°C and 44°C with a maximal frequency at 41°C, whereas cold receptors in the

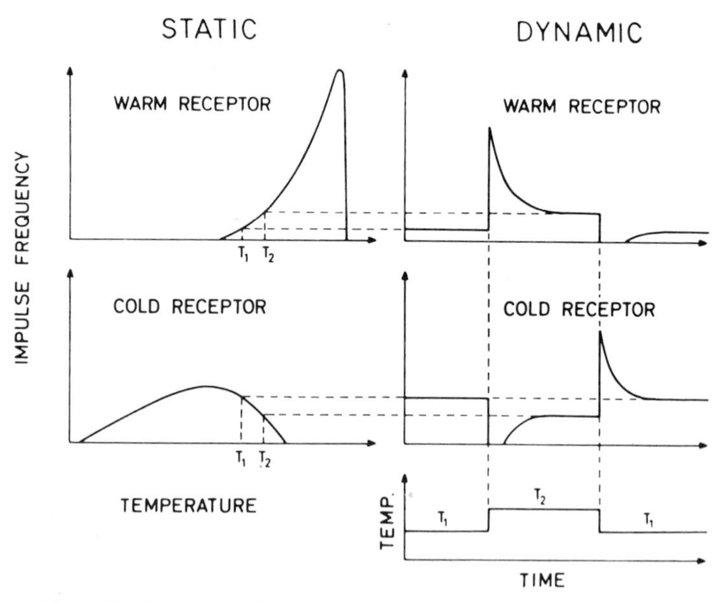

Figure 2. Generalized response of cutaneous single warm and cold receptors to constant temperatures (static response) and to rapid temperature changes (dynamic response). (Reproduced from Hensel, "Temperature Receptors in the Skin," in *Physiological and Behavioral Temperature Regulation*, 1970, (16) by courtesy of Charles C Thomas.)

skin of the rhesus monkey had a maximal frequency of discharge occurring between 24°C and 35°C.

Hensel and Bowman (19) were able to identify by electrophysiological methods cold receptors in human skin. They were unable to identify any warm receptors in the skin of their human subjects. Direct human studies, however, are understandably limited, but it seems quite certain from indirect evidence in man (20, 21) and from the recent direct experiments in infrahuman primates that warm receptors exist in human skin. Although the fundamental mechanisms of thermoreceptor excitation are still unknown, the electrophysiological characteristics of warm and cold cutaneous thermoreceptors in higher primates leave little doubt that they provide not only a source of conscious thermal information, but also a very important peripheral afferent input into the thermoregulatory system of higher vertebrates. Specific examples of the influence of cutaneous thermoreceptors upon effector thermoregulatory responses are discussed under "Thermoregulatory Effector Responses."

Central Nervous System Temperature Sensors

That the hypothalamus plays a major role in the regulation of body temperature has long been suspected and cannot be denied. Numerous studies on a wide variety of species (monkey, dog, cat, rabbit, goat, guinea pig, and others) have demonstrated that localized warming of the hypothalamic region consistently activates effector mechanisms of heat loss, including sweating, vasodilation, and panting (22–25). Conversely, local cooling of the hypothalamus consistently elicits effector mechanisms against cold, including increased heat production (26–28), peripheral vasoconstriction (29, 30), and nonshivering thermogenesis (31). In recent years, as the methods of thermal stimulation of the central nervous system (CNS) have become more refined, the concept has developed that appropriate effector responses can be most consistently elicited by thermal stimulation of the anterior hypothalamus. These same techniques, however, have also provided evidence for the existence of thermal-sensitive areas in other parts of the CNS, including the posterior hypothalamus (32, 33), midbrain (34, 35), and cortex (36). It appears that, in the cat, there are no thermosensitive areas in the medulla oblongata which have a role in thermoregulation because local thermal stimulation of this area produces primarily effects on heart and respiratory rate which are opposite to those induced by thermal stimulation of the hypothalamus (37, 38). Recently, however, Chai and Lin (35) reported that localized cooling and heating of the medulla oblongata initiated appropriate thermoregulatory responses in unanesthetized monkeys. The physiological significance of extrahypothalamic temperature-sensitive areas in the normal control of the thermoregulatory effector responses of primates, however, remains undetermined and deserves further investigation.

Further insight into the functional characteristics of the thermal-sensitive regions in the CNS has been obtained during the past 15 years from electrophysiological studies of individual preoptic anterior hypothalamic (PO/AH) neurons. The earliest attempt to detect and characterize the firing frequency of

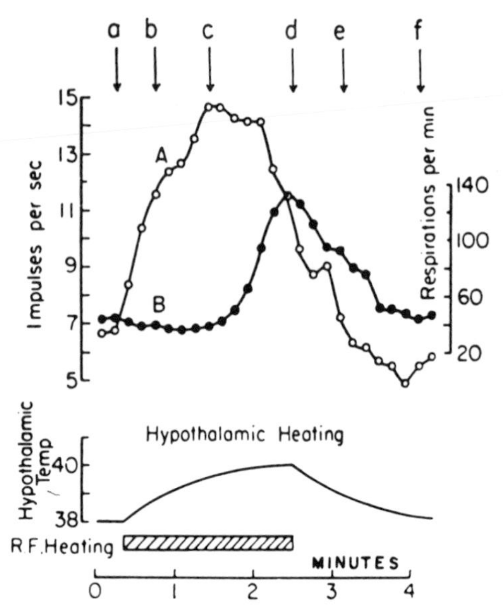

Figure 3. Discharge frequency of a neuron in the preoptic region (*curve A*) and change of respiratory rate (*curve B*) in relation to hypothalamic temperature. (Reproduced from Nakayama et al. (40) by courtesy of Am. J. Physiol.)

individual temperature-sensitive neurons in the PO/AH was made by Nakayama et al. (39). As illustrated in Figure 3, they identified temperature-sensitive neurons in the PO/AH of the cat, whose firing frequency increased with local warming (40). Because the Q_{10} of the response was as high as 10, specific thermal sensitivity of the neuron is strongly suggested. Furthermore, because the increase in firing frequency of the neuron was immediately followed by an increase in respiratory frequency, it has been suggested that these warm-sensitive neurons may function as primary affectors in initiating the appropriate effector responses, in this particular case panting, to a rise in core temperature. Direct physiological evidence of the role of warm neurons as primary affector units, however, remains to be obtained.

In more recent investigations, cold-sensitive units in the PO/AH have also been reported (41–43). Figure 4 demonstrates the changes in firing frequency of a cold neuron as brain temperature was cycled from 35.5–40.5°C. It can be seen that the firing frequency of the cold unit decreased as tissue temperature increased and vice versa. As illustrated in the bottom panel, cold neurons are characterized by a significant negative relationship between activity and local tissue temperature and consequently have Q_{10} values of less than 1.

Cabanac et al. (44) examined the electrical activity of 28 PO/AH thermosensitive neurons of anesthetized rabbits over local temperatures ranging from 32–42°C. They reported both cold and warm neurons which have bell-shaped frequency curves (Figure 5). The mean peak activity of the cold units was obtained at 34–35°C, whereas the mean peak activity of the warm units was

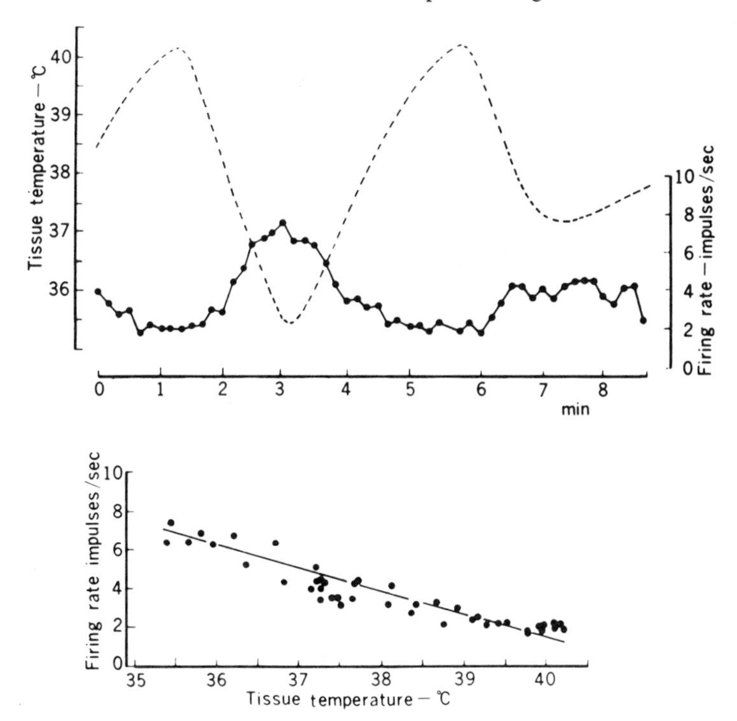

Figure 4. Discharge frequency of a primary cold-sensitive neuron in the preoptic region in relation to tissue temperature. (Reproduced from Ogata and Murakami (137) by courtesy of Igaku Shoin, Ltd.)

obtained at the higher hypothalamic temperature of 38°C. The results from this study, therefore, would suggest that when examined over a fairly wide range of brain temperature, the response characteristics of PO/AH temperature sensors are very similar to those of peripheral sensors, as described above.

In spite of the relatively large amount of recent data on the activity of temperature-sensitive neurons in the CNS, the results are difficult to interpret because a number of basic questions concerning the role of these central neural structures remains unanswered. For example, 1) what is the morphology of temperature-sensitive structures in the CNS? 2) where and how is the primary thermal information from central as well as peripheral temperature sensors integrated? and 3) what are the neural pathways used to send processed information from the CNS to the various effector systems for temperature regulation? Future research must provide answers to some of these questions before the true role of central temperature sensors in the temperature regulation system of higher vertebrates can be effectively conceptualized.

Deep Body Temperature Sensors Outside the Central Nervous System

The idea of extracerebral temperature sensors functionally important in thermoregulation can be traced to the work of Goltz and Edwards (45) published in

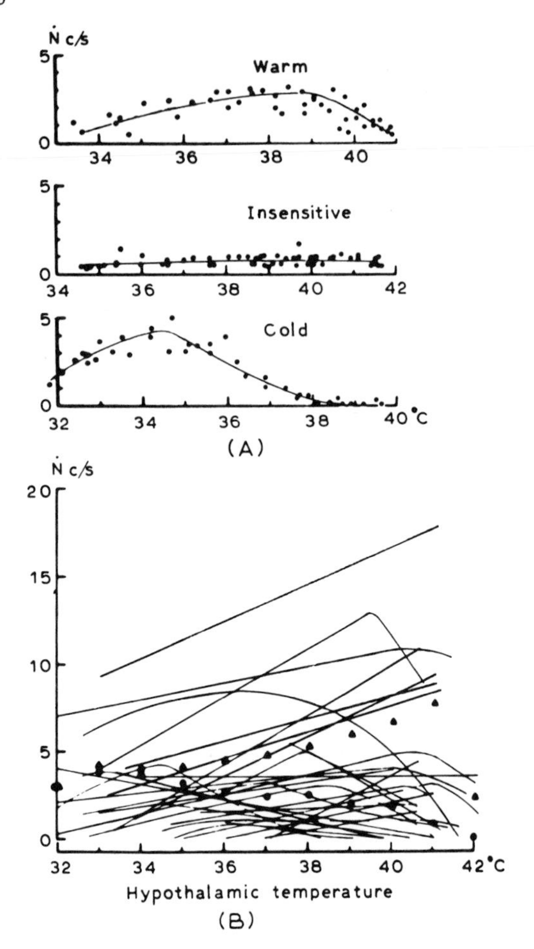

Figure 5. *A,* activity/temperature relationships of a warm-sensitive, a temperature-insensitive, and a cold-sensitive unit in the preoptic region of the hypothalamus of the rabbit. *B,* the activity/temperature relationships of 28 individual units in the preoptic region of the hypothalamus of the rabbit. ▲, mean activity of warm units; ●, mean activity of cold units. (Reproduced from Cabanac et al. (44) by courtesy of J. Appl. Physiol.)

1896. The idea was experimentally supported for the first time by Thauer and Peters (46), who reported that dogs retained limited thermoregulation following chronic high level section of the spinal cord and midbrain transection. The thermal sensitivity of the spinal cord and its significant influence upon thermoregulatory effector responses has recently been demonstrated in a wide variety of mammals, including the dog (47, 48), rat (49), rabbits (50–52), pigs (53), ox (54), and monkeys (55). These studies indicate that any thermoregulatory response initiated by local hypothalamic temperature changes can also be produced by local cooling or heating of the spinal cord in unanesthetized animals. For example, shivering and vasoconstriction were clearly demonstrated in the dog when the spinal cord was locally cooled, whereas brain and skin temperature

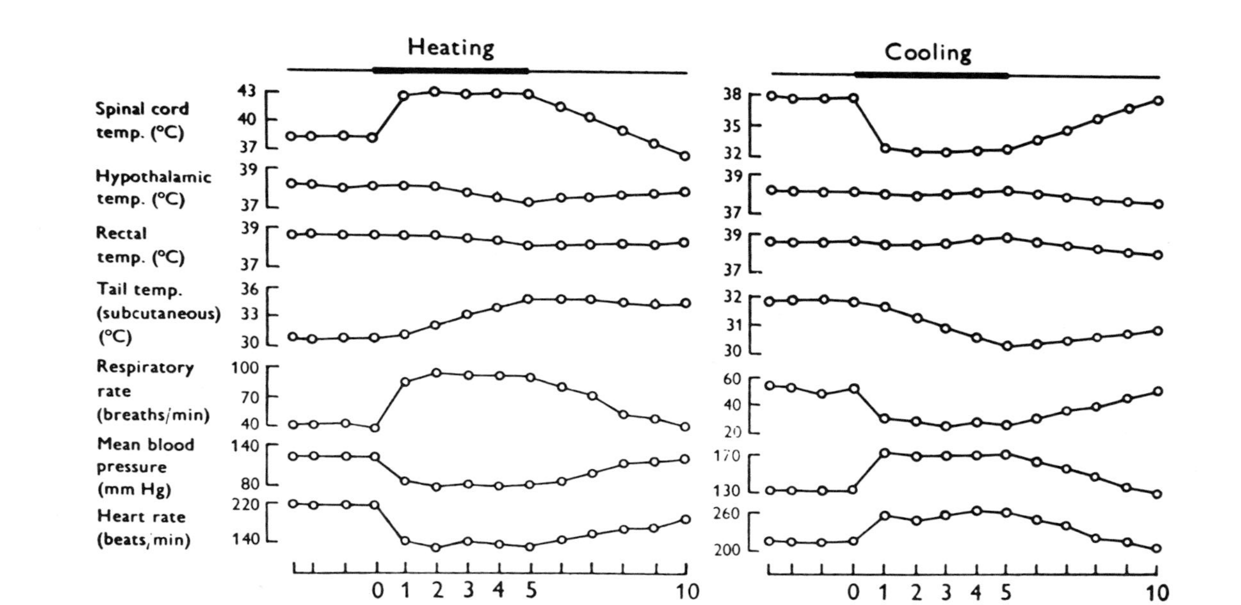

Figure 6. Changes of the temperature of the hypothalamus, rectum, and subcutaneous tissue of the tail and respiratory and cardiovascular reactions during local change of the spinal temperature in an unanesthetized monkey. (Reproduced from Chai and Lin (35) by courtesy of J. Physiol. (Lond.).)

remained essentially constant (56). Similarly, an increase in respiratory rate and vasodilation could be induced by local warming of the spinal cord of the unanesthetized Formosan monkey (*Macaca cyclopis*) (55). As illustrated in Figure 6, heating the spinal cord of this monkey to 43°C resulted in an increase in tail skin temperature of 2.2°C and an increase in respiratory rate of 34 breaths per minute. However, it is not known whether the increase in breathing frequency was in the form of a true panting response. Cardiovascular changes included a decrease in heart rate of 62 beats per minute and a decrease in blood pressure of 45 mmHg. Conversely, when the spinal cord of this primate was cooled to 33°C, the respiratory rate decreased 8 breaths per minute, subcutaneous temperature decreased approximately 1.2°C, whereas heart rate and blood pressure increased 51 beats per minute and 40 mmHg, respectively.

With one exception, the importance of spinal cord temperature in the control of thermoregulatory effector mechanisms seems to have been well established within the last few years. The exception is the role of spinal cord temperature in the control of sweating. In this respect, the influence of spinal cord thermal stimulation has been investigated only in the ox (57). As illustrated in Figure 7, Hales and Jessen (58) found that local spinal cord heating to 44°C resulted in a slight but distinct increase in cutaneous moisture loss and a

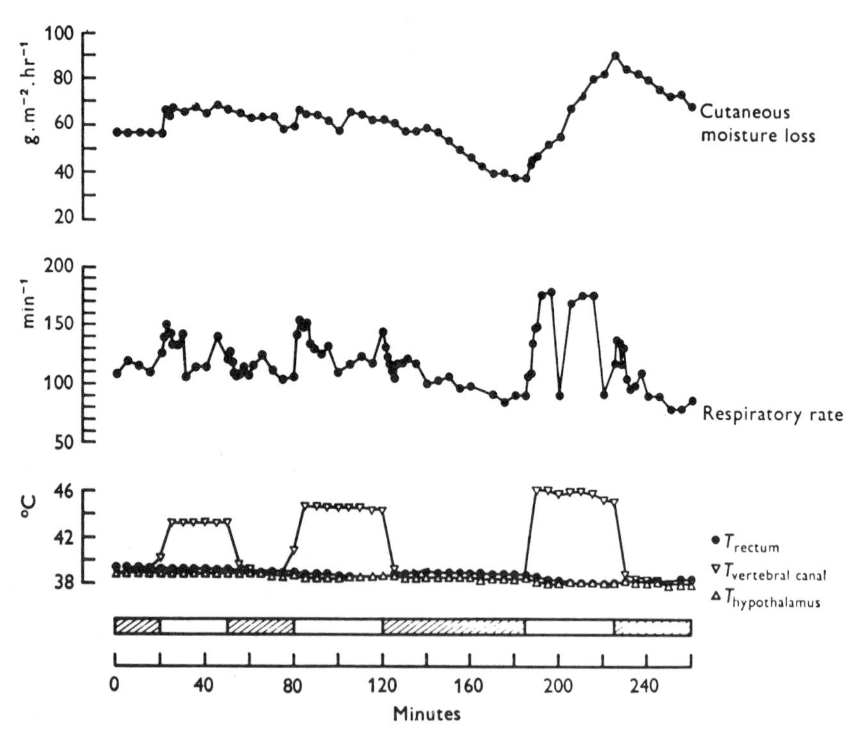

Figure 7. Changes in cutaneous moisture loss and respiratory rate caused by local heating of the spinal cord in the ox. (Reproduced from Hales and Jessen (58) by courtesy of J. Physiol. (Lond.).)

transient rise in respiratory rate. The second period of spinal heating elicited similar but more pronounced effects. The final period of intense spinal cord heating, which resulted in a peridural space temperature of 46°C, doubled the evaporative heat loss due to sweating and also doubled the respiratory rate. Rectal and hypothalamic temperature decreased by approximately 0.6°C. All of the parameters returned to their preheating level when the spinal cord heating was stopped. The authors conclude from these observations that in oxen exposed to moderate external heat stress, heating of the spinal cord results in a marked increase in sweating as well as respiratory water loss. Further studies in other sweating primates such as rhesus monkeys need to be done in an attempt to quantitate the relative importance of spinal cord temperature as compared to skin and hypothalamic temperatures in the control of sweating in higher primates.

Rawson and Quick (59, 60) have recently provided evidence for another site of deep body thermosensitivity outside the central nervous system. They have observed the effects of intra-abdominal thermal stimulation in sheep, with and without splanchnicotomy. These authors reported that intra-abdominal heating of sheep in cold environments depressed shivering, whereas intra-abdominal heating of sheep in a warm environment augmented panting. Furthermore, the splanchnic nerves were indicated as the afferent pathways for these intra-abdominal thermal receptors because unilateral splanchnicotomy abolished the thermoregulatory effector responses to stimulation of the denervated side.

The extensive literature on the spinal cord discussed above and continuous reports of deep body temperature-sensitive structures not only in the abdomen, but possibly in the larger veins and in muscle tissue, suggests that effective thermoregulation in higher primates depends not only upon thermal information from the well established temperature-sensitive receptors in the central nervous system and periphery but also from the spinal cord and other deep body tissues. There are few areas in the study of thermoregulation in homeotherms which show more potential for making a significant contribution than experiments designed to provide quantitative information of the relative contribution of these various thermal receptor inputs in the control of thermoregulatory effector systems of homeotherms under different thermal stress conditions.

THERMOREGULATORY EFFECTOR RESPONSES

The thermoregulatory effector functions of homeotherms can be conveniently classified into two general categories—namely, physiological, which includes autonomic functions which depend upon the integrity of the hypothalamus but not the cortex, and behavioral responses, which require the integrity of the cortex. For more information on the importance of behavioral thermoregulation in higher primates, the excellent recent studies by Adair (26) and Adair and Wright (61) should be consulted. This section reviews the general thermal defense mechanisms of homeotherms with concentration on the physiological responses of higher primates to acute temperature stress.

Homeothermy in Higher Primates

The compensatory physiological responses of higher primates to thermal stress include a) vasomotor adjustments which regulate the rate of heat exchange between the core and peripheral tissues and subsequently between the surface of the body and the environment, b) thermogenesis through shivering and other metabolic processes, and c) evaporative heat loss mechanisms such as sweating and panting.

The thermoregulatory profile for a hypothetical homeotherm is graphically illustrated in Figure 8. Homeothermy in higher primates is characterized by a narrow thermoneutral zone (TNZ), where slight changes in ambient temperature activate few, if any, physiological thermoregulatory effector responses; the TNZ constitutes the ambient temperature range of minimal energy expenditure for thermoregulation. The thermoneutral zone for a resting naked man in the postabsorptive state is very narrow and lies between 25°C and 27°C (62). Man, in this respect, is similar to many tropical species of mammals (63). Arctic mammals, on the other hand, primarily due to better peripheral insulation, tend to have much broader thermal neutral zones, as great as 60°C in some species.

The TNZ also includes the zone of vasomotor control of body temperature and constitutes the ambient temperature range in which any imbalances between heat gain and heat loss by the organism can be compensated by cardiovascular adjustments which modify vascular heat transfer between the core of the body, the skin, and the environment. The lower limit of the vasomotor control zone is characterized by what is termed the "lower critical temperature" (LCT) and constitutes the ambient temperature below which simply minimizing the rates of heat loss by peripheral vasoconstriction is ineffective in maintaining thermal

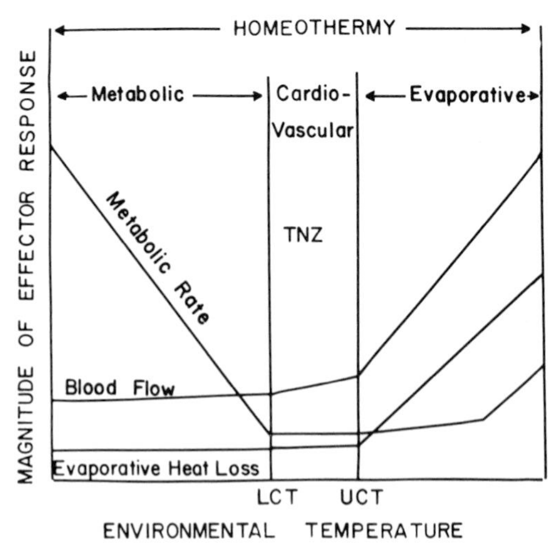

Figure 8. General thermoregulatory profile for a hypothetical homeotherm indicating primary effector responses.

balance. In higher primates, homeothermy at temperatures below the lower critical temperature involves "chemical" control, which provides additional body heat through elevation of whole body metabolic rate and shivering thermogenesis. The upper limit of the vasomotor control zone is characterized by an "upper critical temperature" (UCT) and constitutes the environmental temperature above which simply maximizing the rate of heat transfer from the core to the skin via vasodilation is ineffective in maintaining thermal balance. In higher primates, homeothermy at environmental temperatures above the upper critical temperature depends upon the evaporative heat loss mechanisms of sweating and/or panting, which are used to balance the combined rates of heat gain from the environment and from metabolic processes.

The effects of ambient temperature on the thermoregulatory effector responses of human subjects have been the subject of numerous studies, and the complete thermoregulatory profile for unclothed male subjects resting at ambient temperatures between 12°C and 48°C has recently been reviewed by Hardy et al. (64). Wilkerson et al. (65), using the intercept method of Scholander et al. (66), have recently reported that the average value for the LCT of five human subjects was 25.2°C. The UCT for man, although less critically defined, appears to be 28°C.

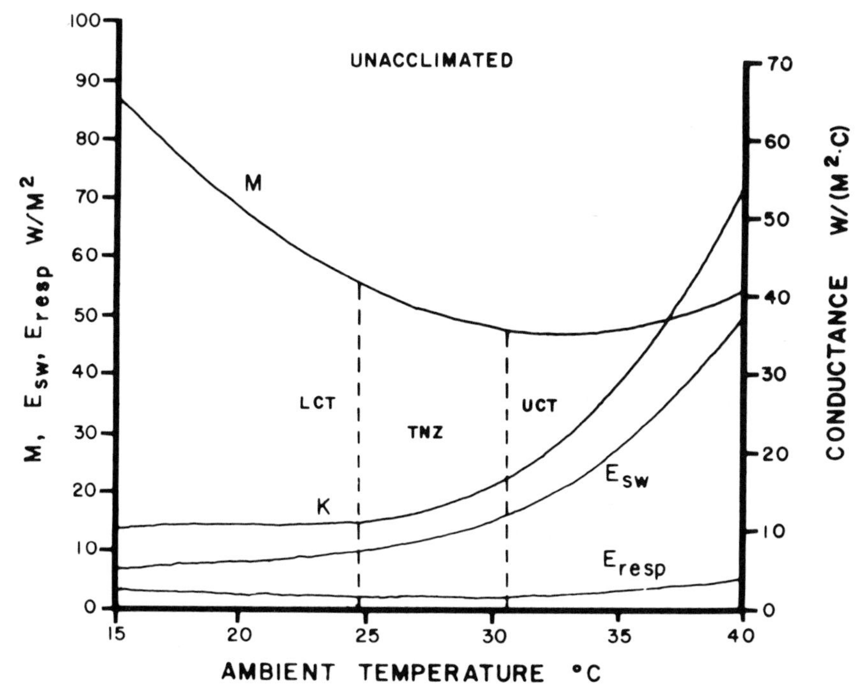

Figure 9. Complete thermoregulatory profile for the unacclimated unanesthetized rhesus monkey equilibrated to ambient temperatures ranging from 15°C to 40°C. (Reproduced from Johnson and Elizondo (104) by courtesy of J. Appl. Physiol.)

Until very recently, no complete thermal balance studies existed in the literature which were designed to characterize the complete thermoregulatory profile of infrahuman primates. The thermoregulatory effector responses of the rhesus monkey (*Macaca mulatta*) exposed to an ambient temperature range of 15–40°C are illustrated in Figure 9. For this infrahuman primate, the thermoneutral range extends from 25–31°C. Below the LCT, metabolic heat production (*M*) increases linearly with decreasing ambient temperature to compensate for the increased heat loss. The metabolic rate of six male rhesus monkeys averaged 52 watts/m^2 (W/m^2) in the TNZ and increased to an average of 91 W/m^2 at 15°C. Chaffee and Allen (67) have shown that cold-acclimated rhesus monkeys possess brown fat; however, it is not known whether nonshivering thermogenesis contributed significantly to increased heat production in the monkeys which were not cold-acclimated. Whole body conductance (*K*) is minimal and constant below the LCT, indicating maximal peripheral vasoconstriction tending to reduce body heat loss. At ambient temperatures above 25°C, whole body conductance increases with increasing ambient temperature, reaching a value of 55 W/m^2 · °C at 40°C. These observations indicate that homeothermy in the rhesus monkey, as in man, is achieved within the TNZ (25–31°C) by vasomotor control. At ambient temperatures above 31°C, the UCT, body heat losses are facilitated by evaporative heat loss due to sweating. Sweating heat loss averaged only 10 W/m^2 in the TNZ and increased to 55 W/m^2 at 40°C. It is interesting to note that the rhesus monkey does not pant, and consequently, evaporative heat loss from the respiratory tract increased from 2.2 W/m^2 in the TNZ to only 4.7 W/m^2 at an ambient temperature of 40°C. At ambient temperatures above the UCT, conductive heat loss progressively increases, but is insufficient to maintain thermal balance. Eccrine sweating, therefore, becomes the major avenue of evaporative heat loss; the role of evaporative heat loss from the respiratory tract in dissipating metabolic and environmental heat gain then becomes insignificant. At ambient temperatures below the LCT, thermal balance is maintained by maximal vasoconstriction to minimize heat loss and an increase in metabolic heat production, presumably due to shivering thermogenesis. These findings in the rhesus monkey are in good agreement with those reported in the squirrel monkey (*Saimiri sciureus*) (68) and the Japanese monkey (*Macaca fuscata*) (69). It remains to be determined whether the thermal balance data obtained in these three species of infrahuman primates are representative of infrahuman primates in general. The significant fact remains, however, that the thermoregulatory profiles of the infrahuman primates examined to date are similar to that of man and can serve as effective animal models for temperature regulation studies which cannot be performed in man.

Cardiovascular Responses As indicated in the previous section, the three most important mechanisms which a homeothermic organism can employ to maintain thermal balance at different ambient temperatures include evaporative heat loss, adjustments in metabolic heat production, and heat loss by convection, conduction, and radiation. Activation of all the above mechanisms is associated to varying extents with adjustments of the circulation. However, only the control of heat exchange by convection, conduction, and radiation is directly and quantitatively dependent upon changes in the circulation.

The skin acts as a thermal insulator, and its effectiveness decreases with increased blood flow and vice versa. Changes in skin blood flow not only affect the rate of heat exchange between the organism and the environment, but also determine the rate of heat transfer from the deep tissues of the body to the periphery. The gross cardiovascular adjustments in man induced during heat stress have recently been reviewed by Rowell (70). This section, therefore, deals principally with a discussion of the factors which are involved in the control of the circulation through the skin of higher primates.

A variety of methods have been used to estimate changes in cutaneous blood flow in man with changing ambient conditions. Changes in skin temperature, venous oxygen tension, thermal conductivity, rates of heat loss by calorimetry, and isotope clearance techniques have all been used. The technical difficulties, theoretical assumptions, and other limitations of these techniques have been reviewed (71, 72). The most direct and extensively used method of measuring blood flow in man is venous occulsion plethysmography. Its use, however, is confined to the extremities, which have been studied much more extensively than other skin areas; thus, most of our knowledge of skin blood flow and its control in man is derived from this technique. It must be emphasized, however, that extrapolation of the results from the extremities to other regions of the body may not be valid because regional differences in normal blood flow, thresholds, and different mechanisms of responses to a given thermal stimulus have been suggested. Extrapolation of results from animal studies, including those in higher primates, is also probably invalid due to the striking differences in the vasomotor innervation and thermal responses of the skin of man and other animals. Stitt and Hardy (28) reported, for example, that the degree of vasodilation is greater in the tail of the squirrel monkey than in the foot and concluded that the tail acts as the principal heat-dissipating organ when this infrahuman primate is exposed to a warm environment (25–30°C), with minimal involvement of other extremities. Recently Wyss et al. (73) reported a lack of significant human-like cutaneous vasodilation in the skin of heated baboons and suggested that this infrahuman primate also is not an appropriate model for vasomotor control in man. It seems clear, therefore, that current understanding of the mechanisms that control peripheral blood flow in human skin is limited by the lack of development of appropriate animal models. Furthermore, relatively noninjurious indirect experimental techniques have been applied predominantly to measure skin blood flow in the extremities.

Due to the limitations outlined above, no direct accurate measurement of total skin blood flow currently exists. Estimates of normal skin blood flow for a resting nude man in a normal indoor environment range from 116 to 289 $ml/m^2 \cdot min$. Estimates for maximal cutaneous blood flow during heat stress range from 2.1 to 3.5 liters. Thus, skin normally has a blood flow which far

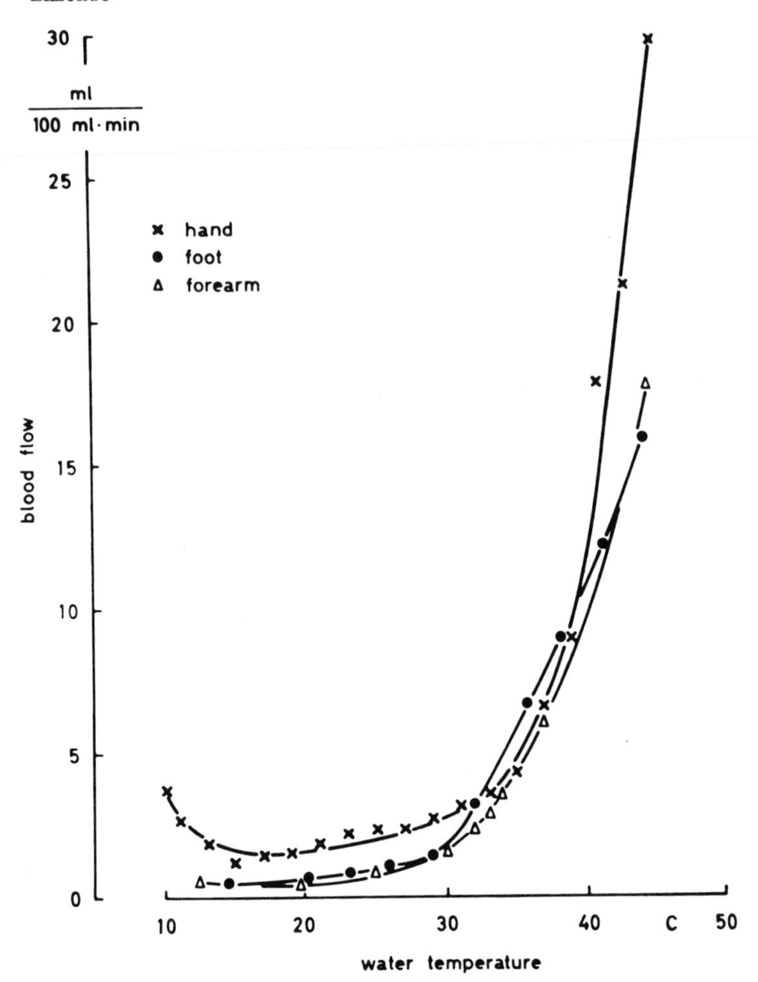

Figure 10. Relationships between blood flow in the human hand, foot, and forearm, and water temperature of the plethysmograph. (Reproduced from Thauer (71) by courtesy of Handbook of Physiology, American Physiological Society.)

exceeds its nutritional needs; when maximally dilated, skin has a total blood flow second only to that for maximally vasodilated skeletal muscle. Therefore, skin blood flow is chiefly determined by thermal considerations.

Numerous investigations have demonstrated that the exposure of any part of the body to a change in temperature induces changes in peripheral circulation not only of the exposed skin area via local temperature effects, but also of other parts of the body through neural reflexes initiated from both peripheral and central thermal receptors.

The effects of local temperature changes on blood flow through the human hand, foot, and forearm are illustrated in Figure 10. It can be seen that hand blood flow at all local temperatures between 15°C and 30°C is greater than through the forearm and foot. At temperatures above 30°C, the regional differences are insignificant, whereas at temperatures above 40°C blood flow through the hand again becomes significantly greater. In general, however, it can be seen that blood flow through the extremities is at its lowest at a local temperature of 15°C (0.3–0.9 ml/100 ml•min). From 15°C to 30°C, there is a gradual rise in skin blood flow with increasing local temperature. With further increases in local temperature from 30°C to 43°C, there is a steep rise to a maximal in hand blood flow of approximately 30 ml/100 ml•min. Although further increases in peripheral blood flow at local temperatures above 43°C have been reported, interpretation of the results is complicated by the fact that temperatures above 43°C stimulate the sensation of pain, which in itself may affect skin temperatures. Because the above responses to local temperature can be demonstrated following sympathectomy, the effect appears to be a truly local one dominated by the physical effects of temperature on the peripheral vasculature (74, 75).

The local effects of temperature on cutaneous blood flow in man depends upon the general thermal state of the body. This is illustrated in Figure 11, which shows the relationship between blood flow in the human hand and local skin temperature as represented by plethysmograph water temperature at three different environmental temperatures. It can readily be seen that at any given local skin temperature the blood flow through the hand is higher as environmental temperature increases. At high local skin temperatures, the absolute effect of local temperature upon cutaneous blood flow is smaller in a cold environment that in a warm one. Conversely, at low local skin temperatures, the absolute effect of local temperature upon blood flow is smaller the higher the environmental temperature. In general, therefore, when vasoconstrictor tone is diminished by total body heating, the direct local temperature effects are potentiated. Similarly, the local effects of cooling are augmented by lowering body temperature.

It is clear that the general pattern of cutaneous vascular responses in man to local changes in skin temperature are well established. The mechanisms, however, which are involved in mediating the local temperature effects remain unknown. The effects of heat on vascular smooth muscle, production of vasodilator metabolites, reduced blood viscosity, and axon reflexes may all contribute, and further experimentation is needed to elucidate the role of all of these mechanisms in the local heating response.

Numerous observations in man have consistently demonstrated that cutaneous vasoconstriction and vasodilation in response to thermal stimuli are not limited to the area of skin heated or cooled. Furthermore, these reflex changes in cutaneous blood flow are mediated by sympathetic nerve fibers to the skin and possibly sweat glands which are activated by changes in the temperature of the blood acting directly on central thermoreceptors and by afferent impulses to the center from peripheral skin receptors.

Controversy continues to exist over the mode of action of vasodilator and vasoconstrictor nerves in the control of peripheral blood flow and their relative

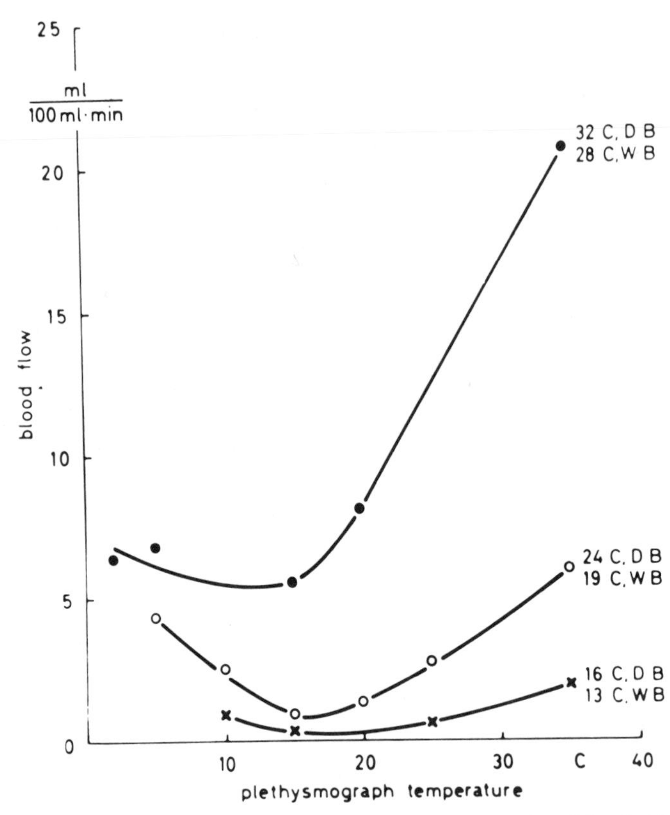

Figure 11. Relationship between blood flow in the human hand and local skin temperatuture as represented by plethysmograph water temperature at three different environmental temperatures. (Reproduced from Thauer (71) by courtesy of Handbook of Physiology, American Physiological Society.)

importance in thermal vasomotor responses in different regions of the body. Experiments designed to determine whether cutaneous vasomotor control is mediated predominantly by changes in sympathetic vasoconstrictor tone or whether vasodilation is also neurogenic indicate that different control mechanisms occur in the hands and the rest of the skin.

Figure 12 illustrates the reflex effects of indirect body heating upon skin blood flow in the normal hand and forearm before (*A*) and after (*B*) sympathetic nerve block. Initially, the subject was placed in a room at 16–18°C to ensure low peripheral blood flow. Within 5 min after the start of indirect body heating, hand blood flow increased rapidly, reaching a level of 30–35 ml/100 ml·min. These results indicate that vasodilation in the hand was mediated entirely by the release of sympathetic vasoconstrictor tone (–VC), as indicated by the observation that nerve block (*B*) alone produced a comparable increase in hand blood flow and subsequent body heating produced no further increase in hand blood flow.

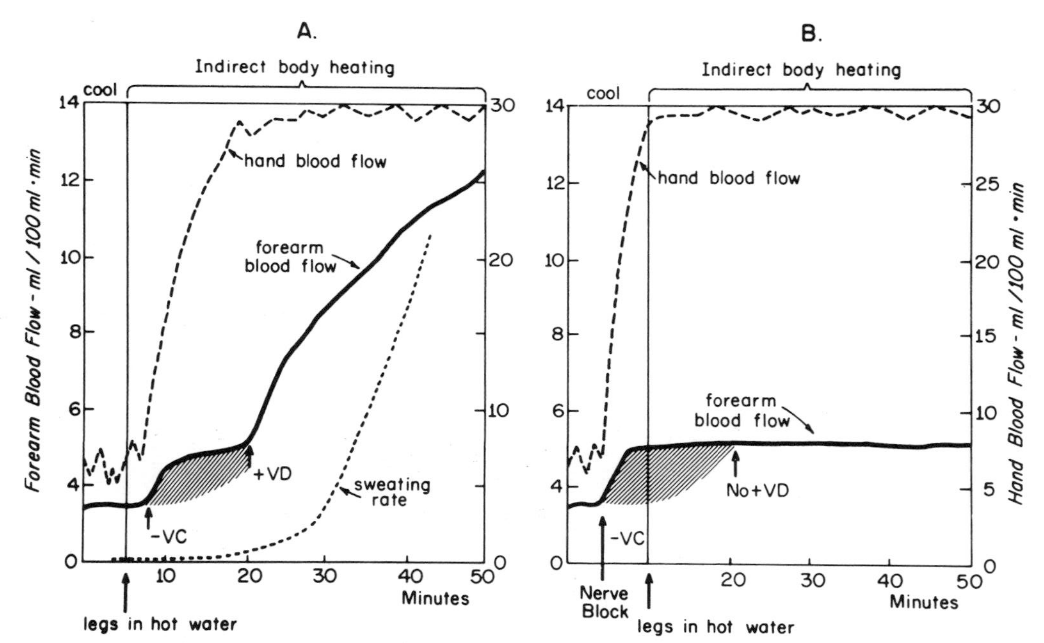

Figure 12. Reflex effects on indirect heating upon skin blood flow in the normal hand and forearm (*A*) and the same regions with sympathetic nerves blocked (*B*). Vasodilation in the hand is mediated only by release of tonic vasoconstrictor tone (–VC). Vasodilation in the forearm results initially from a release of tonic vasoconstrictor tone if the subjects are cooled at the start of heating. After 10–15 min of heating, a second phase of vasodilation (+VD) occurs which is active. (Reproduced from Rowell (138) by courtesy of W. B. Saunders.)

The normal forearm blood flow response (*A*) was characterized by an initial small but definite increase from approximately 3 ml/100 ml·min. The forearm flow remained at this level for 10 min; then there was a sudden further increase which continued until the end of the heating period. It is evident that the initial vasodilation observed in the forearm was due to the release of sympathetic vasoconstriction because nerve blockage produced an equal effect. The second phase of vasodilation, however, is active as indicated by the fact that it was prevented by the blockage of sympathetic nerve transmission (*B*). These observations clearly indicate that although forearm skin, like the hand, also receives tonic sympathetic vasoconstrictor outflow, local nerve block increases flow only if core and skin temperature are relatively low. The increases in flow due to loss of vasoconstrictor tone are small compared to those of the hand in which indirect body heating does not increase it further. The secondary phase of vasodilation observed in forearm skin blood flow with body heating requires sympathetic nerve fibers which in some unknown way induce active vasodilation.

The relationship between active forearm vasodilation and the onset of sweating has received considerable attention since the original suggestion by Fox and Hilton that active vasodilation of forearm cutaneous vessels is secondary to sweat gland activity (76). These authors found a 5-fold increase in bradykinin-like activity in both the sweat and in subdermal perfusates when human sweat glands were stimulated by heating. They, therefore, suggested that the active vasodilation of the forearm during heating is a direct result of the local release of bradykinin from the sweat glands and that only the glands are directly innervated and controlled by sympathetic cholinergic nerve fibers. Although the validity of the bradykinin hypothesis still remains an open question, there are a number of experimental observations which are contradictory to the hypothesis. For example, the onset of sweating and vasodilation do not always coincide and, when they do, cooling a small local area of the skin suppresses vasodilation, but not sweating (77). Sweating rate and the intensity of vasodilation often change in opposite directions during the process of heat acclimation (78). Finally, thermal sweating also occurs in the hands and, as pointed out previously (Figure 12), there is no evidence for an active vasodilator component in the control of peripheral hand blood flow. These and other contradictions suggest that the bradykinin theory does not warrant general acceptance, and final resolution of the controversy depends upon the development of specific inhibitors of the bradykinin mechanism. Until then, it must be concluded that cutaneous vasodilation in response to body heating clearly involves a complex interaction of reduced sympathetic vasoconstrictor tone, direct local effects of heat, and an active cholinergic vasodilator mechanism that requires sympathetic innervation.

The important role of central and peripheral thermoreceptors in the control of skin blood flow has already been indicated and is well documented (79–81). The question of the relative importance of peripheral and central thermal inputs in the control of peripheral vasomotor responses continues to be an area of active investigation; significant advancements in this area have recently been reported. Understandably, the most direct observations have been obtained from animal experiments, and some are reviewed here for the sake of completeness and their possible relevance to the human system.

The role of the receptors in the hypothalamus on peripheral blood flow has been investigated in several species, including the dog (25, 29, 30), cat (82), ox (83), and monkey (26). These studies indicate that, in general, for a given change in hypothalamic temperature, blood flow changes by an amount proportional to the ambient temperature. Furthermore, at any given ambient temperature graded changes in hypothalamic temperature produce graded changes in peripheral blood flow.

Cutaneous blood flow is also influenced by temperature receptors outside the hypothalamus, and it is now clear that heating or cooling of the spinal cord effects cutaneous blood flow in the same way that temperature changes in the hypothalamus do. Recently, it has also been demonstrated that spinal cord heating not only elicits an increase in peripheral blood flow, but also a decrease in the activity of cutaneous sympathetic nerve fibers. The opposite situation was observed in the intestine (84). These neural changes are consistent with the observations on the redistribution of blood flow within the body during external heat stress (70, 85).

Peripheral vasomotor changes that occur in homeotherms in response to changes in ambient temperature are thus controlled by a complex of signals derived from numerous peripheral and central thermoreceptors. A significant advancement in our understanding of the interaction between the various sites of thermoreceptors in the control of vasomotor responses has recently been provided by the studies of Ingram and Legge (86) in the pig. In these studies, blood flow was measured in the tail of unanesthetized pigs by plethysmography. The temperature of the skin on the trunk was controlled independently of ambient temperature by a water perfused coat while the temperature of the head and extremities was controlled by the ambient temperature. Likewise, the temperature of the spinal cord and hypothalamus was locally controlled with water-perfused thermodes. From these studies, it was concluded that peripheral blood flow is influenced by ambient temperature, the temperature of the skin on the trunk, the temperature of the spinal cord, the temperature of the hypothalamus, and a local effect of the temperature of the tail itself. It was further concluded that each temperature appears to exert its influence upon cutaneous blood flow independently of the others. These studies indicate that in a warm environment, when skin temperature is high and body temperature is slightly elevated, cooling of the spinal cord or the hypothalamus or even both simultaneously does not produce complete vasoconstriction. Likewise, in a cool environment, heating of the thermodes may not cause vasodilation. The simplest explanation of these results is that all temperature signals are fed into a central integrating center which, in an as yet unknown manner, determines the appropriate vasomotor effector response. However, because of the apparently large species differences in vasomotor control, extrapolation of these results to other species, including man, should be done with extreme caution. It seems clear that more experiments designed to clarify the precise role of the various thermal inputs in the control of cutaneous blood flow in different species, particularly infrahuman primates, need to be performed.

Metabolic Responses As indicated in the previous section, the ability of higher primates to increase metabolic heat production in response to a decrease in environmental temperature is one of the characteristic effector responses of homeotherms. Thermogenesis from exothermic chemical reactions, especially in skeletal muscle cells, brown adipose tissue, and certain internal organs is one of the variable quantities through which homeothermic organisms can regulate their core temperatures in the cold. The rate at which heat is produced by cellular thermogenic processes is directly related to the total metabolic rate of the organism, as measured by its respiratory exchange. For this reason, the two rates are often considered as relating to the same quantity, namely total oxygen consumption, and are interchangeable, provided the animal is not doing external work.

It is apparent from the data in Figure 9 that even in the absence of external work or cold exposure a relatively large amount of heat is continuously produced in primates. This rate of basal heat production when measured in man under certain standard conditions has been designated as the basal metabolic rate (BMR) and is thought to represent the lowest level of energy utilization compatible with life. The criteria necessary to satisfy this condition in man are that he be in a neutral external temperature, at rest, awake but relaxed, and at least 12 hr postabsorptive. BMR determinations in animals are very difficult, if not impossible to perform, due to the uncertainty that the conditions of rest and psychological relaxation are satisfied. Due to these and other experimental problems inherent in basal heat production measurements in animals, the term "thermal neutral metabolic rate" (TMR), as recently suggested by Webster (87), is preferable.

Thermogenesis in homeotherms is influenced by a large number of factors, such as species, sex, age, size, nutrition, endocrine glands, acclimation, activity, etc. This section, however, focuses primarily on those factors which have a direct significant influence on the control of the effector mechanisms of cold thermogenesis stimulated in higher primates by acute exposure to cold. Additional information on these and other aspects of the mechanisms of heat production in mammals is available in numerous recent reviews (5, 6, 88–91).

It is generally accepted that the activation of cold thermogenesis in homeotherms involves neural information from both peripheral and central thermoreceptors. The functional characteristics of these receptors have been discussed in a previous section, and it is only necessary to emphasize here that cold thermoreceptors have been identified in the skin (92), spinal cord (93), the hypothalamus (94), and the abdomen (60). Although cold receptors from all of these thermal-sensitive areas, and possibly others, play a role in the control of thermogenesis, the relative importance of central and peripheral cold thermosensitivity in the normal control of thermogenesis in homeotherms remains to be elucidated and is an area of current contradictions, some of which have been recently reviewed (12). It seems clear, however, that the afferent neural information from all of the cold thermoreceptors in the body is integrated in the hypothalamus. Furthermore, experimental observations in spinal man (95) and transected animals (96, 97) indicate that some of the efferent pathways from the hypothalamus pass caudally to all levels of the spinal cord, and motor impulses are then transmitted to the effector organs for cold thermogenesis to bring about the appropriate responses (98).

Two principal physiological modes of thermogenesis are used by homeotherms to increase metabolic heat production with decreasing environmental temperature. They include 1) shivering, which involves involuntary tonic or rhythmic muscle activity, and 2) nonshivering thermogenesis (NST), a mechanism for producing heat by means other than shivering. Mammals differ markedly between species and also within certain species in their use of the above thermogenic mechanisms. The thermogenic responses initiated in a particular mammal by a decrease in environmental temperature depend upon the animal's genetic makeup (for example, a hamster will depend upon NST when first exposed to cold, whereas the normal rat will use shivering when initially challenged by a cold stress); its state of postnatal development (NST is an effective mechanism of heat production only in the neonates of large mammals, including man); and its adaptive experience (small mammals such as the rat initially use shivering to increase heat production in response to a cold stress, but as the rat becomes cold acclimated shivering gradually loses importance and is replaced by NST).

It is currently agreed that in adult man and in other large mammals shivering is quantitatively the most important physiological mechanism for thermoregulatory heat production. Recently, however, evidence has been presented which suggests the possibility that adult nonhibernating mammals, including man (99) and infrahuman primates (100), may have the ability to reactivate nonshivering thermogenesis during cold acclimation which presumably has been dormant since neonatal life. At present, however, the role of muscle and brown adipose tissue in mediating nonshivering thermogenic responses in large mammals including man is not understood. These studies have brought into focus the need for further study of the role of NST, not only in long-term cold acclimation, but also in the acute responses of higher primates to cold stress.

As illustrated in Figure 13, there is a clear linear relationship in man between heat production and muscle activity during cold exposure as measured by electromyography (EMG). In these studies, the increase in heat production induced by shivering amounted to approximately 200% of the BMR. Because of the rapidity of the onset of EMG activity upon cold exposure and its rapid cessation when the skin is rewarmed, Davis and Mayer (101) concluded that the observed muscle activity was mediated exclusively by neural mechanisms. The above observations, however, do not exclude the possibility that in higher primates there is also a rapid and significant increase in NST in response to acute cold stress.

The physiological control of metabolic rate in man appears to follow a multiplicative type of regulation in terms of central and peripheral thermal

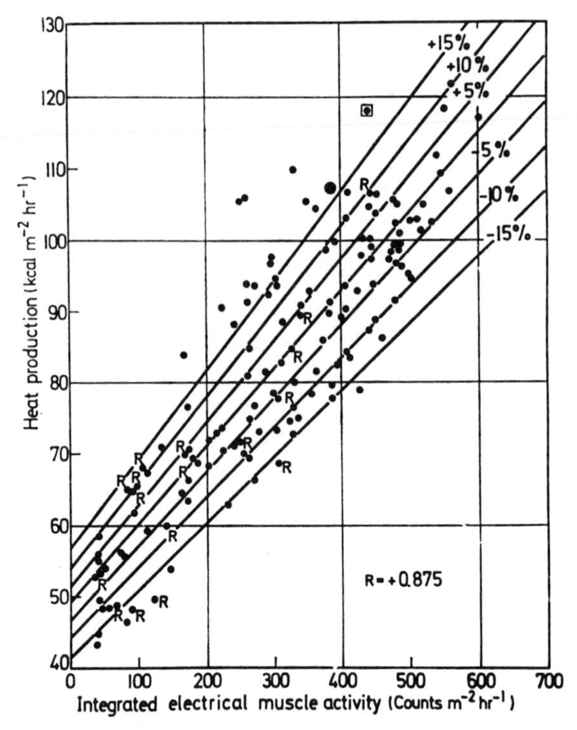

Figure 13. Relationship between 142 individual values for heat production and electrical activity of muscle. One subject stomped his feet for 2.6 min (•) and for 5.15 min (□). R indicates the values attained during 15 min of the 4-hr exposure when the subjects were requested to relax. (Reproduced from Glickman et al. (139) by courtesy of J. Appl. Physiol.)

inputs (102, 103). A similar type of control has recently been reported by Johnson and Elizondo (104) for the metabolic responses of the rhesus monkey. As shown in Figure 9, they found a significant negative linear correlation between metabolic rate and decreasing ambient temperatures below 25°C. As determined by the intercept method of Scholander et al. (66), the lower critical temperature for the initial increase in metabolic rate above the TMR is 24.7°C. In the rhesus monkey, the critical mean skin and rectal temperatures for the initial increase in metabolic rate are 33.5° and 38.1°C, respectively. It is interesting to note that Wilkerson et al. (105) have recently reported that the LCT for unacclimatized male Caucasians averaged 25.2°C and further demonstrated the similarity in the control of thermogenesis in man and this infrahuman primate.

The multiplicative type of regulation of metabolic rate observed in the rhesus monkey and other primates can be modeled by a control systems equation having the following general form.

$$M = \alpha(T_{re} - T_{re_0})\,(T_{\bar{s}} - T_{\bar{s}_0}) + M_0 \ \text{W/m}^2$$

where M equals the metabolic rate at any ambient temperature below the LCT; M_0 is the resting thermoneutral metabolic rate; α is a coefficient describing the gain of the system affecting the metabolic response to cooling of the core and the skin; and T_{re_0} and $T_{\bar{s}_0}$ are the rectal and mean skin temperature thresholds for the onset of the metabolic response to cooling. As previously indicated, the author's group has experimentally determined that the critical $T_{\bar{s}}$ and T_{re} for the rhesus monkey at the LCT of 24.7°C are 33.5°C and 38.1°C, respectively. It is possible to derive α from the data by plotting $(M - M_0)/(T_{re} - 38.1)$ as a function of $T_{\bar{s}}$. The slope of the relationship gives a value for α of 37.8 $W/(m^2 \cdot °C^2)$. Thus, the equation for the control of the metabolic response of the rhesus monkey at ambient temperatures below the LCT is as follows:

$$M = 37.8\,(T_{re} - 38.1)\,(T_{\bar{s}} - 33.5) + 51\ \mathrm{W/m^2}$$

Figure 14 demonstrates a plot of experimentally determined metabolic rate versus predicted metabolic rate calculated from the above model. The *solid line* is the least squares regression line for the data, and the *dashed line* is the line of equality which would result if the model were an exact predictor of the metabolic rate. For a resting monkey, the data points distribute well with

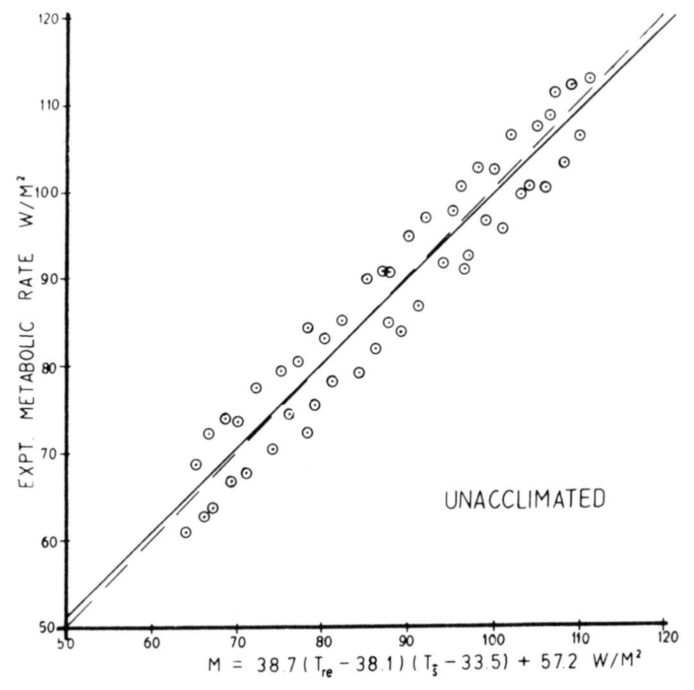

Figure 14. Relationship between experimentally determined metabolic rate for unacclimated rhesus monkey and predicted metabolic rate from a multiplicative model. The *solid line* is the least squares regression line, and the *dotted line* is the line of equality. (R. Elizondo, unpublished data.)

respect to the line of equality, and an analysis of goodness of fit of the model indicated statistical significance and that a predictive model for the control of metabolic rate for the unacclimated rhesus monkey had been obtained.

The importance of such a model in our understanding of the metabolic response of primates to temperature stress is severalfold; 1) it provides a concept of the quantitative interactions of several thermal parameters in the normal control of metabolic rate in a higher primate and 2) it should prove useful in elucidating the nature of the changes in the control of metabolic rate associated with the process of whole body cold acclimation in primates. The significance of such a primate model also lies in the recent observations (106, 107) that prolonged and repetitive local cooling of either the spinal cord or the hypothalamus in the adult normal rat leads to the development of nonshivering thermogenesis and a suppression of shivering thermogenesis. How this shift is controlled in the rat, however, is not known at the present time. Elucidating the significance of local temperature changes and the mechanisms they may activate in higher primates remains an interesting challenge for future research.

Sweating The important role of evaporative heat loss due to sweating in the maintenance of homeothermy in primates was indicated in a previous section. This section, therefore, is primarily concerned with a discussion of our current understanding of the mechanisms that control sweating in man and infrahuman primates.

The sweat glands of higher primates can be classified into two distinct categories based on morphological and physiological characteristics. The apocrine glands (epitrichial), which in man are distributed chiefly in the axillae and pubic regions and in infrahuman primates over the general body surface, secrete relatively small amounts of viscous fluid into the lumen of a hair follicle. They do not become functional until puberty, and there is some decline in activity with age, leading to the hypothesis that they may have a secondary sexual function. The eccrine glands (atrichial), on the other hand, are distributed over the general body surface in both the human and infrahuman primates, secrete relatively large volumes of hyposmotic fluid, and function in heat regulation.

Eccrine sweat glands consists of an irregularly and tightly coiled secretory portion in the dermis, a straight segment, the excretory duct, that extends from the coil to the epidermis, and a superficial spiral segment that lies in the dermis, terminating as an individual pore on the surface of the skin. The secretory coil is composed of a single layer of columnar or pyramidal-shaped cells of two different cell types: smaller ones crowded toward the luminal border which stain with basic dyes, the "dark cells," and larger cells near the basement membrane which stain poorly with basic dyes, the "clear cells." The secretory coil of eccrine sweat glands in primates is innervated by cholinergic sympathetic nerve fibers and is surrounded by blood capillaries.

Recent histochemical studies in man (108) and the rhesus monkey (109) indicate that, after a period of profuse sweating, the glycogen of the clear cells and the acid mucopolysaccharide granules of the dark cells are depleted. It is interesting to note, however, that after recovery from the initial period of profuse sweating both glycogen and the acid mucopolysaccharide granules are replenished in the secretory coil and do not disappear after subsequent periods of profuse sweating. These observations would suggest that in eccrine sweat glands of primates an initial period of profuse sweating imposes such a large energy requirement on the secretory coil that glycogen and mucopolysaccharide stores are decreased. Reappearance of these substances after recovery from the initial period of stress and then their subsequent apparent lack of turnover indicate that an acute adaptive change has taken place at the level of the gland, which probably involves increased utilization of glucose as the major substrate for the energy requirements of the secretory process. Understanding of the physiological and biochemical mechanisms of the secretory process at the level of the gland is very limited and clearly requires further investigation.

Sweat secretion in higher primates is not a continuous function, but is characterized by the cyclical discharge of sweat on the body surface. The technique of resistance hygrometry, which permits continuous measurements of sweating from small areas of the general body surface, has been extensively used to elucidate some of the factors involved in the control of sweating (110, 111).

Figure 15 illustrates a typical continuous sweating record obtained from the forearms of a human subject. This subject had been allowed to equilibrate for 1 hr at an ambient temperature of 43°C and 20% relative humidity. The cyclical nature of the sudomotor mechanism is clearly evident. It is also important to note that the sweating activity on the general body surface of the human is not only cyclical, but is synchronous over the entire body surface. In the human example illustrated in Figure 15, it is evident that the sweating frequency characteristics of the two arms are identical. Figure 16 shows a typical sweating record from the calf and palm of a rhesus monkey after 2 hr of exposure to 42°C and 20% relative humidity. It is interesting to note, that, as in the case of man, both sweating records are cyclical. It is also significant that, although the sweating rate on the palm is approximately twice the rate of the calf, the two

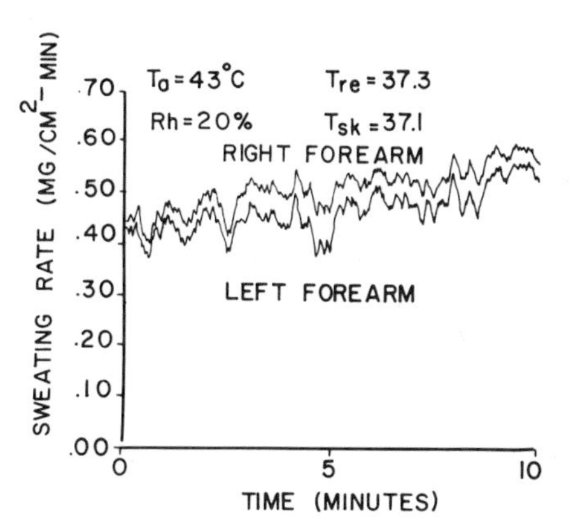

Figure 15. Typical continuous record of sweating activity from the forearms of a human subject in a hot environment (R. Elizondo, unpublished data.)

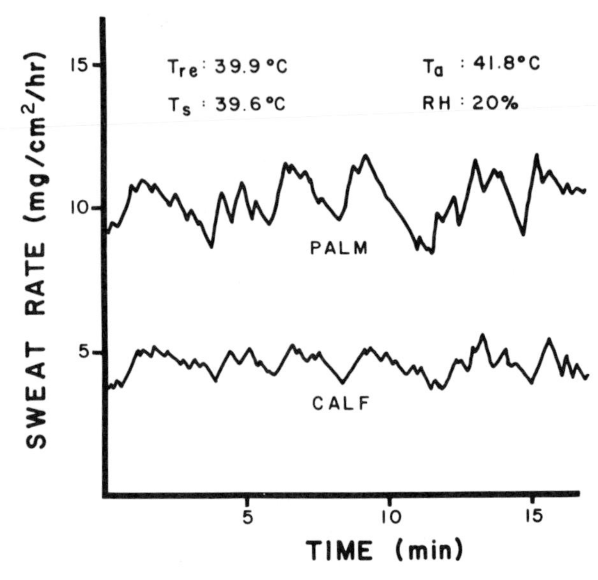

Figure 16. Typical continuous sweating record from the palm and lateral calf of a rhesus monkey in a hot environment. (Reproduced from Johnson and Elizondo (109) by courtesy of J. Appl. Physiol.)

patterns are synchronous. The synchronous discharge of sweat by the eccrine glands over the entire body surface of man and infrahuman primates lends support to the hypothesis that sweating in anthropoid primates is regulated to a large extent by the efferent activity originating in a single control center. The central drive for sweating probably represents an integrated efferent signal determined by numerous relevant thermal inputs (central, peripheral, deep body, etc.) to the thermoregulatory center, presumably in the hypothalamus.

Sweating in resting man has been shown to be controlled by thermal information from both central and peripheral thermal receptors. Benzinger (112, 113) in the late 1960's performed gradient layer calorimeter studies on resting human subjects and reported that, at ambient temperatures in the TNZ and above, the onset of sweating occurred at a predictable cranial temperature (tympanic). As illustrated in Figure 17, increases in skin temperature above 33°C, which would be expected to activate peripheral warm receptors, were reported to have no effect upon the sweat rate. Decreases in skin temperature below 33°C, which would be expected to stimulate peripheral cold receptors, on the other hand, were found to increase the internal temperature at which the onset of sweating occurred and to require higher internal temperatures to maintain equal sweat rates when cold receptors were stimulated. It was concluded, therefore, that the only warm receptor drive involved in the control of sweating was derived from receptors in the central nervous system (hypothalamus), but that a cold receptor drive from cold peripheral receptors introduced

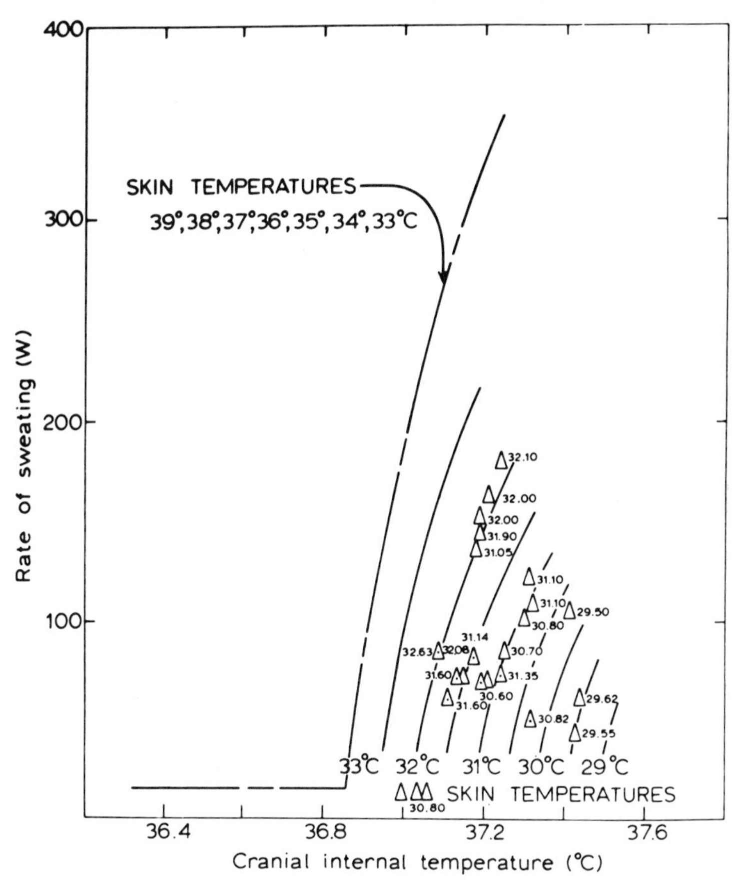

Figure 17. Relationship between the intensity of thermoregulatory sweating in man and tympanic membrane temperature as a function of different skin temperatures. (Reproduced from Benzinger (112) by courtesy of Physiol. Rev.)

an afferent negative feedback signal which reduced the efferent activity from the central controller to the sweat glands.

Banerjee et al. (114), using resistance hygrometry to measure local sweat rates, re-examined the interrelationships between central and peripheral thermal inputs in the control of sweating in man. This continuous technique, with its sensitivity to small immediate changes in evaporative water loss from small skin areas on the general body surface, permitted them to study effectively the dynamic characteristics of the sweating mechanism. In this study, seated male subjects were allowed to equilibrate for 1 hr at ambient temperatures of 37–41°C. The lower leg of each subject was inserted into a water bath in which the level of water, the temperature of the water, and the rate of change of temperature could be independently controlled. Their effects on sweating from

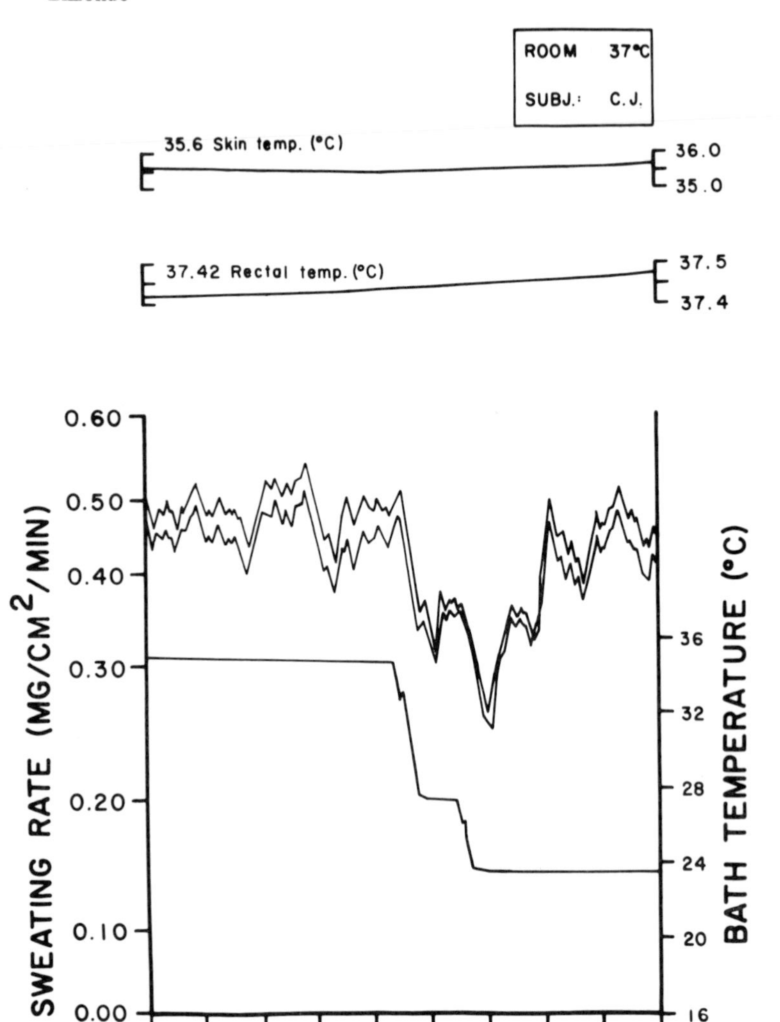

Figure 18. Effects on forearm sweating of a two step decrease in leg skin temperature. In the first step, the decrease was 7°C, and in the second step there was a further decrease of 4°C. The rate of cooling was 13°C/min in both steps. (Reproduced from Banerjee et al. (114) by courtesy of J. Appl. Physiol.)

remote areas on the forearms were evaluated, while central thermal drives remained constant. To establish the neural nature of the afferent input from the peripheral thermal receptors, the circulation to the lower leg was arrested by inflating a cuff on the upper thigh of the immersed leg to supra-arterial pressure. The typical reflex sweat response on the forearms which is associated with a sudden decrease in leg skin temperature is illustrated in Figure 18. In this experiment, the bath temperature (skin temperature) was decreased from 35°C to 24°C in a two step fashion. It can be seen that the initial drop in leg skin temperature to 28°C resulted in a sharp reflex depression in generalized body sweating. This decline was followed by a reversal of the sweating rate toward precooling levels shortly after the bath temperature stabilized at 28°C. A further decrease in the sweat rate was induced by the second step change in skin temperature to 24°C. The second decline was again followed by reversal shortly after the skin temperature of the lower leg stabilized at 24°C. Sweat production on the forearms soon recovered to the initial sweating rate even though the skin temperature of the lower leg was maintained constant at the lower level of 24°C. Identical results were observed whether the skin was decreased from 42°C to 36°C or from 30°C to 24°C, indicating that both a decrease in the firing rate of warm receptors and an increase in the firing rate of cold receptors were involved in the response. These studies, in contrast to the conclusions of Benzinger, clearly established that a rate response from both cold and warm peripheral thermoreceptors provides a significant anticipatory feedback loop to the central control system regulating human eccrine sweating.

Further clarification of the interrelationships between skin temperature, central temperature, and sweat rate in man has been provided by the studies of Nadel et al. (115), which use exercise and radiant heating to induce independent changes in average skin temperature and internal temperature. These studies confirmed the rate of change component of cold and warm peripheral receptors in the control of sweating. As illustrated in Figure 19, they also confirmed earlier observations that when mean skin temperature is constant sweating rate is linearly related to internal (esophageal) temperature. Furthermore, a decrease in mean skin temperature produced a parallel decrease in the sweating versus

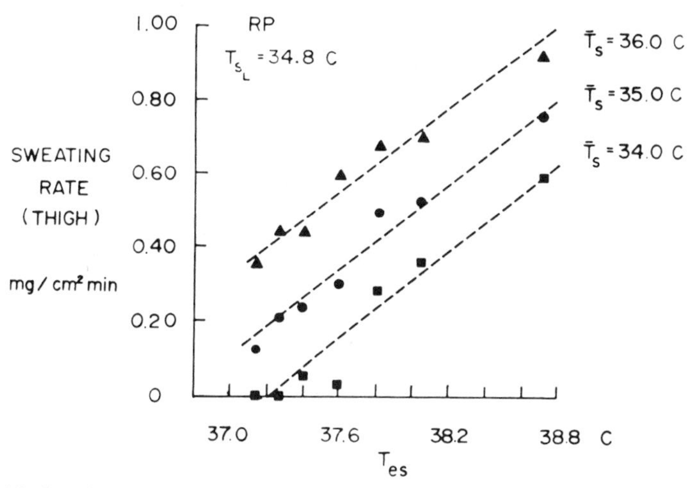

Figure 19. Local sweat rate on the thigh at a constant local skin temperature of 34.8°C as a function of mean skin temperature and esophageal temperature. (Reproduced from Nadel et al. (115) by courtesy of J. Appl. Physiol.)

esophageal temperature relationship, whereas an increase in mean skin temperature produced a parallel increase in the relationship. These results support the concept that the control of sweating in man can be described by a summation model, in which sweat rate is linearly related to core (esophageal) temperature and the level of mean skin temperature shifts the relationship in the appropriate direction.

In 1974, Wyss et al. (81) re-examined the problem of the role of skin versus core temperature in the control of sweating in resting man. In these studies, a water-perfused suit was used to control skin temperature, and both esophageal and arterial blood temperatures were used as indices of core temperature. Sweating was continuously monitored with the use of a sweat capsule sealed to the skin of the forearm. A multiple linear regression analysis of sweat rate against esophageal, arterial blood, and skin temperature revealed that sweat rate was virtually independent of steady state skin temperature when arterial blood was used as a measure of internal temperature. Furthermore, it was concluded that the use of esophageal temperature as a measure of core temperature overestimated the role of skin temperature in the control of sweating and could account for the different conclusions reached by Nadel et al. (115).

Elizondo et al. (116), and Smiles et al. (117) have recently taken a more direct approach than in the previous studies in an attempt to clarify the relative role of central and peripheral temperature inputs in the control of sweating in primates.

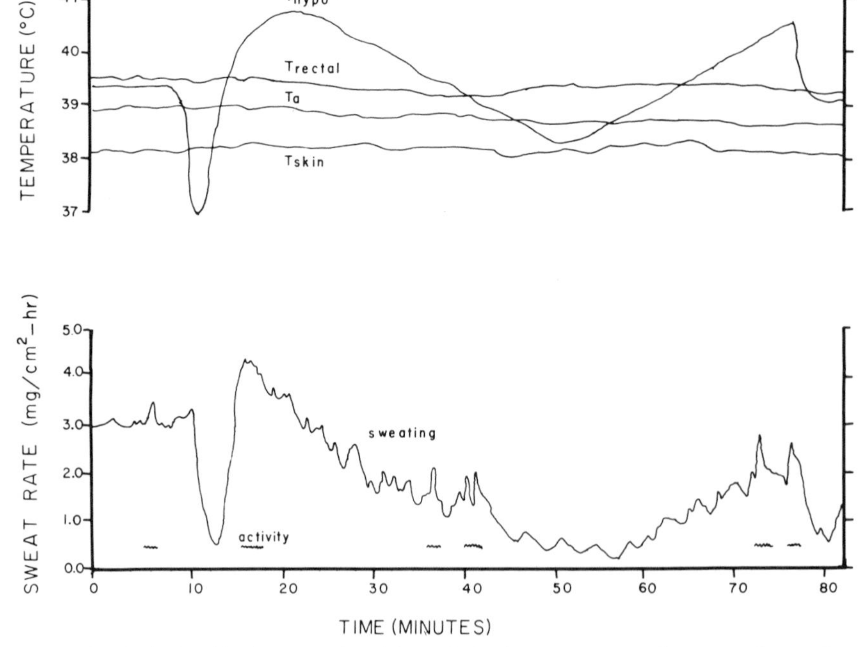

Figure 20. Relationship between local hypothalamic temperature changes and sweating activity on the lateral calf of the rhesus monkey. (R. Elizondo, unpublished data.)

Heat-acclimated unanesthetized rhesus monkeys were used as an animal model for studies which could not be performed in man, in that the monkeys had been implanted with thermodes in the anterior hypothalamic area of the brain. A thermode water perfusion system functionally similar to those of Hammel et al. (118) was used to manipulate hypothalamic temperature independently of skin temperature. The technique of resistance hygrometry was used in conjunction with the thermode perfusion system to quantify the relationships between local sweat rate, hypothalamic temperature, and skin temperature.

Figure 20 illustrates the close coupling observed in this study between sweating and hypothalamic temperature. It is clear that changes in hypothalamic temperature without any significant change in any of the other measured temperatures were followed by the appropriate change in local sweat rates.

Figure 21 summarizes the major findings in this study concerning the interrelationship between sweating on the general body surface of primates, hypothalamic temperature, and skin temperature. It illustrates the relationship between sweating rates as recorded from the lateral calf of a rhesus monkey as a function of different levels of clamped hypothalamic temperature and at two mean skin temperatures. Each data point represents the average sweat rate recorded from the lateral calf during a 3–5 min period associated with a different level of clamped hypothalamic temperature. It can be seen that a significant linear correlation was observed between sweat rate and hypothalamic temperature. A decrease in skin temperature produced a parallel shift to the right, resulting in no change in the slope, but a statistically significant increase in the core (hypothalamic) temperature associated with the first secretion of sweat on the surface of this primate. In other words, an increase in mean skin temperature resulted in an increase in the afferent activity to the central integrator for sweating, resulting in a lower central temperature threshold for sweating with no change in the sensitivity of the sweat gland apparatus itself, as indicated by the parallel shift in the sweating versus hypothalamic temperature relationship. This study provides direct support for the conclusions of Nadel et

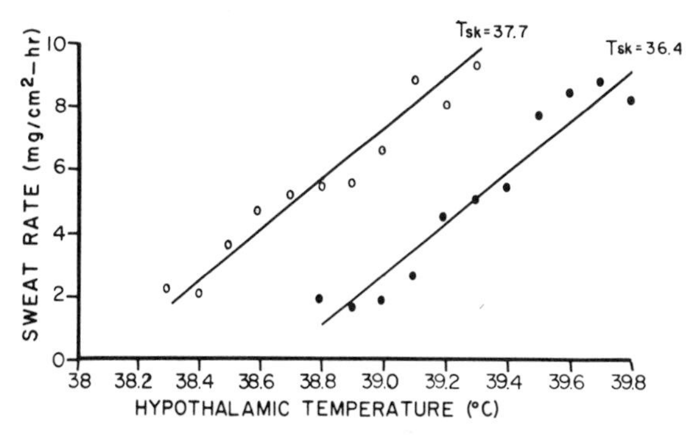

Figure 21. Relationship between hypothalamic temperature and lateral calf sweat rate at two different mean skin temperatures in the same rhesus monkey. (Reproduced from Smiles et al. (117) by courtesy of J. Appl. Physiol.)

al. (115) that central temperature interacts additively with skin temperature in the control of sweating in higher primates.

It seems well established, therefore, that the activity of eccrine sweat glands in primates is under the direct neural control of a single central thermal regulator whose efferent activity is dependent upon integration of both central and peripheral cold and warm thermal receptor inputs. Less well established, however, is the fact that the relationship between the output of the central controller and eccrine sweat gland function in terms of onset and magnitude of sweat secretion on the skin surface can be significantly changed by alterations in the microenvironment at the level of the glands (110, 119). Local factors which have been shown to influence the activity of sweat glands include local skin temperature, the degree of skin hydration, the availability of neurotransmitters, and possibly local training.

Figure 22 illustrates the local factor which has been most extensively studied and demonstrates the effects of a local skin temperature change on sweating

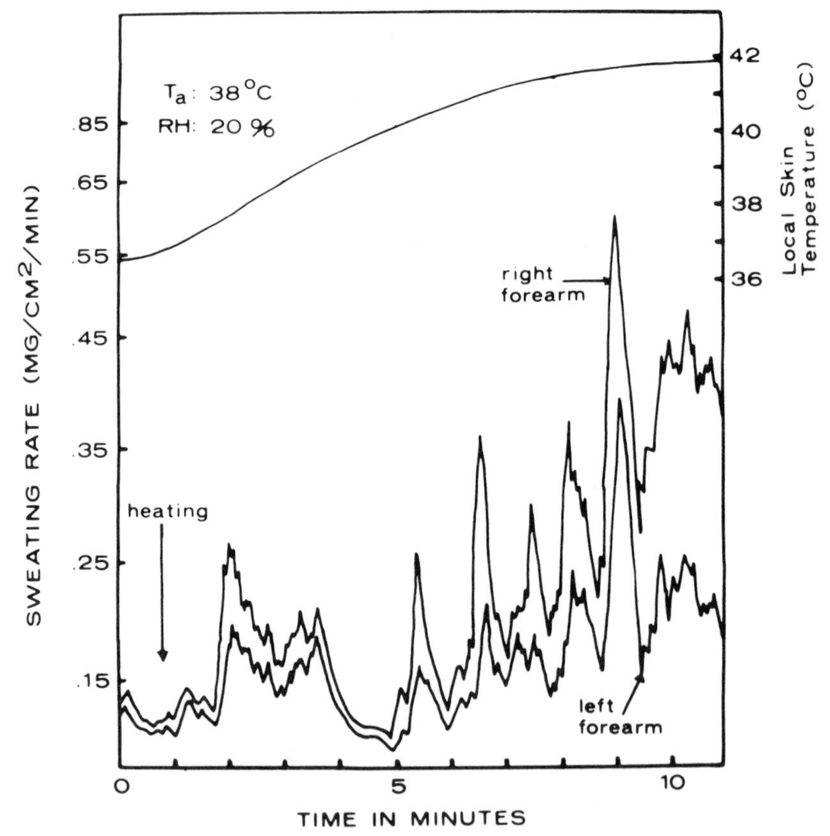

Figure 22. Effect of local skin temperature on the magnitude and cyclical pattern of forearm sweating in the human. (Reproduced from Elizondo (110), Fed. Proc. 32: 1583–1587, by courtesy of Fed. Proc.)

activity. A resting subject was allowed to equilibrate for 1 hr in a warm room maintained at 38°C and 20% relative humidity. At zero time, the area enclosed by the sweat capsule on the right forearm was locally heated. Because the skin area being thermally altered represented approximately 1/1500th of the total body surface, it can be assumed that neither mean skin temperature nor core temperature was significantly affected (119). It can be seen that local heating did not affect the basic synchronous cyclical pattern of the sweating response, but it did significantly increase the amplitude of the sweating cycles of the right arm as compared with the control areas on the contralateral arm. This suggests that for each volley of central efferent impulses arriving at the glands more sweat was produced in the heated area. Furthermore, it has been shown that the local heating response cannot be elicited in the absence of generalized body sweating or after local blocking of neuroglandular transmission in sweating subjects. These observations have led to the conclusion that the local temperature effect on sweating is not simply due to a Q_{10} effect on glandular metabolism of the skin, resulting in increased water evaporation (120). It has been proposed, therefore, that the effect is preglandular and may involve an increase in the amount of transmitter substance released at the neuroglandular junction which acts as a temperature-sensitive amplified or attenuator mechanism at the level of the effector (sweat gland) in the thermoregulatory system of higher primates. Recent studies in man (115, 121) and in the squirrel monkey (122) have provided indirect support for the above hypothesis.

Another factor which can significantly influence sweat gland function is the local availability of neuroglandular transmitter substance. Figure 23 illustrates the effects of an exogenous parasympathomimetic agent on the onset and magnitude of sweating in response to whole body heating. In this experiment, the subject was initially placed in a cold room at 20°C for 30 min to decrease peripheral and core temperature below sweating threshold levels. After 25 min in the cold environment, the subject's right forearm received an intradermal injec-

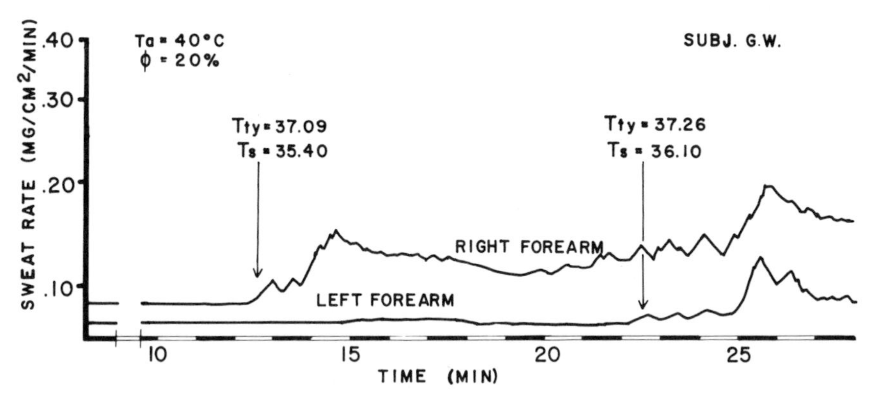

Figure 23. Effect of exogenous pilocarpine treatment of the skin on the latency and threshold temperatures for the onset of forearm sweating in a resting human subject. (Reproduced from Elizondo (110) by courtesy of Fed. Proc.)

tion of pilocarpine (0.1 ml of 10^{-5}). The subject was then transferred into a hot room maintained at 40°C, and hygrometry sweat capsules were attached to contralateral areas on both forearms. It has been consistently observed under these conditions that increased transmitter substance at the level of the sweat glands results in a significant decrease in latency and threshold temperatures for the onset of sweating. In this experiment, it can be seen that the pilocarpine-treated right forearm showed a latency for sweating of only 12 min and was associated with tympanic and skin temperatures of 37.09°C and 35.4°C, respectively. The onset of cyclical sweating in the untreated area had a longer latency for the initiation of sweating of 22 min and was associated with a higher tympanic temperature of 37.26°C and a skin temperature of 36.1°C. It is significant that similar but smaller differences in latencies and threshold temperatures can be induced by simply hydrating one skin area as compared to the control. It seems clear, therefore, that eccrine sweat glands receiving identical integrated central neural inputs can show significantly different latency, magnitude, and threshold temperatures for sweating. Furthermore, these differences are related to the local conditions of the microenvironment (i.e., temperatures, degree of hydration, etc.) at the level of the glands themselves and must be considered an important parameter in the control of eccrine sweat gland function in higher primates. Although the mechanisms by which local factors modify sweat gland function require further study, it seems reasonable to suggest that they may be largely explained by potentiation of both subthreshold and suprathreshold neural impulses arriving at the glands (123).

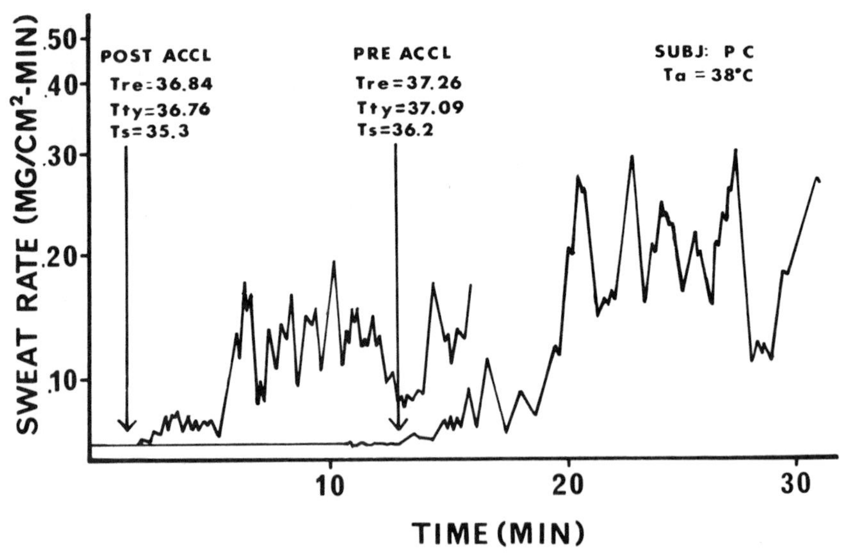

Figure 24. Threshold temperatures and latencies for the onset of cyclical sweating on the forearm of a resting human subject before and after heat acclimation. (Reproduced from Elizondo (111) by courtesy of Int. J. Biometeorol.)

Numerous investigations have reported the common observation that repeated exposure to hot environments leads to a greater sweat production at equivalent central and peripheral thermal drives (124–127). Some of the characteristic functional changes in the sweating mechanism following heat acclimation are illustrated in Figure 24. This figure illustrates the latency and threshold temperatures associated with the onset of sweating activity on the right forearm of the same subject before and after heat acclimation. In each experiment, the subject was initially placed in a cold room at 20°C for 30 min. The subject was then moved into a climatic chamber maintained at 38°C, and a continuous record of sudomotor activity on the forearm was obtained with the use of resistance hygrometry. It can be seen that in the preacclimation experiment no sweating activity was recorded from the forearm during the first 13 min of heat exposure. The body temperatures recorded at the onset of sweating in this experiment are indicated above the *arrow* in the center of the graph marking the onset of forearm sweating. The continuous record on the left illustrates the sweating activity of the right forearm of the same subject during the same experimental conditions following heat acclimation induced by 90 min of work in the heat on 9 consecutive days. It can be seen that the latency for the onset of sweating on the forearm was significantly shorter after heat acclimation. Furthermore, all of the recorded temperatures at the onset of sweating were significantly lower than in the preacclimation experiment. Although not illustrated in Figure 24, the steady state sweat rates on the forearm of 12 different subjects were all consistently higher following heat acclimation.

Although the functional changes induced by heat acclimation in the sweating responses of higher primates have been adequately defined, the physiological mechanisms responsible for the increased responsiveness of the sweating system, however, have not been clearly indicated. Chen and Elizondo (127) recently reported that the increased sweating capacity observed following heat acclimatization was associated with an increase in the responsiveness of the sweat gland apparatus itself. In this study, the steady state sweat rates recorded from the forearm of human subjects during whole body heating (45.5°C) were reproduced in a cool room (24°C) by electrical stimulation of the sweat glands on the same skin area on the forearm. The subjects were then heat acclimatized by a standard 90-min work period in the heat (49°C, 20% relative humidity) performed on 9 consecutive days. After heat acclimatization, the forearm steady state sweat rates for each subject were again determined during identical whole body thermal stress and when stimulated in a cool room with identical electrical stimulation as in the preacclimation measurements.

Table 1 summarizes the forearm sweat rate changes observed under identical thermal and electrical conditions before and after heat acclimation. It can be seen that in the four subjects studied the average sweat rate on the forearm increased from 0.38 mg/cm^2·min to 0.49 mg/cm^2·min after acclimation and represented a 28% increment in forearm sweat rate during whole body heating. Similarly, it can be seen that the average sweat rate produced in a cool room by identical electrical conditions increased from 0.36 mg/cm^2·min prior to acclima-

Table 1. Sweat rate changes observed under identical thermal and electrical conditions before and after heat acclimation

Subject	Condition[a]	Sweat rate ($mg/cm^2 \cdot min$) Pre-acclimation	Post-acclimation	Change (%)
1	45.5°C, 20% rh	0.48	0.70	+46
	1.6 ma at 24°C	0.48	0.70	+46
2	45.5°C, 20% rh	0.51	0.68	+33
	1.4 ma at 24°C	0.49	0.66	+35
3	45.5°C, 20% rh	0.27	0.30	+11
	1.1 ma at 24°C	0.27	0.30	+11
4	45.5°C, 20% rh	0.21	0.26	+24
	1.0 ma at 24°C	0.21	0.27	+29
Group	Thermal stimulation	0.38	0.49	+28
Average	Electrical stimulation	0.36	0.48	+33

[a] rh, relative humidity; ma, milliamperes.

tion to 0.48 $mg/cm^2 \cdot min$ after acclimation, an average increase of 33%. The increase in forearm sweating after acclimation during whole body thermal stress was not significantly different ($p > 0.30$) from the increase observed following local electrical stimulation of the sweat glands in a cool room when no obvious central neural input is involved.

Another series of experiments in this study confirmed the observation of Fox et al. (124) that local cooling of the forearm during the acclimation procedure prevents an increase in sweating capacity. In these experiments, the sweat rate of contralateral areas on both forearms of five unacclimated male subjects was measured by resistance hygrometry. The steady state sweat rates recorded while the subjects were seated in a climatic chamber maintained at 49°C with 20% relative humidity standardized as 100%, to which the sweat rates of the same areas under the same thermal stress were compared after acclimation. During the work period of acclimation, however, the area on the right forearm (treated area) was cooled to an average skin temperature of 5–9°C by circulating ice water through a cuff loosely applied to the forearm. The average skin temperature of the control left forearm was between 35°C and 36°C during the work period. As illustrated in Figure 25, the untreated area on the control arm showed a significant increase in sweating after acclimation, and in five subjects the increase averaged 32%. No increase, however, was observed in the cold-treated contralateral forearm. The central neural drive to both forearm areas is presumably identical, but the sweat output was obviously different. These results are in

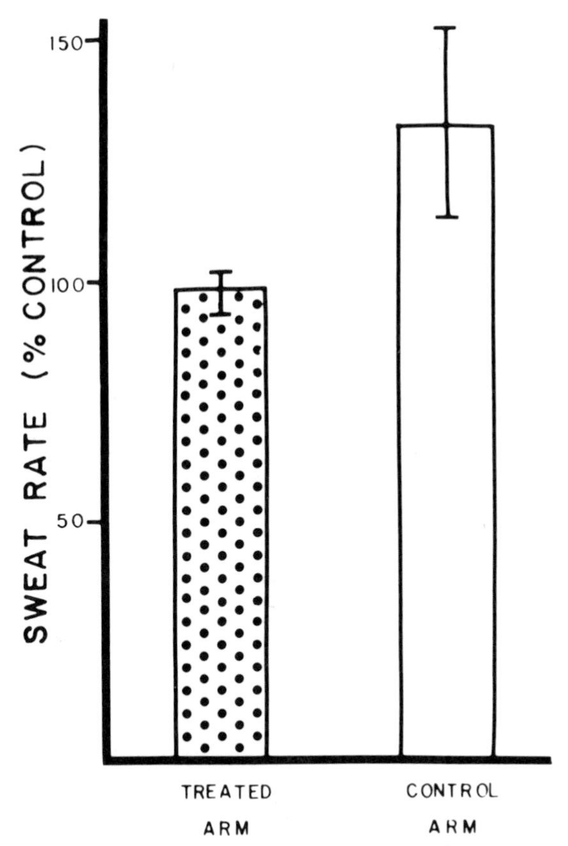

Figure 25. The effects of local skin cooling during the acclimation period on contralateral forearm sweat rates of five human subjects. (Reproduced from Chen and Elizondo (127) by courtesy of J. Appl. Physiol.)

agreement with those of the previous study and support the concept that the greater sweat production following short-term heat acclimation in higher primates is due, in part, to local functional changes resulting in an increase in the responsiveness of the sweat gland apparatus itself (124, 127, 128). These studies, however, are not inconsistent with the currently accepted view that the increased sweating responsiveness also involves significant functional changes in the CNS and modifications in the central drive for sweating following heat acclimatization (125, 129). Future studies of nerve-gland preparations in infrahuman primates should prove very helpful in elucidating the precise mechanisms, both central and peripheral, associated with the increased sweating capacity of higher primates after heat acclimatization.

There now exists a considerable body of evidence which shows that the E prostaglandins are consistently hyperthermic when injected intrahypothalamically or intraventricularly in all placental mammalian species thus far examined

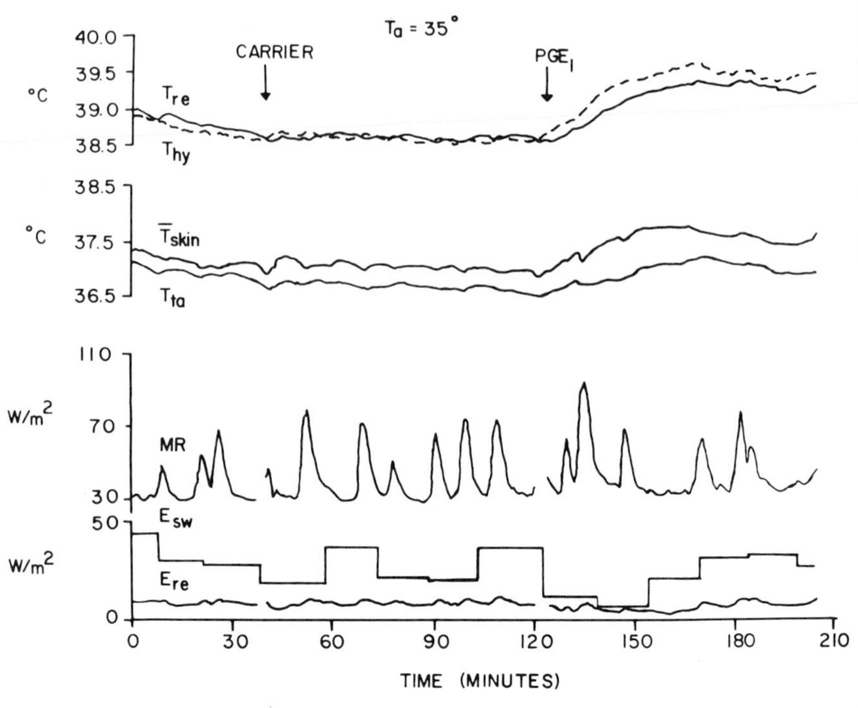

Figure 26. Record of the thermal and metabolic responses of an unanesthetized rhesus monkey to an injection of carrier solution followed by 200 ng of PGE_1 into the same site in the PO/AH at an ambient temperature of 25°C. (Reproduced from Barney and Elizondo (140) by courtesy of J. Appl. Physiol.)

(130). It has been suggested, therefore, that prostaglandins may be the final chemical mediator of pyrogen fevers (131). Although the role of prostaglandins in mediating pryogen fevers remains to be elucidated (132), its hyperthermic effects in the rat (133), cat (134), rabbit (134), sheep (135), and squirrel monkey (136) have been clearly demonstrated. None of the mammals in which the genesis of prostaglandin hyperthermia has been examined, however, use body sweating as the major avenue of evaporative heat loss, and the role of sweating in the development of prostaglandin fever has not been determined.

Recent thermal balance studies in the rhesus monkey (140) indicate that prostaglandin E_1 (PGE_1) has a significant inhibitory affect on general body sweating in higher primates. Figure 26 shows an example of the thermal and metabolic responses of an unanesthetized rhesus monkey to an injection of a carrier solution followed by an injection of 200 ng of PGE_1 into the preoptic anterior hypothalamic region. It can be seen that following the carrier injection neither rectal (T_{re}) nor hypothalamic (T_{hy}) temperature changed significantly. Furthermore, no consistent changes in metabolic heat production (MR), respiratory heat loss (E_{re}), or evaporative heat loss due to sweating (E_{sw}) were observed. Following the PGE_1 injection, there was a maximal spike in metabolic rate and a large decrease in sweating heat loss. These changes consistently led to an increase in both T_{re} and T_{hy}, with the increase in T_{hy} preceeding the T_{re} increase. Mean skin temperature ($\overline{T}_{skin}$) and tail temperature (T_{ta}) also increased following the PGE_1 injection, whereas respiratory heat loss remained essentially unchanged. These results indicate that, in the rhesus monkey equilibrated to a warm environment, PGE_1 fever is mediated primarily by a significant decrease in evaporative heat loss due to sweating and an increase in metabolic heat production. The physiological role of prostaglandins, however, in normal thermal regulation of homeotherms remains totally unknown and remains a significant area of interest for future research in thermoregulation.

REFERENCES

1. Henriques, F. C., Jr. (1947). Studies of thermal injury: the predictability and significance of thermally induced rate leading to irreversible epidermal injury. Arch. Pathol. 43:489.
2. Smith, L. W., and Fay, T. (1939). Temperature factors in cancer and embryonal cell growth. J.A.M.A. 113:653.
3. Bligh, J., and Moore, R. E. (eds.) (1972). Essays on Temperature Regulation. North-Holland, Amsterdam.
4. Cabanac, M. (1975). Temperature regulation. Annu. Rev. Physiol. 37:415.
5. Chaffee, R. R. J., and Roberts, J. C. (1971). Temperature acclimation in birds and mammals. Annu. Rev. Physiol. 33:155.
6. Gale, C. C. (1973). Neuroendocrine aspects of thermoregulation. Annu. Rev. Physiol. 35:391.
7. Hellon, R. (1975). Monoamines, pyrogens and cations: their actions on control of body temperature. Pharmacol. Rev. 26:289.
8. Hensel, H. (1974). Thermoreceptors. Annu. Rev. Physiol. 36:233.
9. Lomax, P., Schonbaun, E., and Jacob, J. (1975). Temperature Regulation and Drug Action. Karger, Basel.
10. Robertshaw, D. (1974). MTP International Review of Science, Physiology Series One, Vol. 7, Environmental Physiology. London, Butterworths.
11. Wyndham, C. H. (1973). The physiology of exercise under heat stress. Annu. Rev. Physiol. 35:193.
12. Bligh, J. (1973). Temperature Regulation in Mammals and Other Vertebrates. North-Holland, Amsterdam.
13. Chowers, I., and Bligh, J. (eds.) (1976). Proceedings of the Jerusalem symposium on temperature regulation. Isr. J. Med. Sci. 12:905.
14. Hensel, H. (1973). Cutaneous thermoreceptors. *In* A. Iggo (ed.), Handbook of Sensory Physiology, Vol. 11, pp. 79–110. Springer, New York.
15. Iggo, A. (1970). The mechanisms of biological temperature reception. *In* J. D. Hardy, A. P. Gagge, and J. A. J. Stolwijk (eds.), Physiological and Behavioral Temperature Regulation, p. 391. Charles C Thomas, Springfield, Illinois.
16. Hensel, H. (1970). Temperature receptors in the skin. *In* J. D. Hardy, A. P. Gagge, and J. A. J. Stolwijk (eds.), Physiological and Behavioral Temperature Regulation, p. 442. Charles C Thomas, Springfield, Illinois.

17. Hensel, H., and Iggo, A. (1971). Analysis of cutaneous warm and cold fibers in primates. Pfluegers Arch. 329:1.
18. Iggo, A. (1969). Cutaneous thermoreceptors in primates and sub-primates. J. Physiol. 200:403.
19. Hensel, H., and Bowman, K. (1960). Afferent impulses in cutaneous sensory nerves in human subjects. J. Neurophysiol. 23:564.
20. Torebjork, H. E., and Halin, R. G. (1972). Activity in C fibers correlated to perception in man. *In* C. Hirsch and Y. Zotterman (eds.), Cervical Pain, p. 171. Pergamon, Oxford.
21. Zenz, M., Fruhstorfer, H., Nolte, H., and Hensel, H. (1973). Dissociated loss of cold and warm sensibility during regional anesthesia. Pfluegers Arch. (Suppl.) 339:171.
22. Smiles, K., Elizondo, R., and Barney, C. (1976). Sweating responses during changes of hypothalamic temperature in the rhesus monkey. J. Appl. Physiol. 40:653.
23. Ingram, D., McLean, J., and Whittow, G. (1963). The effect of heating the hypothalamus and the skin on the rate of moisture vaporization from the skin of the ox (*Bos taurus*). J. Physiol. 169:349.
24. Ingram, D., and Whittow, G. (1962). The effect of heating the hypothalamus on respiration in the ox (*Bos taurus*). J. Physiol. 163:200.
25. Schonüng, W., Wagner, H., Jessen, C., and Simon, E. (1971). Differentiation of cutaneous and intestinal blood flow during hypothalamic heating and cooling in aneasthetized dogs. Pfluegers Arch. 328:145.
26. Adair, E. R. (1976). Autonomic thermoregulation in squirrel monkey when behavioral regulation is limited. J. Appl. Physiol. 40:694.
27. Gale, C. C., Jobin, M., Proppe, D. W., Notter, D., and Fox, H. (1970). Endocrine thermoregulatory responses to local hypothalamic cooling in unanesthetized baboons. Am. J. Physiol. 219:193.
28. Stitt, J. T., and Hardy, J. D. (1971). Thermoregulation in the squirrel monkey (*Saimiri sciureus*). J. Appl. Physiol. 31:48.
29. Schonüng, W., Jessen, C., Wagner, H., and Simon, E. (1971). Regional blood flow antagonism induced by central thermal stimulation in the conscious dog. Experientia 27:1291.
30. Iriki, M., Riedel, W., and Simon, E. (1971). Regional differentiation of sympathetic activity during hypothalamic heating and cooling in anesthetized rabbits. Pfluegers Arch. 328:320.
31. Bruck, K., and Schwennicke, H. P. (1971). Interaction of superficial and hypothalamic thermosensitive structures in the control of non-shivering thermogenesis. Int. J. Biometerol. 15:156.
32. Hardy, J. D. (1973). Posterior hypothalamus and the regulation of body temperature. Fed. Proc. 32:1564.
33. Wunnenberg, W., and Hardy, J. D. (1972). Response of single units of the posterior hypothalamus to thermal stimulation. J. Appl. Physiol. 33:547.
34. Cabanac, M. (1970). Interaction of cold and warm temperature signals in the brain stem. *In* J. D. Hardy, A. P. Gagge, J. A. J. Stolwijk (eds.), Physiological and Behavioral Temperature Regulation, p. 549. Thomas, Springfield, Illinois.
35. Chai, C. Y., and Lin, M. T. (1972). Effects of heating and cooling the spinal cord and medulla oblongata on thermoregulation in monkeys. J. Physiol. 225:297.
36. Baker, J. L., and Carpenter, D. O. (1970). Thermosensitivity of neurons in the sensorimotor cortex of the cat. Science 169:597.
37. Tabatabai, M. (1972). Respiratory and cardiovascular responses resulting from heating the medulla oblongata in cats. Am. J. Physiol. 222:1558.

38. Tabatabai, M. (1972). Respiratory and cardiovascular responses from cooling the medulla oblongata in cats. Am. J. Physiol. 223:8.
39. Nakayama, T., Eisenman, J. S., and Hardy, J. D. (1961). Single unit activity of anterior hypothalamus during local heating. Science 134:560.
40. Nakayama, T., Hammel, H. T., Hardy, J. D., and Eisenman, J. S. (1963). Thermal stimulation of electrical activity of single units of the preoptic region. Am. J. Physiol. 204:1122.
41. Eisenman, J. S., and Jackson, D. C. (1967). Thermal response pattern of septal and preoptic neurons in cats. Exp. Neurol. 19:33.
42. Hellon, R. F. (1970). Hypothalamic neurons responding to changes in hypothalamic and ambient temperatures. *In* J. D. Hardy, A. P. Gagge, J. A. J. Stolwijk (eds.), Physiological and Behavioral Temperature Regulation, p. 463. Thomas, Springfield, Illinois.
43. Guieu, J. D., and Hardy, J. D. (1971). Integrative activity of preoptic units. I. Responses to local and peripheral temperature changes. J. Physiol. (Paris) 63:253.
44. Cabanac, M., Stolwijk, J. A. J., and Hardy, J. D. (1968). Effect of temperature and pyrogens on single unit activity in the rabbits brain stem. J. Appl. Physiol. 24:645.
45. Goltz, F., and Edwards, I. R. (1896). Der hund mit verkurztem ruckenmark. Pfluegers Arch. 63:362.
46. Thauer, R., and Peters, G. (1937). Warmergulation nach operativer ausschaltung des warmezentrums. Pfluegers Arch. 239:483.
47. Jessen, C., and Simon, E. (1971). Spinal cord and hypothalamus as core sensors of temperature in the conscious dog. III. Identity of functions. Pfluegers Arch. 324:217.
48. Jessen, C., and Ludwig, O. (1971). Spinal cord and hypothalamus as core sensors of temperature in the conscious dog. II. Addition of signals. Pfluegers Arch. 324:205.
49. Lin, M. T., Yin, T. H., and Chai, C. Y. (1974). Independence of spinal cord and medulla oblongata on thermal activity. Am. J. Physiol. 226:1066.
50. Guieu, J. D., and Hardy, J. D. (1970). Effects of preoptic and spinal cord temperature in control of thermal polypnea. J. Appl. Physiol. 28:540.
51. Riedel, W., Siaplauras, G., and Simon, E. (1973). Intraabdominal thermosensitivity in the rabbit as compared with spinal thermosensitivity. Pfluegers Arch. 340:59.
52. Kosaka, M., Simon, E., Walther, O. E., and Thauer, R. (1969). Response of respiration to selective heating of the spinal cord below partial transection. Experientia 25:36.
53. Ingram, D. L., and Legge, K. F. (1972). The influence of deep body temperature and skin temperatures on respiratory frequency in the pig. J. Physiol. 220:283.
54. Jessen, C., McLean, J. A., Calvery, D. J., and Findlay, J. D. (1972). Balanced and unbalanced temperature signals generated in spinal cord of the ox. Am. J. Physiol. 222:1343.
55. Chai, C. Y., and Lin, M. T. (1972). Effects of heating and cooling the spinal cord and medulla oblongata on thermoregulation in monkeys. J. Physiol. 225:297.
56. Jessen, C., Simon, E., and Kullmann, R. (1968). Interaction of spinal and hypothalamic thermodetectors in body temperature regulation of the conscious dog. Experientia 24:694.
57. McLean, J., Hales, J., Jessen, C., and Calvert, D. (1970). Influences of spinal cord temperature on heat exchange of the ox. Aust. Physiol. Pharmacol. Soc. 1 2:32.

58. Hales, J., and Jessen, C. (1969). Increase of cutaneous moisture loss caused by local heating of the spinal cord in the ox. J. Physiol. 204:40.
59. Rawson, R. D., and Quick, K. P. (1971). Unilateral splanchnicotomy: its effect on the response to intra-abdominal heating in the ewe. Pfluegers Arch. 930:362.
60. Rawson, R. O., and Quick, K. P. (1972). Localization of intra-abdominal thermoreceptors in the ewe. J. Physiol. 222:665.
61. Adair, E. R., and Wright, B. A. (1976). Behavioral thermoregulation in the squirrel monkey when response effort is varied. J. Comp. Physiol. Psychol. 90:197.
62. Erikson, H., and Krog, J. (1956). Critical temperature in naked man. Acta Physiol. Scand. 37:35.
63. Scholander, P. F., Hock, R., Walters, V., and Irving, L. (1950). Heat regulation in some arctic and tropical mammals and birds. Biol. Bull. 99:237.
64. Hardy, J. D., Stolwijk, J. A., and Gagge, A. P. (1971). Man. *In* G. C. Whittow (ed.), Comparative Physiology of Thermoregulation, Vol. II, p. 327. Academic Press, New York.
65. Wilkerson, J. E., Raven, P. B., and Horvath, S. M. (1972). Critical temperature of unacclimated male Caucasians. J. Appl. Physiol. 33:451.
66. Scholander, P. F., Walters, V., Hock, R., and Irving, L. (1950). Body insulation of some arctic and tropical mammals and birds. Biol. Bull. 99:224.
67. Chaffee, R. R. J., and Allen, J. R. (1973). Effects of ambient temperature on the resting metabolic rate of cold and heat acclimated *Macaca mulatta*. Comp. Biochem. Physiol. 44a:1215.
68. Stitt, J. T., and Hardy, J. D. (1971). Thermoregulation in the squirrel monkey (*Saimiri sciureus*). J. Appl. Physiol. 31:48.
69. Nakayama, T., Hori, T., Nagasaka, T., Tokura, H., and Tadaki, E. (1971). Thermal and metabolic responses in Japanese monkey at temperatures of 5–38°C. J. Appl. Physiol. 31:332.
70. Rowell, L. B. (1974). Human cardiovascular adjustments to exercise and thermal stress. Physiol. Rev. 54:75.
71. Thauer, R. (1963). Circulatory adjustments to climatic requirements. Handbook of Physiology, Vol. III, p. 1921. American Physiological Society, Washington, D.C.
72. Greenfield, A. D. M., Whitney, R. J., and Mowbray, J. F. (1963). Methods for the investigation of peripheral blood flow. Br. Med. Bull. 19:101.
73. Wyss, C. R., Rowell, L. B., and Feigl, E. O. (1976). Lack of significant humanlike cutaneous vasodilation in heat-stressed baboons. J. Appl. Physiol. 41:528.
74. Freeman, N. E. (1935). The effect of temperature on the rate of blood flow in the normal and in the sympathectomized hand. Am. J. Physiol. 113:384.
75. Shepherd, J. T. (1963). Physiology of the Circulation in Human Limbs in Health and Disease, p. 91. W. B. Saunders, Philadelphia.
76. Fox, R. H., and Hilton, S. M. (1958). Bradykinin formation in human skin as a factor of heat vasodilitation. J. Physiol. 142:219.
77. Senay, C. C., Prokop, L. D., Caneu, L., and Hertzman, A. B. (1963). Relation of local skin temperature and local sweating to cutaneous blood flow. J. Appl. Physiol. 18:781.
78. Shvartz, E., Sarr, E, Meyerstein, N., and Benor, D. (1973). A comparison of three methods of acclimatization to dry heat. J. Appl. Physiol. 34:214.

79. Bligh, J. (1973). Temperature Regulation in Mammals and Other Vertebrates, p. 94. North Holland, Amsterdam.
80. Shepherd, J. T., and Webb-Peploe, M. M. (1970). Cardiac output blood flow distribution during work in the heat. *In* J. D. Hardy, A. P. Gagge, J. A. J. Stolwijk (eds.), Physiological and Behavioral Temperature Regulation, p. 237. Thomas, Springfield, Illinois.
81. Wyss, C. R., Brengelman, G. L., Johnson, J. M., Rowell, L. B., and Niederberger, M. (1974). Control of skin blood flow, sweating, and heart rate: role of skin vs. core temperature. J. Appl. Physiol. 36:726.
82. Strom, G. (1950). Influence of local thermal stimulation of the hypothalamus of the cat on cutaneous blood flow and respiratory rate. Acta Physiol. Scand. 20 (Suppl. 70):47.
83. Whittow, G. C. (1968). Cardiovascular response to localized heating of the anterior hypothalamus. J. Physiol. 198:541.
84. Walther, O. E., Iriki, M., and Simon, E. (1970). Antagonistic changes of blood flow and sympathetic activity in different vascular beds following central thermal stimulation. II. Cutaneous and visceral sympathetic activity during spinal cord heating and cooling in anesthetized rabbits and cats. Pfluegers Arch. 319:162.
85. Hales, J. R. S., and Dampney, R. A. L. (1975). The redistribution of cardiac output in the dog during heat stress. J. Therm. Biol. 1:29.
86. Ingram, D. L., and Legge, K. F. (1971). The influence of deep body temperatures and skin temperatures on peripheral blood flow in the pig. J. Physiol. 215:693.
87. Webster, A. J. F. (1974). Physiological effects of cold exposure. *In* D. Robertshaw (ed.), MTP International Review of Science, Physiology Series One, Vol. 7, Environmental Physiology, p. 48. Butterworths, London.
88. Jansky, L. (1973). Non-shivering thermogenesis and its thermoregulatory significance. Biol. Rev. 48:85.
89. Hochachka, P. W. (1974). Regulation of heat production at the cellular level. Fed. Proc. 33:2162.
90. Himms-Hagen, J. (1976). Cellular thermogenesis. Annu. Rev. Physiol. 38:315.
91. Edelman, I. S. (1976). Transition from the poikilotherm to the homeotherm: possible role of sodium transport and thyroid hormones. Fed. Proc. 35:2184.
92. Keatinge, W. R., and Nadel, J. R. (1965). Immediate respiratory response to sudden cooling of the skin. J. Appl. Physiol. 20:65.
93. Simon, E., Rautenberg, W., and Jessen, C. (1965). Initiation of shivering in unanesthetized dogs by local cooling within the vertebral canal. Experientia 21:476.
94. Bligh, J. (1972). Neuronal models of mammalian temperature regulation. *In* J. Bligh and R. E. Moore (eds.), Essays on Temperature Regulation, pp. 105–120. North Holland, Amsterdam.
95. Downey, J. A., Miller, J. M., and Darling, R. C. (1969). Thermoregulatory responses to deep and superficial cooling in spinal man. J. Appl. Physiol. 27:209.
96. Kosaka, M., and Simon, E. (1968). Der Zentralnervose spinale mechanismus des kaltezitterns. Pfluegers Arch. 302:357.
97. Stuart, D., Ott, K., Ishikawa, K., and Eldred, E. (1966). The rhythm of shivering: I. General sensory contributions. Am. J. Phys. Med. 45:61.
98. Bligh, J. (1966). The thermosensitivity of the hypothalamus and thermoregulation in mammals. Biol. Rev. 41:317.

99. Itoh, S., and Kuroshima, A. (1972). Lipid metabolism of cold-adapted man. *In* S. Itoh, K. Ogata, and H. Yoshimura (eds.), Advances in Climatic Physiology, pp. 260–277. Igaku Shoin, Tokyo.
100. Chaffee, R. R. J., Allen, R. M., Arine, J. F. Rochelle, R. H., and Rosander, J. (1975). Studies on thermogenesis in brown adipose tissue in temperature acclimatized *Macaca mulatta.* Comp. Biochem. Physiol. 50:303.
101. Davis, T. R. A., and Mayer, J. (1955). Nature of the physiological stimulus for shivering. Am. J. Physiol. 181:669.
102. Nadel, E. R., Horvath, S. M., Dawson, C. A., and Tucker, A. (1970). Sensitivity to central and peripheral thermal stimulation in man. J. Appl. Physiol. 29:603.
103. Stolwijk, J. A. J., and Hardy, J. D. (1966). Temperature regulation in man. Pfluegers Arch. 291:129.
104. Johnson, S. G., and Elizondo, R. S. Thermoregulation in *Macaca mulatta*: a thermal balance study. J. Appl. Physiol., in press.
105. Wilkerson, J. E., Raven, P. B., and Horvath, S. M. (1972). Critical temperature of unacclimatized male caucasians. J. Appl. Physiol. 33:451.
106. Banet, M., and Hensel, H. (1976). Non-shivering thermogenesis induced by repetitive hypothalamic cooling in the rat. Am. J. Physiol. 250:522.
107. Banet, M., and Hensel, H. (1976). Non-shivering thermogenesis induced by repetitive cooling of spinal cord in rat. Am. J. Physiol. 230:720.
108. Dobson, R. (1965). The human eccrine sweat gland: structural and functional inter-relationships. Arch. Environ. Health 11:423.
109. Johnson, G. S., and Elizondo, R. S. (1974). Eccrine sweat gland in *Macaca mulatta*: physiology, histochemistry and distribution. J. Appl. Physiol. 37:814.
110. Elizondo, R. S. (1973). Local control of eccrine sweat gland function. Fed. Proc. 32:1583.
111. Elizondo, R. S., and Bullard, R. W. (1971). Local determinants of sweating and the assessment of set point. Int. J. Biometeorol. 15:259.
112. Benzinger, T. H. (1969). Heat regulation: homeostasis of central temperature in man. Physiol. Rev. 49:671.
113. Benzinger, T. H. (1964). The thermal homeostasis of man. Symp. Soc. Exp. Biol. 18:49.
114. Banerjee, M., Elizondo, R., and Bullard, R. (1969). Reflex responses of human sweat glands to different rates of skin cooling. J. Appl. Physiol. 26:787.
115. Nadel, E. R., Bullard, R. W., and Stolwijk, J. A. J. (1971). Importance of skin temperature in the regulation of sweating. J. Appl. Physiol. 31:80.
116. Elizondo, R., Smiles, K., and Barney, C. (1976). Effects of local hypothalamic heating and cooling on the sweat rate in the rhesus monkey. Isr. J. Med. Sci. 12:1026.
117. Smiles, K., Elizondo, R., and Barney, C. (1976). Sweating responses during changes of hypothalamic temperature in the rhesus monkey. J. Appl. Physiol. 40:653.
118. Hammel, H. T., Hardy, J. D., and Fusco, M. M. (1960). Thermoregulatory responses to hypothalamic cooling in unanesthetized dogs. Am. J. Physiol. 198:481.
119. Ogawa, T. (1972). Local determinants of sweat gland activity. *In* S. Itoh, K. Ogata, and H. Yoshimura (eds.), Advances in Climatic Physiology, pp. 92–108. Igaku Shoin, Tokyo.
120. MacIntyre, B., Bullard, R., Banerjee, M., and Elizondo, R. (1968). Mecha-

nism of enhancement of eccrine sweating by localized heating. J. Appl. Physiol. 25:255.

121. Mitchell, D., and van Rensburg, A. (1972). Thermoregulatory significance of the effect of local skin temperature on local sweat rate. Proc. Int. Union Physiol. Sci. 9:369.
122. Nadel, E., and Stitt, J. (1970). Control of sweating in the squirrel monkey. Physiologist 13:267.
123. Ogawa, T., and Bullard, R. (1972). Characteristics of subthreshold sudomotor neural impulses. J. Appl. Physiol. 33:300.
124. Fox, R. H., Goldsmith, R., Hampton, I. F. G., and Lewis, H. E. (1963). Acclimation to heat in man by controlled elevation of body temperature. J. Physiol. 166:530.
125. Wyndham, C. H. (1967). Effect of acclimation on sweat rate, rectal temperature relationship. J. Appl. Physiol. 22:27.
126. Mitchell, D., Senay, C., Wyndham, C. H., van Rensburg, A. J., Rogers, G. G., and Strydom, N. B. (1976). Acclimatization in a hot, humid environment: energy exchange, body temperature, and sweating. J. Appl. Physiol. 40:768.
127. Chen, W. Y., and Elizondo, R. S. (1974). Peripheral modification of thermoregulatory function during heat acclimation. J. Appl. Physiol. 37:367.
128. Hanane, R., and Valatx, J. L. (1973). Thermoregulatory changes induced during heat acclimation by controlled hyperthermia in man. J. Physiol. 230:255.
129. Nadel, E. R., Randolf, K. B., Roberts, M. F., and Stolwijk, J. A. J. (1974). Mechanisms of thermal acclimation to exercise and heat. J. Appl. Physiol. 37:512.
130. Cammock, S., Dascombe, M. J., and Milton, A. S. (1976). Prostaglandins in thermoregulation. *In* B. Samuelsson and R. Paoletti (eds.), Advances in Prostaglandin and Thromboxane Research, Vol. 1, pp. 375–380. Raven Press, New York.
131. Feldberg, W., and Saxena, P. N. (1971). Fever produced by prostaglandin E_1. J. Physiol. 217:547.
132. Cranston, W. I., Hellon, R. F., and Mitchell, D. (1975). A dissassociation between fever and prostaglandin concentration in cerebrospinal fluid. J. Physiol. 253:583.
133. Lipton, J. M., Welch, J. P., and Clark, W. G. (1973). Changes in body temperature produced by injecting prostaglandin E_1 EGTA and bacterial endotoxin into the PO/AH and the medulla oblongata of the rat. Experientia 29:806.
134. Milton, A. S., and Wendlandt, S. (1971). Effects on body temperature of prostaglandins of the A, E and F series on injection into the third ventricle of unanesthetized cats and rabbits. J. Physiol. 218:325.
135. Hales, J. R. S., Bennett, J. W., Baird, J. A., and Fawcett, A. A. (1973). Thermoregulatory effects of prostaglandins E_1, E_2, F_1, and F_2 in the sheep. Pfluegers Arch. 339:125.
136. Crawshaw, L. I., and Stitt, J. (1975). Behavioral and autonomic induction of prostaglandin E_1 fever in squirrel monkeys. J. Physiol. 244:197.
137. Ogata, K., and Murakami, N. (1972). Neural factors affecting the regulatory responses of body temperature. *In* S. Itoh, K. Ogata, and H. Yoshimura (eds.), Advances in Climatic Physiology, pp. 50–67. Igaku Shoin, Ltd., Tokyo.

138. Rowell, L. B. (1974). The cutaneous circulation. *In* T. C. Ruch and H. D. Patton (eds.), Physiology and Biophysics, Vol. II, pp. 185–199. W. B. Saunders, Philadelphia.
139. Glickman, K., Mitchell, H., Keeton, R., and Lambert, E. (1967). Shivering and heat production in men exposed to intense cold. J. Appl. Physiol. 22:1.
140. Barney, C., and Elizondo, R. The effect of ambient temperature on the development of prostaglandin E_1 hyperthermia in the Rhesus monkey. J. Appl. Physiol. Submitted for publication.

International Review of Physiology
Environmental Physiology II, Volume 15
Edited by David Robertshaw
Copyright 1977 University Park Press Baltimore

4
Exercise and Environmental Heat Loads: Different Mechanisms for Solving Different Problems?

C. R. TAYLOR
Concord Field Station,
Harvard University, Bedford, Massachusetts

DIFFERENCES BETWEEN EXERCISE AND ENVIRONMENTAL HEAT LOADS 121
Origin of Heat 121
Comparison of Magnitudes of Two Types of Heat Loads 122

SALIVA SPREADING AS EVAPORATIVE COOLING MECHANISM 125
Evidence that Salivation is Evaporative Cooling Mechanism 125
Situations Requiring Heat Dissipation by Salivation 125

THERMAL PANTING 126
Recondensation of Water from Exhaled Air 126
Nasal Mucosa as Surface for Evaporative Cooling 128
Steno's Gland, Source of Water for Evaporative Cooling from Nasal Mucosa 132

EXERCISE PANTING–BUCCAL AND TONGUE SURFACES AS SITES FOR EVAPORATIVE COOLING 133
Apparent Paradox–Heat Lost from Buccal Surfaces and Tongue 133
Buccal Cavity and Tongue as Evaporative Surfaces During Exercise 135

SWEATING: ADDITIONAL EVAPORATIVE SURFACE FOR DISSIPATING LARGE EXERCISE HEAT LOADS 137
Animals That Sweat 137
Size Dependency for Sweating 137
Reasons for Dependence on Body Size 138

INTERACTION OF DIFFERENT EVAPORATIVE COOLING MECHANISMS 138

CONCLUSIONS 140
Advantages and Disadvantages of Different Evaporative Cooling Mechanisms 140
Some Unanswered Questions 143

In the first series on environmental physiology of the MTP International Review of Science, this author wrote a chapter entitled "Exercise and Thermoregulation" (1) and J. R. S. Hales wrote a chapter entitled "Physiological Responses to Heat" (2). In this contribution to the second series, I should like to propose and develop a hypothesis which helps to pull together information presented in both of these chapters.

It will be argued that exercise and environmental heat loads pose different problems for homeotherms and that different mechanisms have been adopted for dealing with these two types of heat loads during the evolution of homeothermy. The review of the literature will not be exhaustive, nor will all the excellent papers which have appeared during the last 4 years be mentioned.

The chapter focuses on the discussion of evaporative cooling. Evaporative cooling is generally considered in terms of three separate mechanisms (2, 3):

1. Salivation and grooming: the water for evaporation is produced by salivary glands and spread over the surface of the animal, particularly the sparsely furred forelimbs, where it evaporates and cools the animal.
2. Panting: respiratory ventilation is increased, usually without alkalosis, and evaporation takes place from the moist respiratory surfaces.
3. Sweating: specialized glands on the body surface produce the water for evaporation, which takes place from the generalized body surface.

It is useful to define four evaporative cooling mechanisms, rather than the usual three. Panting can be divided into two separate mechanisms, each having a different evaporative surface, different structures for providing the water for evaporation and responding to different types of thermal loads. These two panting mechanisms are as follows:

1. Shallow thermal panting: the lateral nasal glands provide much of the moisture for evaporation, and evaporation takes place from the nasal mucosa.
2. Deep exercise panting: the salivary glands provide most of the moisture for evaporative cooling, and evaporation takes place from the buccal surfaces and the tongue.

It is hoped that this approach to thermoregulation will help answer some old questions (e.g., why do some animals pant while others sweat to dissipate excess heat?), pose some new questions, point out areas where experiments should be carried out to fill the voids, and give some insights into the evolution of thermoregulatory mechanisms.

DIFFERENCES BETWEEN EXERCISE AND ENVIRONMENTAL HEAT LOADS

Origin of Heat

At first, it might appear that the most basic difference between an exercise and an environmental heat load is the site at which the heat originates (i.e., inside versus outside the animal). The rate at which an animal consumes energy as it contracts its muscles can increase dramatically during exercise. Because exercising animals usually do little work on the external environment, most of the energy which their muscles use appears as heat within the body. This heat must be dissipated if the animal is to maintain a constant body temperature. In the case of an exercise heat load, it is clear that the heat originates inside, rather than outside, the animal's body.

The situation with environmental heat loads, however, is not so clear-cut. Only in the case of the most extreme environmental heat loads does the heat originate outside the animal (i.e., there is a net inward flow of heat from the hot environment into the animal by nonevaporative means). In nature, this situation probably only arises in a few hot deserts, and then only at midday or when an animal stands exposed to the sun. Even though this is a somewhat unusual situation, it merits careful consideration. It represents the maximal environmental heat load which an animal might encounter in nature. This situation also has the advantage that the magnitude of the heat load can be easily quantified. Both the heat which the animal produces and the heat which flows into it from the hot environment must be lost by evaporation. The rate of heat loss from the animal can be measured by weighing the animal at timed intervals, because under these conditions almost all of the weight loss is due to evaporation of water. If the animal's metabolic rate is also measured, then the net inward flow of heat from the environment is simply the difference between the rate at which metabolic heat is produced and the rate at which the animal loses heat by evaporation.

It is unreasonable to limit our definition of an environmental heat load to only the situation in which there is a net inward flow of heat into an animal by nonevaporative means. This author has decided, therefore, to consider any environment in which resting humans or animals must sweat or pant (i.e., use evaporative cooling) to keep cool as a hot environment which confronts the animal with an "environmental heat load." In the vast majority of these situations, even when sweat is literally dripping from the body or an animal is panting vigorously, body temperature exceeds the temperature of the surrounding environment and there is a net outward flow of heat by nonevaporative means. It is simply that the gradient for nonevaporative heat loss has become too small to dissipate all the metabolic heat which is being produced. Evaporative cooling mechanisms must be recruited to make up the difference. The heat, however, originates inside, rather than outside, the body.

Comparison of Magnitudes of Two Types of Heat Loads

The magnitude of exercise heat loads was discussed in the previous MTP series on environmental physiology (1). It was shown that the heat production of running quadrupedal animals can be estimated fairly well from their body weight and their speed by using the following equation:

$$M_{run} = 40.6\ W^{-0.4}\ (Vel) + (29\ W^{-0.25})$$

where W is the body weight in grams, M_{run} is metabolic rate in cal $(g \cdot hr)^{-1}$, and Vel is running speed in $km \cdot hr^{-1}$. By using the top speeds of animals reported in the literature (4, 5) and an average weight, it was calculated that a 30-g mouse running at its top speed would produce heat at a rate 10–15 times its resting level; intermediate-sized animals, weighing from 10–100 kg (such as the Thompson's gazelle, cheetah, pronghorn, or dog), running at their reported top speeds would produce heat at 50–80 times their resting levels; whereas a very large 1,000-kg mammal such as a horse or an elephant would increase its metabolism by only about 15 times resting levels at its reported top speed (1). Looked at in this way, it seems safe to conclude that when mammals run at their top speed, they are faced with the problem of dissipating metabolic heat at rates 10–80 times higher than at rest in order to maintain a constant body temperature.

Because this discussion of heat loads is focused on evaporative cooling, it is useful to know how much of the heat which is produced during exercise must be lost by evaporation, and how much can be lost without recruiting evaporative cooling mechanisms. Estimating the magnitude of nonevaporative heat loss in natural environments is very complex, and to make things still more difficult, it changes rapidly. Mitchell (6) provides an excellent discussion of the physical parameters involved in nonevaporative heat exchange in his chapter in the previous series on environmental physiology. Virginia Finch, in a series of elegant experiments, has accomplished the seemingly impossible task of quantifying the various avenues of heat exchange of some East African ungulates (7). As a result of these complexities, it is useful to concentrate this discussion on relatively simple environmental situations (e.g., experiments in an environmental room in which radiation temperatures are uniform and equal to skin or air temperature) and to make simplifying assumptions.

Operationally, the easiest way to consider nonevaporative heat loss is to lump the various avenues of nonevaporative heat loss (radiation, convection, and conduction) into a simple term—conductance. As in my previous chapter (1), Tucker's definition of conductance (8) will be used because it is concerned only with nonevaporative heat loss and it corrects for work which is done against the environment:

$$\text{Conductance (cal cm}^{-2}\ {}^\circ\text{C}^{-1}\ \text{hr}^{-1}) = \frac{\begin{bmatrix}\text{heat produced}\\ (\text{cal hr}^{-1})\end{bmatrix} - \begin{bmatrix}\text{evaporative heat loss}\\ (\text{cal hr}^{-1})\end{bmatrix} \pm \begin{bmatrix}\text{heat stored}\\ (\text{cal hr}^{-1})\end{bmatrix} \pm \begin{bmatrix}\text{work}\\ (\text{cal hr}^{-1})\end{bmatrix}}{\begin{bmatrix}\text{surface area}\\ (\text{cm}^2)\end{bmatrix} \begin{bmatrix}T_{\text{body}} - T_{\text{air}}\\ ({}^\circ\text{C}) \quad ({}^\circ\text{C})\end{bmatrix}}$$

Conductance defined in this way expresses the rate of nonevaporative flow of heat across each square centimeter of surface for each °C gradient between core temperature and air temperature. It has the advantage that all of the terms, with the exception of surface area, can be easily measured. The surface area of animals of similar shape is proportional to (body weight)$^{2/3}$ and can be estimated from body weight (9).

The relative importance of evaporative and nonevaporative cooling mechanisms for dissipating exercise heat loads depends to a large extent on an animal's size. Small animals have not only more surface area per unit of mass, but also more surface area per calorie of heat produced at rest. As a result, mouse- or rat-sized animals (<1,000 g) are able to dissipate most of the heat which they produce during exercise without having to recruit evaporative cooling mechanisms (at least as long as their body temperature is 10–20°C above ambient temperature). Large animals (>10 kg) cannot rely on nonevaporative cooling, and they must either increase evaporative cooling or store heat to cope with moderate exercise heat loads. This can be clearly demonstrated by making some reasonable assumptions and doing some simple calculations. Let us assume that there is a 20°C temperature gradient for nonevaporative heat loss (i.e., ambient temperature is about 20°C) and conductance is 1 cal $(\text{cm}^2 \cdot \text{hr}^\circ\text{C})^{-1}$ (the highest reported for running mammals (1). Surface area and resting metabolism is calculated from body weight. It is then possible to plot the ratio of the rate of maximal nonevaporative heat loss to the rate of resting metabolic heat production as a function of body size (Figure 1). Small mouse- to rat-sized animals are able to dissipate heat nonevaporatively at rates equal to about 8–10 times their resting metabolic heat production. This decreases to only about 3–4 times for a horse-sized animal. Thus, small animals are able to dissipate almost all of the heat they would produce while running at top speed without resorting to evaporative cooling, but intermediate and large animals must rely mainly on evaporative cooling mechanisms (or store the heat and dissipate it after the run) even when they run at relatively slow speeds.

The maximal heat load from a hot environment will be proportional to the surface area which is exposed to this environment. Schmidt-Nielsen made this

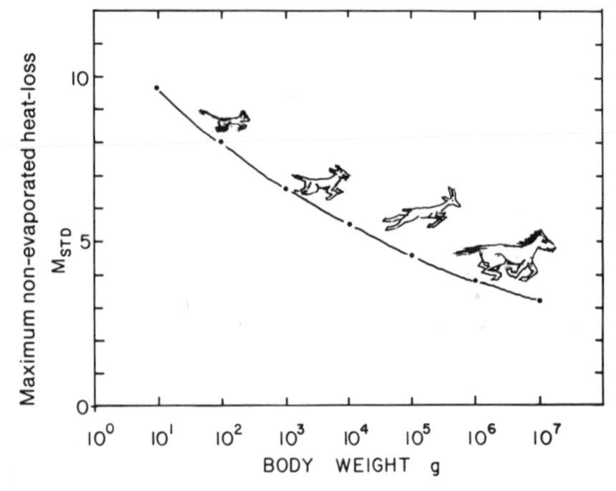

Figure 1. Interdependence of evaporative cooling for maintaining a constant body temperature and body size. Plotting the ratio of the maximal rate of nonevaporative heat loss to the predicted rate of resting metabolic heat production clearly demonstrates that the potential for nonevaporative heat loss decreases with increasing body size. Large animals become more reliant on evaporative cooling to dissipate the large amount of heat produced during exercise.

point very clearly in his book *Desert Animals* (10). He used data that D. B. Dill and his collaborators had collected in the deserts of the Southwestern United States during the construction of the Boulder Dam. Dill and his colleagues took a donkey and a dog on hikes in the desert. They measured the amount of water which the man and animals lost by evaporation by weighing the animals and recording the time intervals between weighings. Normalizing their data for surface area, Schmidt-Nielsen showed that each animal evaporated water at a rate of about 0.6 $kg \cdot m^{-2} \cdot h^{-1}$. This figure includes both the heat which the animals produced during the walk and heat which flowed into the animals from the hot environment. When the heat which was produced by the animals' metabolism is estimated and subtracted from the total evaporative heat loss, then the inward heat flow from a hot environment under these most severe desert conditions amounts to only 2–4 times the animals' predicted resting metabolic heat production. Borut et al. (11) and Adolph (12) have reported similar values for resting black Bedouin goats and nude men exposed to sun at midday in hot deserts.

Three important points emerge from this discussion:

1. The maximal exercise heat loads which intermediate- and large-sized animals (>10 kg body weight) encounter are nearly an order of magnitude greater than the maximal environmental heat which they might encounter in nature.
2. These animals have to rely almost entirely on evaporative cooling mechanisms in order to dissipate this heat and maintain a constant body temperature.
3. Small animals (<1 kg) can dissipate exercise heat loads without recruiting evaporative cooling mechanisms, so long as their body temperature is 10–20°C above air temperature. It should also be noted that the only option open to small animals in a hot environment is escape. They simply do not contain

enough water to use evaporative cooling on a sustained basis. They would dehydrate very quickly if they tried to dissipate the heat they produced by evaporative cooling mechanisms. Most small animals escape from hot environments by retreating into cool burrows. The evaporative cooling mechanisms which have been adopted by small animals are not geared for sustained heat dissipation, but they give animals which are caught in a hot environment more time to find shelter.

SALIVA SPREADING AS EVAPORATIVE COOLING MECHANISM

Evidence that Salivation is Evaporative Cooling Mechanism

Schmidt-Nielsen (10, 13) pointed out that salivation might be an important evaporative cooling mechanism for rodents, cats, and marsupials. He noted that when these animals encountered lethal and near lethal temperatures, large portions of their fur became completely soaked with moisture, the result of saliva being spread on the fur. It increased evaporation and prolonged survival in these hot environments.

Hainsworth (14), in studies of white rats, demonstrated that saliva spreading was of critical importance for survival of these animals in hot environments. He found that evaporative cooling could be increased by this means to rates high enough to allow rats to maintain a constant body temperature in situations in which ambient temperature exceeded body temperature by 3°C. Thus, evaporation was sufficient to dissipate not only the resting metabolic heat production, but also a significant nonevaporative inward flow of heat from the hot environment. Removing the submaxillary-sublingual salivary glands abolished the rat's ability to maintain a constant body temperature under these conditions. Other studies by Hainsworth and his colleagues (15, 16) and by Elmer and Ohlin (17) have shown that salivation from these glands is under hypothalamic control and that parasympathetic innervation to the glands is of major importance in controlling rates of secretion.

Saliva spreading also occurs in a number of marsupials (10). Perhaps the most convincing studies are those of Needham et al. (18), who have shown that an elaborate vascular network exists in the arms and legs of the big red kangaroo. This animal licks these areas when it gets hot. This vascular network maximizes the effectiveness of evaporative heat loss from these regions.

Situations Requiring Heat Dissipation by Salivation

It is obvious that saliva spreading is not very effective as an evaporative cooling mechanism during exercise. The animal would need to stop running to lick its fur. It is also clear from earlier studies that salivation does not enable small rodents to withstand extremely hot desert environments without rapidly dehydrating. In these situations, it is useful in allowing time to escape to a cooler environment. It might also be useful for withstanding short exposures to high temperatures, or even for dealing with a situation in which the temperature gradient is not adequate for all of the metabolic heat to be lost by nonevaporative means over a longer period of time.

THERMAL PANTING

Recondensation of Water From Exhaled Air

It is obvious that inhaled air is warmed to deep body temperature and humidified before it reaches the lungs. It had been generally believed that this air was exhaled still saturated with water vapor at deep body temperature until the experiments of Jackson and Schmidt-Nielson (19) showed that the exhaled air temperature of two small mammals was almost the same as the temperature of inhaled air. These authors explained their unexpected results by proposing a nasal countercurrent heat exchange mechanism, where the heat exchange is separated in time rather than in space (i.e., a "temporal" countercurrent heat exchange as contrasted with a "spatial" exchange). This "nasal countercurrent exchanger" traps both heat and water inside the animal. An understanding of how, when, and where (i.e., in what animals) it operates is basic to an understanding of thermal panting mechanisms, for this exchange must be circumvented in order to have effective evaporative cooling from the respiratory tract.

First, how does the temporal countercurrent heat exchanger work? Let us assume for the purpose of this discussion that dry air is inhaled at 22°C and that the core body temperature (i.e., the temperature of the lungs) is 38°C. In order for the temporal countercurrent exchanger to operate, the inspired air and the expired air both must pass across the exchanger (i.e., both inhalation and exhalation must take place through the nose). The exchange of heat and water takes place between the surfaces of the nasal mucosa and the air. Let us consider the case of a single respiratory cycle (Figure 2). Prior to inhalation, a temperature gradient exists along the nasal mucosa. The surface of the nasal mucosa is just slightly warmer than the inspired air at the point where the inspired air first reaches it. Nasal mucosal temperature increases along the pathway that the air is to follow until it is approximately equal to deep body temperature before the air arrives at the lungs. Both heat and water are lost from the nasal mucosa to the air as it flows past the nasal mucosa. During the time interval between inhalation and exhalation, the temperature gradient between the nose and the lungs is maintained. Little heat or water is added to the nasal mucosa by the animal. On exhalation, the expired air leaves the lung saturated with water vapor at 38°C. As it passes across the nasal mucosa in the opposite direction it has taken during inhalation, it encounters surface temperatures which are progressively cooler. The temperature and vapor pressure gradient are reversed, and heat is lost from the exhaled air to the cooler nasal mucosa along this continuous gradient until, just before the exhaled air leaves the nose, it has returned to approximately the same temperature as that of the air that was inhaled initially. As the air is cooled, it becomes supersaturated with water vapor, and water then recondenses on the surfaces of the nasal mucosa. When the exhaled air leaves the nose, it is

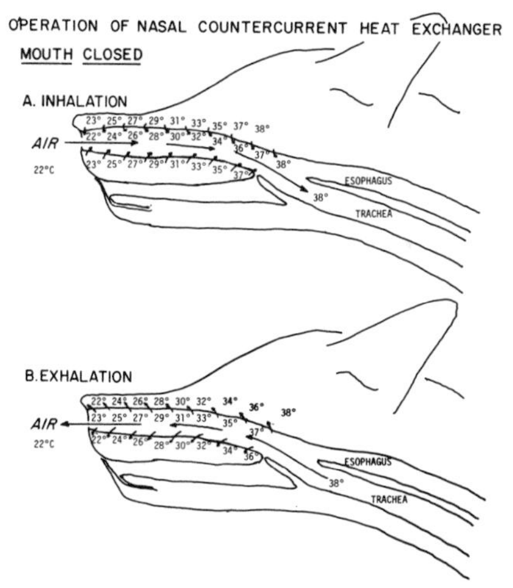

Figure 2. Operation of the nasal countercurrent heat exchanger. See text for explanation. *A*, inhalation. Air entering the nose at 22°C gains heat and water as it passes across the nasal mucosa. It arrives at the trachea saturated with water vapor at body temperature (38°C). The *small arrows* indicate transfer of heat from the mucosa to the air. *B*, exhalation through the nose. The air leaves the lungs saturated with water vapor at 38°C. It loses heat to the nasal mucosa (which has been cooled on inhalation) and water recondenses onto the mucosal surface. The air exits through the nose saturated with water vapor at about 22°C. The *small arrows* indicate transfer of heat from the air to the mucosa. Cooling of the nasal mucosa has been disregarded in order to simplify the discussion. Schmidt-Nielsen (53) gives a clear discussion of the exchanger and presents a mathematical model for heat exchange.

still saturated with water vapor, but at a temperature approximately equal to that of the inhaled air rather than to that of the lungs. At a temperature of 22°C, each liter of air contains approximately 20 mg of water, whereas at 38°C it contains approximately 50 mg of water. Thus, 60% of the water which evaporated from the nasal mucosa as the inspired air was humidified on its way to the lungs has been recondensed on exhalation. This treatment of the nasal heat exchanger is greatly simplified. It neglects the loss of heat from the mucosa as a result of evaporation. In reality, the exhaled air may be cooler than the inhaled air (19). The general principles are the same, however, and the arguments presented here are not altered by neglecting the cooling by evaporation of the mucosa.

This temporal heat exchanger minimizes the loss of heat and water from the respiratory tract of small mammals (19), birds (20), and even from ectothermic reptiles with a high body temperature (21). From our earlier discussion of the large surface to volume ratio of small animals, it is obvious that it is advantageous for resting small animals to minimize the loss of heat when confronted with the problem of maintaining a constant high body temperature in the face of a large temperature gradient for nonevaporative cooling. But small animals are not always resting in a cool environment.

What happens to the nasal countercurrent exchanger when these animals increase their heat production during exercise? Raab and Schmidt-Nielsen (22) ran white rats and kangaroo rats on treadmills at speeds up to 1.2 km hr^{-1} and found that the nasal heat exchanger operated just as effectively at metabolic rates 4–5 times resting levels as it did at rest in these small rodents.

What happens when the gradient for nonevaporative loss between the animal and the environment decreases and the animal must increase heat loss evaporatively in order to maintain constant body temperature? Does the existence of the nasal mucosal heat exchanger eliminate the possibility of using the respiratory tract as an evaporative cooling surface in these animals?

On reflection, it is obvious that there is a certain degree of autoregulation within this type of countercurrent system. The temperature and the vapor pressure of the exhaled air will automatically increase as air temperature increases. For example, as inhaled air temperature increases from 22°C to body temperature of 40°C, so does exhaled air temperature, and the heat exchanger is effectively abolished. The exhaled air is always saturated with water vapor, and when its temperature is 38°C, the exhaled air will be saturated with water vapor at 38°C and will contain about two and a half times as much water as it did at 22°C (50 mg of H_2O/liter of air versus 20 mg of H_2O/liter of air). Approximately 580 cal of heat are bound with each gram of water which evaporates. Thus, as air temperature increases from 22°C to 40°C, maximal evaporative heat loss (maximal because this assumes that inspired air temperature contains no water) increases from ≅ 12 cal for each liter of air moved across the system to ≅ 29 cal.

If it is assumed that the oxygen content of the respired air is reduced from 21% on inhalation to 16% in the lungs, and that dead space makes up an insignificant part of the tidal volume, then the percentage of the metabolic heat produced by the body which can be lost by respiratory evaporation can be calculated. Fifty ml of oxygen would be used from each liter of air which the animal breathed, and assuming a caloric equivalent of 4.8 cal/ml of oxygen consumed, 240 cal would be produced. In this extreme case (i.e., no dead space ventilation), respiratory evaporation could account for 5% of heat production when air was inspired at 22°C. This would increase to 12% when dry air was inspired at 38°C. Evaporative heat loss would, of course, decrease as the water content (relative humidity) of the inspired air was increased. At 50% relative humidity, for example, the loss at 38°C would be halved. It seems clear that if the nasal mucosa is to become an important evaporative surface, specialized mechanisms must be adopted. Autoregulation, in itself, is not sufficient to dissipate the heat which the animal produces as air temperature approaches body temperature and the gradient for nonevaporative heat loss diminishes.

Nasal Mucosa as Surface for Evaporative Cooling

It is obvious that autoregulation of the nasal heat exchanger will not provide rates of evaporation high enough to deal with either environmental or exercise heat loads. The most obvious way to increase evaporative cooling from the nasal mucosa is to increase the amount of air which is moved across it. Increasing the volume of air that flows in and out of the lungs, however, poses a severe problem for the animal–respiratory alkalosis. It has already been mentioned that the oxygen content of the lungs is 5% lower than that of inspired air. This oxygen has been replaced with CO_2. The bicarbonate buffer system of the blood is in equilibrium with the 5% CO_2 in the alveoli of the lungs. If the alveolar ventilation is increased without a simultaneous increase in the rate of oxygen consumption, then the concentration of CO_2 in the lungs will fall, which will lead to a decrease in the bicarbonate in the blood and an increase in blood pH (i.e., respiratory alkalosis). In order for the nasal mucosa to serve as an effective evaporative surface, total ventilation must be increased without increasing alveolar ventilation. This can be accomplished by taking advantage of the fact that part of the inspired air never reaches the lungs. The air in the respiratory tract between the point of inspiration and the alveoli moves back and forth with each breath. It never comes in contact with respiratory exchange surfaces and is not involved in gas exchange. This ventilation of nonrespiratory surfaces is referred to as the dead space ventilation. Evaporation from the respiratory tract could be increased without causing alkalosis if the ventilation of this dead space were increased without increasing alveolar ventilation. This is indeed what happens. It

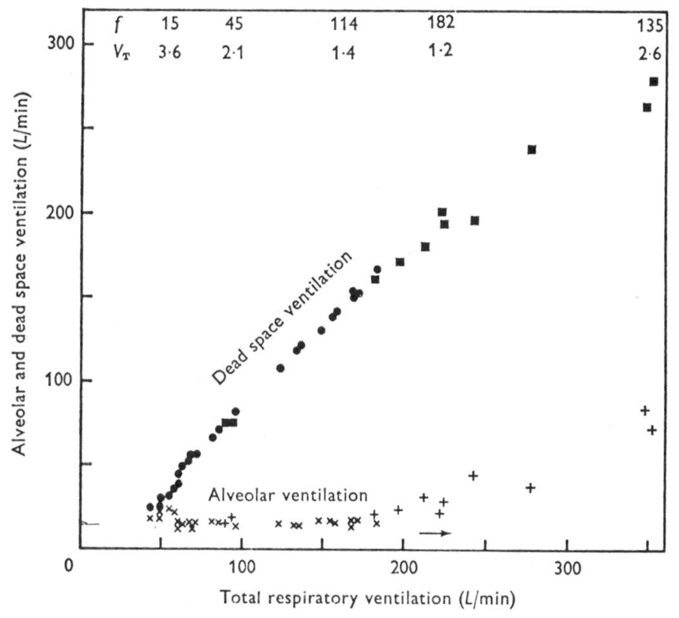

Figure 3. Increase of dead space ventilation in the panting ox without any increase in alveolar ventilation until total ventilation exceeds 200 liters/min. Alveolar ventilation increases markedly when total respiratory ventilation exceeds 200 liters/min. This is accompanied by an increase in total volume (V_t) and a decrease in respiratory frequency (f). (Tidal volumes and frequency are given at the top of the graph.) (Reproduced from Hales (23) with permission.)

is accomplished by increasing the frequency of respiration while decreasing the amplitude. As the amplitude decreases, proportionally less air from each breath reaches the lungs, whereas the amount of air which moves in and out without reaching the lungs (the dead space) remains constant. Thus, dead space ventilation constitutes a larger and larger fraction of total ventilation.

Hales (23) measured both dead space and alveolar ventilation in the ox as it was exposed to progressively hotter environments and began to use thermal panting. Figure 3 shows his results. Alveolar and dead space ventilation in the thermal neutral environment were approximately the same. As the ox was exposed to hotter and hotter environments, shallow thermal panting was initiated and ventilation increased. Dead space ventilation, however, increased by approximately 10-fold before there was any increase in alveolar ventilation. Thus, evaporation from the nasal mucosa in the ox could increase by 10-fold before respiratory alkalosis would become a problem. If dry air was inspired at 40°C and our previous calculations hold true for the ox, then it could maintain a constant body temperature and lose all of its metabolic heat by evaporation from the nasal mucosa. This would work only if the inhaled air were absolutely dry.

It seems obvious that in order for ventilation of the nasal mucosa to be an effective evaporative cooling mechanism there must be some way of bypassing the countercurrent exchanger. Heat loads are not imposed simply by changes in air temperature. A simple method to bypass this exchanger would be to increase the blood flow to the nasal mucosa and warm the mucosal surface between inhalation and exhalation. Murrish (54) has shown that this mechanism exists in penguins. In other experiments in which it might have been observed, Raab and Schmidt-Nielsen (22) found that the exchanger of small rodents was always operative.

Most panting animals probably have adopted a simple mechanical device for bypassing the heat exchanger similar to that which has been found in dogs, which can flip a valve between inspiration and expiration and vent air which was inhaled through the nose out through the mouth, thereby bypassing the nasal heat exchanger (Figure 4). Schmidt-Nielsen et al. (24) have shown that this is exactly what dogs do during rapid thermal panting, when they need to dissipate large amounts of heat (Figure 5). It seems reasonable to assume that other panting animals can use a similar physical bypass during shallow, open-mouth panting. This valving mechanism provides a means of modulating heat loss without changing either respiratory rate or volume. The air can either vent across the exchanger, reducing loss of heat and water, or vent out the mouth, maximizing the loss of heat and water.

Thermal panting as an evaporative cooling mechanism poses still another problem. Animals must expend energy to move the air across the evaporative surfaces. This produces additional heat which must also be lost by evaporation. Crawford (25) has proposed that animals avoid this problem and minimize the energy expenditure by panting at a resonant frequency, allowing much of the energy required for moving the air to be stored in and recovered from elastic elements. Crawford estimated that if such a resonant system did not exist, dogs would produce more heat than they could dissipate. Recently, Crawford (26) has also provided evidence which suggests that birds as well as mammals make use of resonant panting to minimize energy expenditure during thermal panting.

Hales (2) has reviewed the oxygen cost of shallow thermal panting. It is clear that there is little additional energy expended when an animal pants vigorously. In fact, the amount is so small that it cannot be measured in the African hunting dog (27), sheep (2), oxen (2), goat (28), or gazelle (29). However, this situation is not quite so simple. Thiele and Albers (30) found that dogs required a considerable increase in energy expenditure for moving the air during panting. They reported that as much as 40% of the increased evaporation resulting from the increased ventilation during shallow panting is required to dissipate the additional heat generated by the panting itself. Hales (31, 32) has offered a plausible explanation for this apparent discrepancy. He finds in sheep that the increase in metabolism as the result of panting is offset by a decrease in the metabolism of other tissues.

Whatever the explanation, it is clear that there is no measureable net increase in metabolic heat production during shallow panting in most animals that use this mechanism for evaporative cooling, regardless of the cost of the panting mechanism itself. It also seems clear that there is some reduction in energy expenditure by exploiting a resonant frequency for panting.

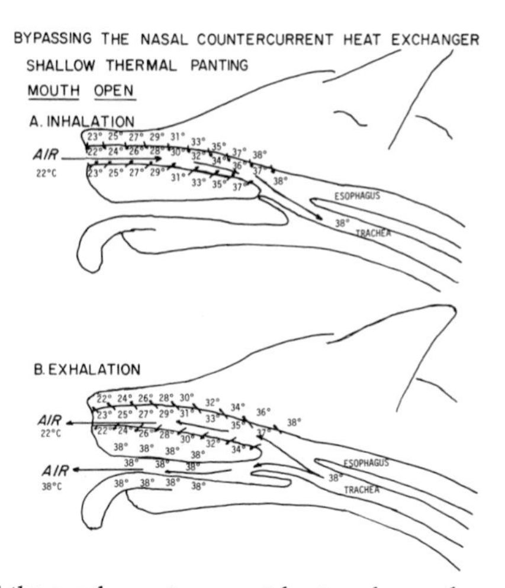

Figure 4. Bypassing of the nasal countercurrent heat exchanger by opening the mouth. See text for explanation. *A*, inhalation air entering the nose at 22°C gains heat and water as it passes across the nasal mucosa. A one-way valve prevents air entering through the open mouth. The air reaches the trachea saturated with water at body temperature (38°C). *B*, exhalation through the mouth and nose. The air leaves the lungs saturated with water vapor at 38°C. The mouth is open, allowing the positive air pressure to open the one-way valve. Air exits through both the nose and the mouth. The air that exits through the mouth bypasses the heat exchanger and remains saturated with water vapor at body temperature. The small arrows indicate flow of heat from the mucosa to the air on inhalation and from the air to the mucosa on exhalation.

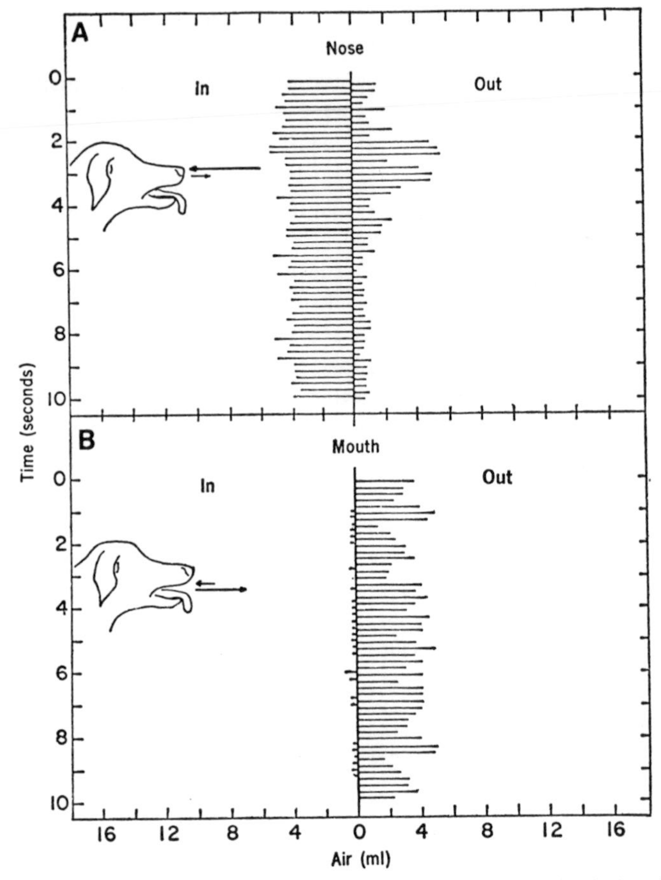

Figure 5. Volume of air entering and leaving nose and mouth during "phase 1" panting. *A*, the volume of air entering and leaving the nose of a dog during shallow thermal panting. Each line extending to the left of zero (the midline of the graph) represents the volume inspired in a single breath, and each line extending to the right of zero represents the volume expired in a single breath. *B*, the volume of air entering and leaving the mouth at inhalation through the mouth is virtually zero, but air can exit through either the nose or the mouth. Mean inspired and expired volumes for 10 s are indicated by vectors adjacent to the dog head. (Reproduced from Schmidt-Nielsen, Bretz and Taylor (24) with permission. Copyright 1970 by the American Association for the Advancement of Science.)

Steno's Gland, Source of Water for Evaporative Cooling from Nasal Mucosa

It has been seen that it is possible to increase the evaporation from the nasal mucosa by nearly 100-fold, both by increasing respiratory frequency and by exhaling the air through the mouth. The rate of evaporative heat loss from this surface can exceed the resting heat production of a panting animal. This immediately raises a question: where does the water come from? In 1664, Nicolaus Steno (33), a Danish anatomist, described a large serous type of gland found in the nasal cavities of dogs. Since Steno first described this gland, it has been found in a variety of animals which use thermal panting for evaporative cooling (dog, cat, pig, sheep, goat, and small antelope (34)). In animals that increase evaporative cooling primarily by sweating (man, horse, and cattle), the gland is either absent or of microscopic proportions (34). This certaintly suggests that this gland might produce the water for evaporation from the nasal mucosa. Blatt et al. (35) canulated this gland and found that it secreted a watery, hypotonic salt solution in response to heat stress. These authors found that a large part of the water for the evaporative cooling which takes place from the nasal mucosa could be supplied by these glands. Their function seems somewhat analogous to those of sweat glands of man. In the dog, each gland drains through a single duct which opens about 2 cm inside the opening of the nostril. Thus, the water which the gland produces is carried back over the nasal mucosa with the inspired air. This location may be essential to avoid dessication of the mucosa during thermal panting.

EXERCISE PANTING–BUCCAL AND TONGUE SURFACES AS SITES FOR EVAPORATIVE COOLING

Apparent Paradox–Heat Loss from Buccal Surfaces and Tongue

One has only to look at a panting dog with its tongue hanging out to be convinced that the tongue and buccal cavities are important surfaces for evaporative cooling. Not only does our intuition tell us that this is the case, but there are anatomical and functional specializations to facilitate heat loss from these surfaces. The tongue is richly supplied with arteriovenous anastomoses and is capable of very high blood flow (36), and the salivary glands can supply large amounts of water for evaporative cooling (37). However, a paradox is immediately evident. In the discussion of shallow thermal panting, it has been seen that there is only an outward flow of air from the mouth. Air enters the nose, is warmed and humidified, and then exits either through the nose or the mouth. Therefore, the air which passes over the tongue and buccal surfaces is saturated with water vapor at body temperature–a situation which seems to rule out either evaporative or nonevaporative heat loss from these surfaces. In order for the tongue and buccal surfaces to be involved in evaporative cooling, inhalation must take place through the mouth. Negus, in his book *The Comparative Anatomy and Physiology of the Larynx* (38), states that only man and anthropoid apes can inhale through their mouths, because a unidirectional valving mechanism is present in all other mammals.

There are several experiments which make this paradox even more puzzling. Pleschka and his colleagues (36, 39) have shown that the blood flow to the tongue of the dog increased dramatically in response to heating and that there was a large temperature difference between the arterial blood supplying the tongue and buccal surfaces and the venous blood draining these areas. When they multiplied the temperature drop by the blood flow, they found that a significant fraction of the total heat loss from the dog had taken place from the tongue. Hammel and his colleagues (37, 40) have also shown that saliva production is a thermoregulatory response. They measured rates of saliva production of a dog running at 6.4 km hr^{-1} on a treadmill while they changed the temperature of the hypothalamus by using thermodes. Increasing the temperature of the hypothalamus caused a dramatic increase in salivation from the parotid glands, whereas a decrease in hypothalamic temperature causes a sudden decrease. Both of these experiments certainly support the idea that the buccal cavity and the tongue are important surfaces for evaporative cooling.

How does one reconcile these observations with unidirectional outward flow of warm saturated air across these surfaces during thermal panting? There are two different types of panting, and it is in this difference that the answer to our paradox can be found. In the rapid, shallow thermal panting which has been considered to this point ("phase 1 panting"), there is a unidirectional flow of air out of the mouth, and little evaporation takes place from the buccal surfaces and tongue. This type of panting occurs when animals are exposed to warm environments with which they can cope (i.e., they are able to increase evaporation to the extent that they maintain a constant body temperature). When panting animals are confronted with extreme thermal loads and body temperature begins to rise, a different pattern of panting is initiated. Respiratory rate changes from the very rapid, shallow pattern to a slower, deeper pattern. This second phase of thermal panting ("phase 2 panting") has been observed in dogs (41), sheep (42), and cattle (43). Total ventilation, alveolar ventilation, and respiratory evaporation all increase dramatically, but respiratory alkalosis develops. Thus, even

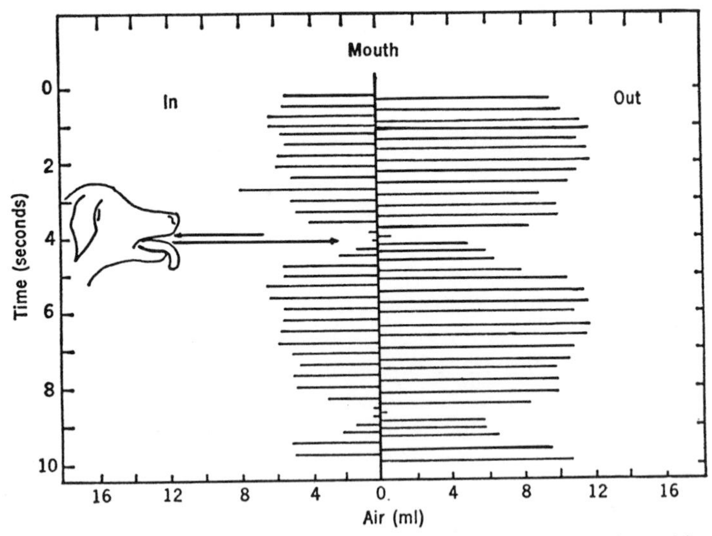

Figure 6. Inhalation through the mouth of a dog. The volume of air entering and leaving the mouth during deep, "phase 2" panting is indicated in the same manner as in Figure 5. More than half of the inspired air enters the mouth under these conditions. Mean inspired and expired volumes for 10 s are indicated by vectors adjacent to the dog head. (Reproduced from Schmidt-Nielsen, Bretz and Taylor (24) with permission. Copyright 1970 by the American Association for the Advancement of Science.)

though this type of panting provides for additional evaporation in response to extreme environmental heat loads, it can be used for a very limited period of time before the animal develops severe respiratory alkalosis.

Schmidt-Nielsen et al. (24) also observed two types of thermal panting when they measured the patterns of air flow in the panting dog. As might be anticipated from the previous discussion, they found that during the slower, deeper "phase 2" panting (in response to extreme thermal loads) the air entered through both the nose and the mouth (Figure 6). Therefore, during this "phase 2" of thermal panting, cool, dry air would pass over the tongue and buccal surfaces, and they could become important sites for evaporative cooling. Although this clearly shows that a mechanism exists for inspiration through the mouth, under these conditions alkalosis develops. Thus, it certainly would not seem a very important mechanism for the animal in response to heat, except perhaps as a last resort for keeping cool while it escaped to a cooler environment. Is there any situation in which this mechanism might be important?

Buccal Cavity and Tongue as Evaporative Surfaces During Exercise

Now, a new problem has arisen. Although the increased rates of evaporation are desirable during extreme heat loads, the ensuing respiratory alkalosis cannot be tolerated for any prolonged period of time. How then can the use of the buccal surfaces and the tongue to be used for sustained evaporative cooling be explained? The answer is obvious. One simply has to look for a situation in which alveolar ventilation is increased without producing respiratory alkalosis, which is exactly what happens during exercise, when O_2 consumption and CO_2 production are increased, and alveolar ventilation therefore also must increase. The experiments in which Hammel and his colleagues (37, 40) found that saliva production was a thermoregulatory response were carried out on exercising dogs. The rates of saliva production were already high while the dog ran at 6.4 $km \cdot hr^{-1}$. They could then be increased by raising hypothalamic temperature or reduced by lowering hypothalamic temperature. It seems entirely plausible that the second pattern of panting, with inhalation through the mouth, normally occurs during exercise.

Is there any evidence to support this hypothesis? Recently, in unpublished studies in our laboratory, it was found that dogs normally inhale through their mouths when they run. Masks were built which fit over the mouth, but not the nose. A differential pressure transducer was then used to measure the pressure in the mask. Positive pressure indicated outward flow of air from the mouth into the mask, and a negative pressure indicated an inward flow of air from the mask to the lungs. During rapid, shallow thermal panting, only a positive pressure pulse associated with each breath was observed. When the animal began to run, however, both positive and negative pressures were recorded, indicating a bidirectional flow through the mouth during exercise. This change in the valving pattern was confirmed with the use of high speed x-ray movies of animals panting at rest and while running on a treadmill. During running, the valve described by Negus (38) is simply moved out of the way (Figure 7).

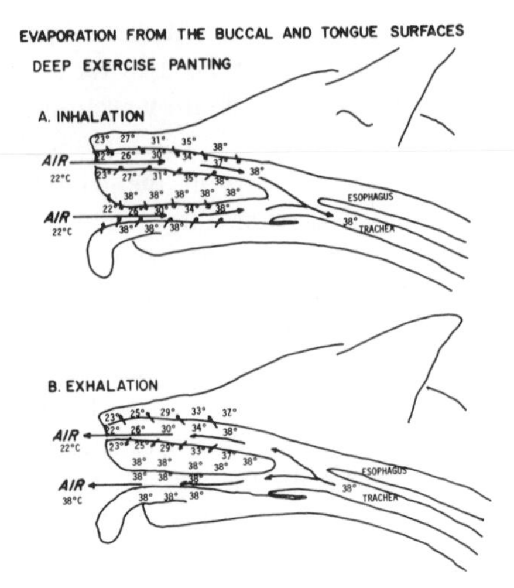

Figure 7. Air flow in and out of the mouth of the running dog. The one-way valve is pulled forward and lies flat against the tongue allowing the air to enter the trachea through the mouth in both directions. Under these circumstances, evaporation takes place from both the nasal mucosa and the buccal-tongue surface. The *small arrows* indicate flow of heat from the surfaces to the air and from the air to the surfaces. These schematics have been made from x-ray movies taken at 200 frames/s of a small dog running on a treadmill in this laboratory.

Recruiting the buccal and tongue surfaces for evaporative cooling during exercise should result in higher rates of respiratory evaporation during exercise than in response to environmental heat loads in steady state experiments (i.e., no respiratory alkalosis). Is this the case? There are data available for dog (27), goat (28), and gazelle (29) in which maximal rates of evaporative cooling from the respiratory tract have been measured at the highest ambient temperatures which the animals could tolerate before they change from rapid, shallow panting to deep, second-phase panting. Rates of evaporative cooling from the respiratory tract are also available from these animals while they ran on

Table 1. Maximal rates of respiratory evaporation during rapid thermal panting in a hot environment compared with the high rates of respiratory evaporation reported for these animals while they ran on a treadmill (not necessarily maximal)[a]

	Hot environment (EHL_R/max M_{STD})	During running (EHL_R/M_{STD})
Dog (27) (3 kg)	1.97	3.64
Goat (28) (31 kg)	0.75	1.82
Gazelle (29) (23.5 kg)	1.15	2.42

[a]The data are normalized by presenting the ratio of evaporative heat loss from the respiratory tract (EHL_R) to the predicted resting metabolic rate (M_{STD}).

a treadmill. Table 1 shows that these animals can evaporate at least twice as much water from the respiratory tract during exercise as they can during maximal, rapid, shallow steady state thermal panting. This supports the contention that an additional evaporative surface is recruited in the respiratory tract during exercise.

SWEATING: ADDITIONAL EVAPORATIVE SURFACE FOR DISSIPATING LARGE EXERCISE HEAT LOADS

Animals That Sweat

Many animals, humans perhaps being the most familiar example, sweat in response to both exercise and environmental heat loads. The water evaporates from the skin, binding approximately 580 cal of heat with each gram that vaporizes. Two types of sweat glands are found in mammals, one which opens directly onto the skin surface and another which opens into the lumen of a hair follicle. Although these have commonly been referred to as "eccrine" and "apocrine" glands (44), Bligh (3) has suggested the more appropriate terms "atrichial" (without hair) and "epitrichial" (with hair). There does not appear to be any clear functional distinction between the two types. There are examples of both types which can produce large quantities of sweat for sustained periods of time (3). The horse, with its epitrichial sweat glands, seems just as capable of dissipating large amouts of heat by cutaneous evaporation as humans with their atrichial sweat glands.

Sweat glands serve different functions: 1) thermoregulation, perhaps the most obvious function; 2) an increase in friction between the foot pad and the substratum (45, 46) due to secretion from sweat glands on the paws and palms; and 3) social and sexual signals (47) produced by the odor of some sweat glands.

In this chapter, only the thermoregulatory function of sweat glands is discussed. Weiner and Hellman (47) and Robertshaw (48) have surveyed the literature and they have concluded that general body sweat glands are found in many groups of mammals: primates, equidae, rhinocerotidae, bovidae, suidae, camelidae, canidae, and felidae, but not in rodentia. General body sweating also has important thermoregulatory functions in a marsupial, the big red kangaroo (49). In many animals, a thermoregulatory function of the sweat glands has not yet been demonstrated. It seems clear, however, that many distantly related mammals use generalized body sweating as an evaporative cooling mechanism and that it has evolved and been lost more than once in the evolution of mammals. Because phylogeny does not provide an answer, is there another explanation for the adoption of sweating as an evaporative cooling mechanism by different mammals?

Size Dependency for Sweating

While considering the different ways in which large and small animals have dealt with hot desert environments, Schmidt-Nielsen pointed out that large desert animals sweat to dissipate heat, but small ones do not (50). This size dependency does not seem to be restricted just to desert animals. Whenever an attempt is made to generalize about the presence or absence of sweating as an evaporative cooling mechanism among mammals, the conclusions seem to be that 1) larger animals sweat, 2) smaller animals rely on respiratory more than cutaneous evaporation, and 3) there is no phylogenetic pattern which explains whether or not a particular species will sweat.

The size dependency is found even within a single family of closely related species. For example, Robertshaw and Taylor (51) examined the importance of sweating among eight species of wild bovids ranging in size from the duiker (6-kg) to the eland and buffalo (1,000-kg) and three species of domestic bodid: sheep, goats, and cattle. They concluded that there is indeed a correlation between the rate of cutaneous evaporation and the adult size of an animal. Smaller animals relied primarily on the respiratory tract for evaporative heat dissipation, whereas larger animals relied more on cutaneous evaporation.

Reasons for Dependence on Body Size

In an earlier section, the relative magnitude of heat loads which an animal might encounter during exercise or in a hot environment was considered. Several generalizations emerged: 1) exercise heat loads can be nearly an order of magnitude larger than environmental heat loads; 2) small animals (<1,000 g) can dissipate most of the heat that they produce during exercise by nonevaporative means at the ambient temperatures in which they usually operate; and 3) under the same ambient temperature conditions, large animals (>10 kg) must resort to evaporative cooling or store heat in order to cope with moderate exercise heat loads. It seems that the maximal capacity for respiratory cooling is insufficient to maintain a constant body temperature during exercise in larger animals and that they must recruit an additional evaporative surface—the skin—in order to be capable of sustained exercise.

What evidence is available to support these speculations? With the same respiratory cooling, large sweating animals are able to achieve much higher rates of evaporative cooling than large animals that do not sweat. Unfortunately, measurements of maximal rates of evaporative cooling from large animals are practically nonexistent. Much more information is needed on maximal rates of cutaneous and respiratory evaporation in animals of different sizes, particularly in large animals like the eland and horse. If our hypothesis is correct, then the maximal rate of cutaneous evaporation should increase with body size to compensate for the decreased ability to dissipate heat by nonevaporative means.

INTERACTION OF DIFFERENT EVAPORATIVE COOLING MECHANISMS

Four evaporative cooling mechanisms have been shown and it has been suggested that each operates in response to a particular type of heat load, i.e., either environmental or exercise. If different evaporative cooling mechanisms deal with two different types of heat loads, then the question is, what happens when the animal encounters both types of heat loads simultaneously? How do the mechanisms interact?

Saliva spreading is primarily a mechanism which small mammals use for dealing with environmental heat loads, and it obviously is impractical for sustained evaporative cooling during exercise. Small rodents use neither "phase 1" nor "phase 2" panting, nor do they sweat; thus there is no possibility of an interaction of evaporative cooling mechanisms. Saliva spreading appears to be an emergency cooling mechanism in most of the animals that employ it. Although it may be used by some mammals together with another mechanism, it seems unlikely that this would happen on a sustained steady state basis.

Let us next consider the two respiratory mechanisms. Is there an interaction between rapid thermal panting and deep exercise panting when animals that possess both mechanisms run in a warm environment? Can the rapid nasal panting be adjusted to compensate for lower rates of nonevaporative heat loss from an exercising animal as air temperature increases? Experiments need to be specifically designed to answer this question.

What about animals that sweat? Do they use either rapid thermal or deep exercise panting in addition to sweating? Humans and anthropoid primates have apparently lost the ability to use the nasal mucosa as an evaporative cooling surface. Figure 8*A* demonstrates that they do not increase respiratory evaporation in response to heat. Respiratory evaporation accounts for only 5–10% of their total evaporation at 45°C. This contrasts with the situation in those animals that do use thermal panting—the dog and goat—which increase respiratory evaporative heat loss from less than 10% of heat production at 22°C to more than 120% at 45°C. By respiratory evaporation, they are able to dissipate both the heat which they are producing and the heat flowing into them from the environment.

Some sweating animals, however, do make use of evaporation from the buccal and tongue surfaces as a mechanism for dissipating heat during exercise. During strenuous exercise, these animals are able to dissipate heat by evaporation from the respiratory tract at a rate equal to the rate at which they produce heat at rest (Figure 8*B*). Under these conditions, however, it accounts for only 10% of their total heat production.

Many thermal panting animals such as the dog, goat, kangaroo, and gazelle possess sweat glands over most of their body surface. If our hypothesis is correct, it might be expected that these sweat glands would function during exercise, but not in the heat. Dawson et al. (49) found that this happened in the big red kangaroo, which pants in response to hot environments and sweats to dissipate the large heat loads encountered during exercise. Taylor et al. (27) postulated that the African hunting dog might sweat in response to exercise, but they found no significant sweating even when the animal ran at ambient temperatures above 40°C. Taylor et al. (29) also looked for exercise sweating in the gazelle, but found none, even when the gazelle ran at high speeds at ambient temperatures higher than body temperature. This leaves unresolved the question as to whether or not the sweat glands of panting animals serve an important role

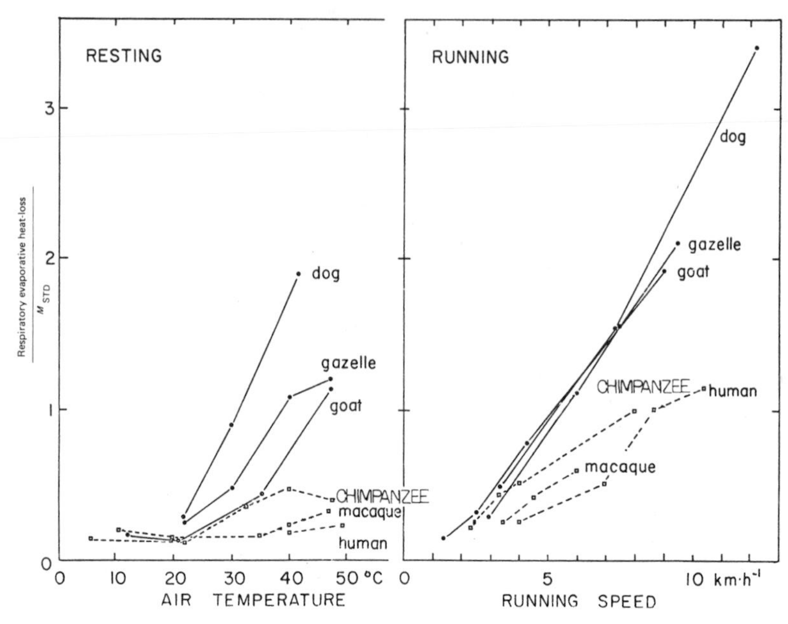

Figure 8. The ratio of respiratory evaporative heat loss to predicted metabolic heat production (M_{STD}) in panting and sweating animals as a function of air temperature and running speed. *A,* the panting animals (dog, goat, and gazelle) increased respiratory evaporation by as much as 10-fold as air temperature increased from 22–47°C. The sweating primates, however, did not increase respiratory evaporation with increasing temperature. *B,* both panting and sweating animals increase respiratory evaporation during exercise. In chimpanzees and humans, the rate of respiratory evaporative heat loss during exercise can equal the predicted resting metabolic heat production. Data from the goat (28) and gazelle (29) are from the literature, whereas data from the dog and primates have been collected in this laboratory and are as yet unpublished.

in heat dissipation. Recent experiments by A. Shkolnik and his colleagues (personal communication) indicate that sweating can become an important avenue of heat loss when gazelles or goats run exposed to the hot sun. Evidently, direct solar radiation is needed together with the exercise to trigger this sweating. This suggests that the general body surface sweat glands of many panting animals may operate in this manner. However, this situation can only be understood through more experiments.

CONCLUSIONS

Advantages and Disadvantages of Different Evaporative Cooling Mechanisms

Many biologists and physiologists have considered the advantages and disadvantages of sweating or panting for evaporative cooling (1–3, 10, 25, 27). It seems worthwhile to review some of these arguments in the light of the previous discussion.

Table 2. Sweating animals gain heat from a hot environment at about 3 times the rate of panting animals[a]

Panters			Sweaters		
	% Evaporation from respiratory tract	Conductance cal$\cdot$(cm^2 $\cdot$ hr $\cdot$ $^\circ$C)$^{-1}$		% Evaporation from skin	Conductance cal$\cdot$(cm^2 $\cdot$ hr $\cdot$ $^\circ$C)$^{-1}$
Dog (3 kg)	77	0.38	Spider monkey (3 kg)	91	0.93
Gazelle (24 kg)	79	0.09	Stump-tail macaque (4 kg)	88	0.65
Goat (29 kg)	69	0.41	Baboon (7 kg)	71	0.61
Cheetah (44 kg)	79	0.08	Chimpanzee (18 kg)	89	0.70
	$\overline{m}$	0.24		$\overline{m}$	0.72

From Taylor and Rowntree (52).

[a]Measurements of conductance were made at an ambient temperature of 47°C. They express nonevaporative heat flow per unit of surface area for each °C of temperature gradient per hour.

In a very hot environment, evaporative cooling from the respiratory tract appears to have the following advantages over evaporative cooling from the skin:

1. Panting animals should have a slower inward flow of heat from the hot environment than sweating animals. Panting animals should have both a higher skin surface temperature and a lower rate of blood flow to the skin. Taylor and Rowntree (52) have quantified different rates at which heat flowed into sweating and thermal panting animals from a hot environment of 47°C. They found that heat flowed into the sweating animals at about twice the rate per unit of surface area per °C temperature gradient that it did into thermal panting animals (Table 2). Thus, when ambient temperature exceeds body temperature, respiratory evaporation offers a considerable advantage over cutaneous evaporation. It must be remembered, however, that in most hot environments body temperature still exceeds air temperature, and there is still a net nonevaporative heat loss. In these situations, high rates of blood flow to the skin is an advantage. Sweating animals still have some disadvantage in that their skin would usually be cooler (as a result of the evaporation taking place from the skin) than the skin of a panting animal. Thus, their gradient for nonevaporative heat loss is smaller.
2. Panting provides forced convection across the respiratory surfaces, whereas sweating animals at rest must rely on free convection if there is no wind.
3. Sweating animals may encounter large salt losses, which can pose a serious problem in hot environments.

Evaporative cooling from the skin offers the following advantages over panting:

1. It should require the expenditure of less energy for dissipating heat. It has been argued that since energy is required to transport salts in producing sweat, and panting animals minimize the energy expenditure for moving air by panting at a resonant frequency, the total energy expenditure might be about the same for both evaporative cooling mechanisms. This argument neglects the fact that salt must also be transported to produce the water for respiratory evaporation (i.e., the secretion from the Steno's gland or the salivary glands). Therefore, panting probably requires a greater expenditure of energy than does sweating, although the difference may be small.
2. Sweating provides a greater maximal rate of evaporative cooling than panting.
3. Sweating does not pose the danger of respiratory alkalosis.
4. Sweating animals can eat and cool evaporatively simultaneously, while animals using open-mouthed panting cannot.

If exercise is considered rather than environmental heat loads, the situation changes somewhat.

a. The lack of forced convection across the skin no longer is a problem for a sweating animal.
b. Second-phase panting may allow increased rates of respiratory evaporation without alkalosis.

c. Air has to be moved across the respiratory surface to provide for the increased rates of gas exchange in any case.

Some Unanswered Questions

A good case has been made for the existence of different thermoregulatory mechanisms for dealing with exercise and environmental heat loads. This approach helps to explain some perplexing findings, such as the anatomical and physiological adaptations for evaporative cooling from the surfaces of the buccal cavity and the tongue and the distribution of general body sweat glands among mammals. The approach has also left many unanswered questions. Only when the answers to these questions are available can the validity of this hypothesis be evaluated. Therefore, rather than end this chapter with conclusions, it will end with questions which need to be answered experimentally:

1. How important is exercise panting as a mechanism for evaporative cooling, and under what circumstances is it used? This can only be answered by partitioning the evaporation between the nasal mucosa and the buccal-tongue surfaces.
2. How do the maximal rates of nonevaporative heat loss, respiratory evaporation, and cutaneous evaporation change as a function of body size.
3. Is exercise sweating (as has been observed in the big red kangaroo, goat, and gazelle) a general phenomenon among thermal panting animals (other than humans and anthropoid apes), and if so, what triggers it?

REFERENCES

1. Taylor, C. R. (1974). Exercise and thermoregulation. *In* MTP International Review of Science, Physiology Series I, Vol. 7, Environmental Physiology, pp. 163–184. D. Robertshaw (ed.), Butterworths, London.
2. Hales, J. R. S. (1974). Physiological responses to heat. *In* MTP International Review of Science, Physiology Series I, Vol. 7, Environmental Physiology, pp. 107–162. D. Robertshaw (ed.), Butterworths, London.
3. Bligh, J. (1973). Temperature Regulation in Mammals and Other Vertebrates. North Holland Publishing Company, Amsterdam.
4. Howell, A. B. (1944). Speed in Animals. Chicago University Press, Chicago.
5. Layne, J. N., and Benton, A. H. (1954). Some speeds of small mammals. J. Mammal. 35:103.
6. Mitchell, D. (1974). Physical basis of thermoregulation. *In* MTP International Review of Science, Physiology Series I, Vol. 7, Environmental Physiology, pp. 1–32. D. Robertshaw (ed.), Butterworths, London.
7. Finch, V. A. (1972). Thermoregulation and heat balance of the East African eland and hartebeest. Am. J. Physiol. 222:1374.
8. Tucker, V. A. (1972). Metabolism during flight in the laughing gull, *Larus atricilla.* Am. J. Physiol. 222:237.
9. Kleiber, M. (1961). The Fire of Life. An Introduction to Animal Energetics. Wiley and Sons, New York.
10. Schmidt-Nielsen, K. (1964). Desert Animals. Physiological Problems of Heat and Water. Oxford University Press, New York.

11. Borut, A., Dmi'el, R., and Shkolnik, A. (1974). Heat balance of resting and walking goats exposed to natural desert conditions. Abstracts of Jerusalem Satellite Symposium on Environmental Physiology, XXVI International Congress of Physiological Sciences, p. 7.
12. Adolph, E. A. (1947). Physiology of Man in the Desert. Interscience Publishing Company, Inc., New York.
13. Schmidt-Nielsen, K. (1964). Terrestrial animals in dry heat: desert rodents. *In* Handbook of Physiology, Vol. 4, Adaptation to the Environment, pp. 493–507. D. B. Dill, E. F. Adolph and C. G. Wilber (eds.), American Physiological Society, Washington, D.C.
14. Hainsworth, F. R. (1967). Saliva spreading, activity and body temperature regulation in the rat. Am. J. Physiol. 212:1288.
15. Hainsworth, F. R., and Stricker, E. M. (1969). Evaporative cooling in the rat: effects of partial desalivation. Am. J. Physiol. 217:494.
16. Rodland, K. D., and Hainsworth, F. R. (1973). Peripheral neural control of thermoregulatory salivary secretion in the rat. Can. J. Physiol. Pharmacol. 51:213.
17. Elmer, M., and Ohlin, P. (1971). Salivary secretion in the rat in a hot environment. Acta Physiol. Scand. 83:174.
18. Needham, A. D., Dawson, T. J., and Hales, J. R. S. (1974). Forelimb blood flow and saliva spreading in the thermoregulation of the red kangaroo (*Megaleia rufa*). Comp. Biochem. Physiol. 49A:555.
19. Jackson, D. C., and Schmidt-Nielsen, K. (1964). Countercurrent heat exchange in the respiratory passages. Proc. Natl. Acad. Sci. USA 51:1192.
20. Schmidt-Nielsen, K., Hainsworth, F. R., and Murrish, D. E. (1970). Countercurrent heat exchange in the respiratory passages: effect on water and heat balance. Respir. Physiol. 9:263.
21. Murrish, D. E., and Schmidt-Nielsen, K. (1970). Exhaled air temperature and water conservation in lizards. Respir. Physiol. 10:151.
22. Raab, J., and Schmidt-Nielsen, K. (1972). Effect of running on water balance of the kangaroo rat. Am. J. Physiol. 222:1230.
23. Hales, J. R. S. (1966). The partition of respiratory ventilation of the panting ox. J. Physiol. (Lond.) 188:45P.
24. Schmidt-Nielsen, K., Bretz, W. L., and Taylor, C. R. (1970). Panting in dogs: unidirectional air flow over evaporative surfaces. Science 169:1102.
25. Crawford, E. C., Jr. (1962). Mechanical aspects of panting in dogs. J. Appl. Physiol. 17:249.
26. Crawford, E. C., Jr., and Kampe, G. (1971). Resonant panting in pigeons. Comp. Biochem. Physiol. 40A:549.
27. Taylor, C. R., Schmidt-Nielsen, K., Dmi'el, R., and Fedak, M. (1971). Effect of hyperthermia on heat balance during running in the African hunting dog. Am. J. Physiol. 220:823.
28. Taylor, C. R., and Roundtree, V. J. (1973). Temperature regulation and heat balance in running cheetahs: a strategy for sprinters? Am. J. Physiol. 224:848.
29. Taylor, C. R., Dmi'el, R., Shkolnik, A., Baharav, D., and Borut, A. (1974). Heat balance of running gazelles: strategies for conserving water in the desert. Am. J. Physiol. 226:439.
30. Thiele, P., and Albers, C. (1963). Die Wasserdampfabgabe durch die Atemwege und der Wirkungsgrad des Wärmehechelns beim wachen Hung. Pfluegers Arch. 278:316.
31. Hales, J. R. S. (1973). Effects of exposure to hot environments on the regional distribution of blood flow and on cardiorespiratory function in sheep. Pfluegers Arch. 344:133.

32. Hales, J. R. S. (1973). Effects of heat stress on blood flow in respiratory and nonrespiratory muscles in the sheep: an explanation of the apparent high efficiency of panting. Pfluegers Arch. 345:123.
33. Steno, N. (1664). De musculi et glandulis. Amstelodami. (Quoted by Broman, I. (1921). Z. Ges. Anat. 60:439.)
34. Nickel, R., Schummer, A., and Seiferle, E. (1960). Lehrbuch der Anatomie dar Haustiere, Vol. 2. Parley, Berlin.
35. Blatt, C. M., Taylor, C. R., and Habal, M. B. (1972). Thermal panting in dogs: the lateral nasal gland, a source of water for evaporative cooling. Science 177:804.
36. Kindermann, W., and Pleschka, K. (1973). Local blood flow and metabolism of the tongue before and during panting in the dog. Pfluegers Arch. 340:251.
37. Hammel, H. T., and Sharp, F. (1971). Thermoregulatory salivation in the running dog in response to preoptic heating and cooling. J. Physiol. (Paris) 63:260.
38. Negus, V. E. (1949). The Comparative Anatomy and Physiology of the Larynx. Grune and Stratton, New York.
39. Pleschka, K., and Krönert, H. (1974). Thermoregulatory adjustment of lingual blood flow in the conscious dog at high ambient temperature. Abstracts of Jerusalem Satellite Symposium on Temperature Regulation, XXVI International Congress of Physiological Sciences, p. 27.
40. Hammel, H. T. (1972). The set point in temperature regulation: analogy or reality. *In* Essays on Temperature Regulation, pp. 121–137. J. Bligh and R. Moore (eds.), North-Holland Publishing Company, Amsterdam.
41. Hales, J. R. S., and Bligh, J. (1969). Respiratory responses of the conscious dog to severe heat stress. Experientia 25:818.
42. Hales, J. R. S., and Webster, M. E. D. (1967). Respiratory function during tachypnoea in sheep. J. Physiol. (Lond.) 190:241.
43. Hales, J. R. S., and Findlay, J. D. (1968). Respiration of the ox: normal values and the effects of exposure to hot environments. Respir. Physiol. 4:333.
44. Schiefferdecker, P. (1971). Die Hautdrusen des Menschen und der Säugetiere, ihre biologische und rassenana tomische Bedentung, so wie die Muscularis sexualis. Biol. Zentralbl. 37:534.
45. Adams, T., and Hunter, W. S. (1969). Modification of the mechanical properties by eccrine sweat gland activity. J. Appl. Physiol. 26:417.
46. Adelman, S., Taylor, C. R., and Heglund, N. (1975). Sweating on paws and palms: what is its function? Am. J. Physiol. 229:1400.
47. Weiner, J. S., and Hellman, K. (1960). The sweat glands. Biol. Rev. 35:141.
48. Robertshaw, D. (1975). Catecholamines and the control of sweat glands. *In* Handbook of Physiology, Endocrinology, Vol. 7, Adrenal Gland, pp. 591–603. R. O. Greep, E. B. Aftwood, H. Blaschko, G. Sayers, A. D. Smith, and S. R. Geiger (eds.), American Physiological Society, Washington, D.C.
49. Dawson, T. J., Robertshaw, D., and Taylor, C. R. (1974). Sweating in the kangaroo: a cooling mechanism during exercise, but not in the heat. Am. J. Physiol. 227:494.
50. Schmidt-Nielsen, K. (1954). Heat regulation in small and large desert mammals. *In* Biology of Deserts, pp. 182–187. J. L. Cloudsley-Thompson (ed.), Institute of Biology, London.
51. Robertshaw, D., and Taylor, C. R. (1969). A comparison of sweat gland activity in eight species of East African bovids. J. Physiol. (Lond.) 203:135.

52. Taylor, C. R., and Rowntree, V. J. (1974). Panting vs. sweating: optimal strategies for dissipating exercise and environmental heat loads. Proceedings of the International Union of Physiological Science, XXVI International Congress, New Delhi, XI:348.
53. Schmidt-Nielsen, K. (1972). How Animals Work. Cambridge University Press, New York.
54. Murrish, D. E. (1973). Respiratory heat and water exchange in penguins. Resp. Physiol. 19:262–270.

International Review of Physiology
Environmental Physiology II, Volume 15
Edited by David Robertshaw

5
Thermoregulation During Sleep and Hibernation

H. C. HELLER AND S. F. GLOTZBACH

Stanford University,
Stanford, California

DAILY CYCLE OF BODY TEMPERATURE: INDIRECT EFFECT OF THE SLEEP/ACTIVITY CYCLE? 148

THERMOREGULATORY CHANGES SPECIFICALLY RELATED TO SLEEP 149

CHANGES IN BRAIN AND BODY TEMPERATURES AND THERMOREGULATORY RESPONSES ASSOCIATED WITH STATES OF SLEEP 151
- Electrophysiological Correlates of Sleep 151
- Changes in Brain and Skin Temperatures During Sleep 151
- Changes in Thermoregulatory Responses During Sleep 153
- Changes in Hypothalamic Thermosensitivity During Sleep 155

EFFECTS OF HYPOTHALAMIC AND AMBIENT TEMPERATURES ON SLEEP STATES 159
- Effect of Skin Temperature on EEG of Immobilized Cats 159
- Influence of Ambient Temperature on Distribution of Sleep States 160
- Influence of Hypothalamic Temperature on Sleep 161
- Combined Influence of Hypothalamic and Ambient Temperatures on Distribution of Sleep States 162

HIBERNATION AS AN EXTENSION OF SLEEP: ELECTROPHYSIOLOGICAL EVIDENCE 165

THERMOREGULATION DURING HIBERNATION 168
- Is T_b Regulated During Hibernation? 168
- Thermosensitivity During Hibernation 171
- The Nature of Regulation During Hibernation 173

CONCLUSIONS 179

This chapter is a review and synthesis of information pertaining to the regulation of body temperature during sleep and hibernation. In remaining true to this topic, many interesting, related studies in the separate fields of sleep, thermoregulation, and hibernation research have been omitted. The general outline of the chapter is as follows. First, studies are reviewed which demonstrate fluctuations in various body temperatures and changes in the activities of thermoregulatory effector organs in humans and other mammals during sleep and during specific sleep states. Second, a general, functional explanation for these diverse observations based on recent research on sleep-state-dependent changes in the characteristics of the central nervous thermoregulatory system is offered. Next, studies on the influence of ambient temperature on total sleep time and percent occurrence of sleep states are reviewed, and some new information on the combined influence of ambient temperature and brain temperature on these parameters is presented. In light of the changes occurring in the thermoregulatory system during sleep states, an adaptive explanation for the influence of ambient temperature on sleep time and percent occurrence of sleep states is offered.

The transition from sleep studies to hibernation studies is made through a discussion of the possibility that these two phenomena are homologous. The evidence brought to bear on this question are several electrographic studies of animals entering hibernation. Electrographic records from animals entering hibernation are compared with the classic electrographic patterns associated with sleep states. Thermoregulatory homologies between sleep and hibernation are also sought. The evidence that body temperature is regulated during hibernation is reviewed, and then recent research on the central nervous thermoregulatory system during hibernation is discussed. Because both lines of evidence suggest physiological homologies between sleep and hibernation, this chapter is concluded speculating on the evolution of these two phenomena.

DAILY CYCLE OF BODY TEMPERATURE: INDIRECT EFFECT OF SLEEP/ACTIVITY CYCLE?

Daily rhythms of body temperature and metabolism associated with daily sleep/activity rhythms have been described for a wide variety of animals, including humans (1–3). The constant association of higher body temperatures and metabolic rates with activity and of lower body temperatures and metabolic rates with inactivity led to early hypotheses that the body temperature/metabolism rhythm was the indirect effect of muscular activity, food intake, and/or sleep. Even though these factors normally do influence body temperature (T_b) and metabolic heat production, they are not sufficient explanations for the T_b/metabolism rhythm (3). This rhythm persists in food-deprived humans and animals (4–7). It also persists in humans deprived of sleep (8–11), in humans kept inactive in bed (12), and even in totally paralyzed humans (9). A daily rhythm of T_b partially independent of activity has been demonstrated in the rat by Heusner (13, 14) and in three species of hummingbirds by Morrison (15). In these studies, T_b during day and night was plotted as a function of activity. For the rat, the night curves were significantly higher than the day curves, and vice versa for the hummingbirds, thereby revealing an underlying difference in T_b between day and night, independent of level of activity. The most convincing evidence for a daily T_b rhythm that is independent of the sleep/activity cycle is the observation that the rhythms of activity and T_b of humans can become desynchronized and free run, with different periodicities in the absence of environmental time cues (16–18). It seems likely that the central thermoregulatory system is under the influence of a circadian oscillator which is separate from, but normally synchronized with, one or more oscillators driving the sleep/activity rhythm (18).

Hammel et al. (19) proposed that the daily fluctuation of T_b was due to a changing "set point" of the hypothalamic proportional controller. The concept of a set point for T_b has been supplanted by the concept of specific threshold temperatures for individual thermoregulatory responses. There is now a substantial amount of evidence that temperature thresholds for thermoregulatory responses fluctuate in a daily rhythm even in awake subjects (20–25).

THERMOREGULATORY CHANGES SPECIFICALLY RELATED TO SLEEP

The existence of a daily T_b rhythm independent of the sleep/activity cycle does not exclude sleep-related thermoregulatory adjustments which are normally superimposed on the T_b rhythm. However, this possibility was rejected by Geschickter et al. (26), based on their measurement of T_b, rate of sweating, and electroencephalograms (EEG) in 8 subjects over 14 nights of sleep. Because a decline in T_b always began prior to the onset of sleep (mean Δt = 144 min), and a rise in T_b always began prior to awakening (mean Δt = 88 min), and drops in T_b were not observed during several episodes of afternoon napping, they concluded: "Reduction of body temperature did not seem directly dependent on the act of sleep . . . " (26).

Contrary to the conclusions of Geschickter et al. (26), many studies have produced evidence for thermoregulatory adjustments specifically associated with the transition from wakefulness to sleep. Day (27) measured rectal and skin temperatures and evaporative water loss during afternoon naps in nine children between the ages of 5 months and 4 years. He consistently observed a decline in rectal temperature (mean of 0.55°C) beginning with the onset of sleep. The beginning of the decline of rectal temperature was coincident with increased evaporative water loss and a rise in skin temperature, indicating that the T_b decline was a regulated phenomenon. The end of the decline in rectal temperature was abrupt and was coincident with decreases in evaporative water loss and skin temperatures. Geschickter et al. dismissed the discrepancy between the findings of Day and themselves by assuming that the diurnal T_b rhythms of the children were not yet firmly established. However, there is clear evidence for diurnal T_b rhythms in infants as young as 3½ months old, and such rhythms are firmly established during the 2nd year of life (9). It is clear that Day demonstrated a consistent readjustment of the thermoregulatory system coincident with the onset of sleep.

Thermoregulatory adjustments associated with sleep have also been observed in human adults. In neutral or cool environments, skin temperatures rise while rectal temperature falls at the onset of sleep (19, 28–30). In neutral or warm environments, sweating was seen to increase at the onset of sleep (26, 31, 32).

Animal studies have also produced evidence for thermoregulatory adjustments coupled to the onset of sleep. Euler and Söderberg (33) induced episodes of sleep in cats, evidenced by a synchronization of the cortical EEG, by gently stroking the animals' backs. The onset of the synchronized EEG was concomitant with a cessation of shivering, vasodilation in the ear pinnae, and a subsequent fall in T_b. In a study of hypothalamic temperature (T_{hy}) of cats, Adams (34) found that T_{hy} always fell during behavioral sleep. Abrams and Hammel (35) continuously measured T_{hy} in unrestrained, unanesthetized albino rats. Their results showed that behavioral sleep was always associated with a fall in brain temperature, and the magnitude of the fall was proportional to the duration of the sleep episode.

Interesting experiments demonstrating both circadian and sleep-related thermoregulatory changes were reported by Hammel et al. (19) on a rhesus monkey confined to a primate chair. Hypothalamic temperature was recorded continuously over many days and showed a diurnal rhythm, the magnitude of which depended upon ambient temperature (T_a). The fact that T_{hy} fell at night, even in a 35°C environment, indicated that the nocturnal drop in body temperature represents an active decrease in the regulated body temperature. In a 7-hr experiment, ear pinna temperature and T_{hy} were recorded, while activity and periods of eye closure were noted by an observer (Figure 1). It was assumed that

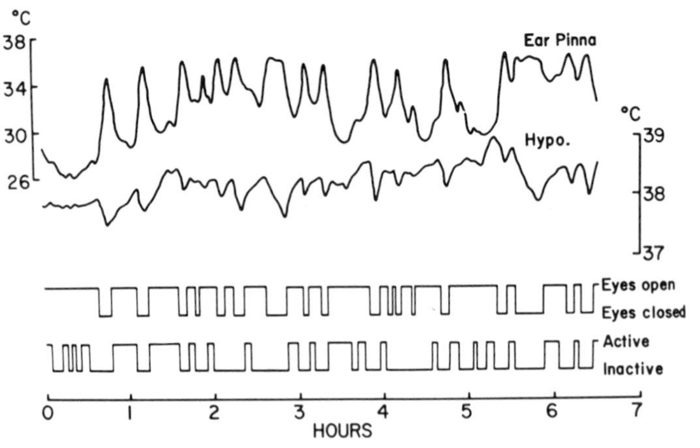

Figure 1. Continuous recordings of hypothalamic and ear pinna temperatures of a rhesus monkey confined to a primate chair in a closed chamber maintained at 22–24°C. An observer noted activity and whether the eyes were open or closed. (Reproduced from Hammel et al. (19), with permission.)

eye closure signified drowsiness or sleep. Changes in T_{hy} and ear temperature did not coincide with gross body movements, but they were highly correlated with eye closure. Whenever the animal closed its eyes, ear pinna temperature rose, indicating vasodilation, and T_{hy} fell. The reverse happened when the eyes opened.

In summary, evidence strongly suggests the existence of thermoregulatory events specifically related to sleep which are independent of and normally superimposed upon a separate 24-hr rhythm of T_b.

CHANGES IN BRAIN AND BODY TEMPERATURES AND THERMOREGULATORY RESPONSES ASSOCIATED WITH STATES OF SLEEP

Electrophysiological Correlates of Sleep

An implicit assumption in many of the studies reviewed in the previous section was that sleep is a unitary phenomenon; this is definitely not the case. Since the classic studies of Kleitman and associates in the 1950s (36, 37), sleep researchers have defined distinctive sleep states primarily on the basis of electrophysiological criteria. One state has been variously referred to as rapid eye movement sleep (REM), active sleep, desynchronized sleep, fast wave sleep, or paradoxical sleep. For consistency, we shall use the term "paradoxical sleep" (PS) in this paper. PS is generally characterized by a low voltage (<50 μV), fast (>8 cps), desynchronized, cortical EEG pattern similar to wakefulness (hence "paradoxical" sleep); atonia of the skeletal muscles; irregularities in respiratory rate and heart rate; and occasional rapid eye movements (38). The other stages of sleep are generally referred to as synchronous sleep, non-REM sleep (NREM), or slow wave sleep (SWS). In mammals other than primates, NREM sleep and SWS are considered synonymous, whereas in humans and other primates NREM sleep has been classified into four stages, two of which are called SWS. In this chapter, we are dealing almost exclusively with studies of nonprimates and shall use the term "SWS" throughout. SWS is characterized by a high voltage (>75 μV), slow (<3 cps), synchronous, cortical EEG; reduction in frequency in electrical activity relative to wakefulness in many subcortical regions; and decreased muscle tonus (38).

Changes in Brain and Skin Temperatures During Sleep

A large number of workers have described consistent changes in deep brain temperatures as a function of sleep state. Most of these studies do not report the ambient temperature (T_a) at which the experiments were conducted. Because SWS is almost always the sleep state entered from wakefulness, older literature discussing the transition from wakefulness to sleep is interpreted as describing the transition from wakefulness to SWS, even though EEG was not recorded. In general, a fall in brain temperature occurs during the transition from wakefulness to SWS: cat (33, 34, 39–46), rat (47), rabbit (48), monkey (19, 45, 49), dog

(19, 50), the armadillo (51), and the opossum (52). In contrast, the transition from SWS to PS is usually accompanied by a rise in deep brain temperatures: cat (40, 41, 44–46, 53), dog (54), rat (47), rabbit (41, 48, 55), sheep (45), opossum (52), and mole (56). A decrease in cortical and deep brain temperatures during PS has been observed in the monkey (45, 57). The hypotheses offered to explain these brain temperature changes include changes in metabolic rate of the neural tissue and changes in cerebral blood flow. The most convincingly documented and unifying explanation, however, is that changes in peripheral vasomotor activity occur as a function of sleep states (41, 45). Although Baker and Hayward studied cats, monkeys, rabbits, dogs, and sheep, we shall briefly discuss their study on rabbits (41) as an example of their approach, techniques, and conclusions. Thermocouples were implanted in the midline of the brainstem on three frontal planes. Thermocouples were also implanted at the base of the brain near the circle of Willis and in the aorta. Skin temperatures were measured with thermocouples taped to the ear and the back. EEG was recorded from two bone screws over the parietal cortex. There was a close correlation between intracranial temperatures and aortic blood temperature. Back skin temperature also showed a constant relationship to blood temperature. The important finding was that ear skin temperature was inversely related to blood temperature. Increases in ear skin temperature reflected vasodilation and high heat loss resulting in decreases of blood and brain temperatures. Decreases in ear skin temperature reflected vasoconstriction and decreases in thermal conductance resulting in increases of blood and brain temperatures. At ambient temperatures of 25°C and 32°C, the entrance into PS was always accompanied by vasoconstriction in the ear and increases of brain and blood temperatures (Figure 2). At 15°C ambient temperature, the ear was already vasoconstricted during SWS, and therefore, no changes in ear skin, blood, or brain temperatures were observed during the transition to PS. Baker and Hayward (41) conclude that the brain temperatures are mainly influenced by the temperature of the perfusing blood which rapidly reflects changes in peripheral vasomotor tone. In animals that lack internal carotid arteries but have carotid retes (e.g., cats, dogs, and sheep), brain temperatures are buffered from changes in the temperature of the blood arriving in the external carotid artery, but are strongly influenced by the temperatures of venous blood returning from such peripheral structures of the head as skin, nasal mucosa, and horns (41, 45, 58).

The studies of Parmeggiani and colleagues (46, 59) on cats must be mentioned here because they appear to conflict with the findings of Baker and Hayward on the rabbit at low ambient temperatures. At all T_a values from 32°C to −15°C, the cat showed a rise in T_{hy} during the transition from SWS to PS. It is difficult to imagine that the peripheral vessels of the cat are actively dilated at −15°C and return to a passive state of relative vasoconstriction during PS. However, Parmeggiani et al. (59) report a fall in a subcutaneous temperature (inguinal) coincident with the transition into PS even at a T_a of −15°C. Of course, the relevant skin temperatures to measure would be on the head and not

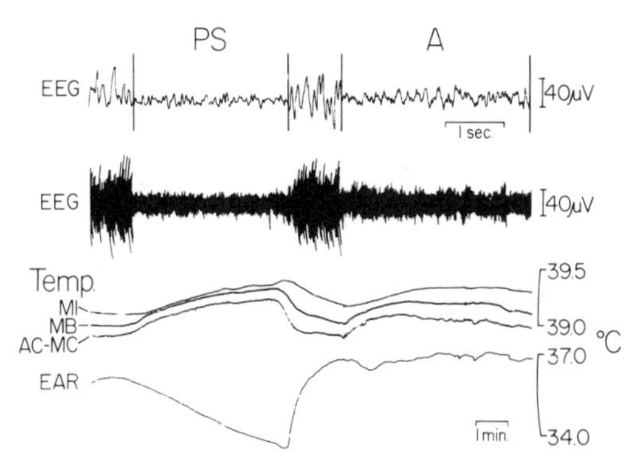

Figure 2. Recordings of central and peripheral body temperatures and biparietal cortical electroencephalogram during PS, SWS, (unlabeled), and wakefulness (*A*) in a freely moving rabbit at a T_a of 25°C. *Lower EEG trace* was recorded simultaneously with the temperatures, and *upper EEG* are representative samples recorded at a faster speed from the same animal during similar arousal states. *MI,* massa intermedia of the thalamus; *MB,* midmammillary body; *AC-MC,* cerebral arterial blood at junction of the anterior and middle cerebral arteries in the basal subarachnoid space; *ear,* skin of the ear. (Reproduced from Baker and Hayward (41), with permission. Copyright 1967 by the American Association for the Advancement of Science.)

the body, because the cat has a carotid rete that buffers the temperature of the cerebral blood from changes in arterial blood temperature.

There are a variety of other hypotheses purporting to explain changes in the temperature of the cerebral arterial blood during sleep stages. These include changes in local blood flow rates (60–62) and changes in local tissue metabolic rates due to altered patterns of neuronal activity (63, 64). Changes in autonomic responses other than vasomotor activity could also cause changes in cerebral blood temperature. Alterations in depth or frequency of respiration could influence the temperature of the venous blood from the tongue and the nasal mucosa which comes into intimate contact with the carotid rete. Sufficient data do not exist to enable us to resolve the question of the cause or causes of changes in cerebral blood temperature during sleep stages. However, it does seem clear that deep brain temperatures are sensitive indices of autonomic adjustments, some of which are most certainly under the influence of the thermoregulatory system.

Changes in Thermoregulatory Responses During Sleep

Consistent brain temperature changes during sleep stages, even if shown to be due to peripheral vasomotor activity, are not necessarily reflecting changes in the central nervous regulation of body temperature. Additional evidence pointing to changes in the central nervous system (CNS) regulator of T_b during sleep is derived from studies which demonstrate coordinated changes in thermoregulatory responses other than vasomotor activity at the onset of sleep—for example, cessation of shivering in a cool environment (65, 66) and increased evaporative water loss in a neutral or warm environment (17, 26, 31, 32, 67). Parmeggiani and Sabattini (68) measured shivering and panting in cats sleeping at high, neutral, and low temperatures. At high ambient temperatures (30–35°C), they observed an increase in the rate of panting during SWS in comparison to wakefulness. This observation is compatible with the body of evidence cited above in suggesting that a regulated decrease in body temperature occurs at the onset of SWS. At low ambient temperatures (0–10°C), they observed an increase of shivering during SWS relative to wakefulness which appears to be incompatible with such an hypothesis. No temperature records were presented, and it is possible that the shivering was caused by an excessive drop in core temperature. In a separate study, Parmeggiani et al. (59) measured inguinal subcutaneous temperature of cats during sleep at different T_a values. At low T_a values, inguinal subcutaneous temperatures increased during SWS, indicating vasodilation and, therefore, an increased thermal conductivity which would cause a fall in T_b. All evidence to date supports the view that T_b is regulated during SWS, but at a lower level than during wakefulness.

In spite of the increase in brain temperatures commonly seen during PS in most species studied, thermoregulatory responses do not show consistent coordinated changes which would be indicative of a shift in the regulated T_b during this sleep state. Measurements of thermoregulatory responses during PS at different T_a values suggest an inhibition of thermoregulatory functions. Parmeggiani and Rabini (69), working on cats, were the first to observe a cessation of shivering at low T_a values and a cessation of panting at high T_a values during the

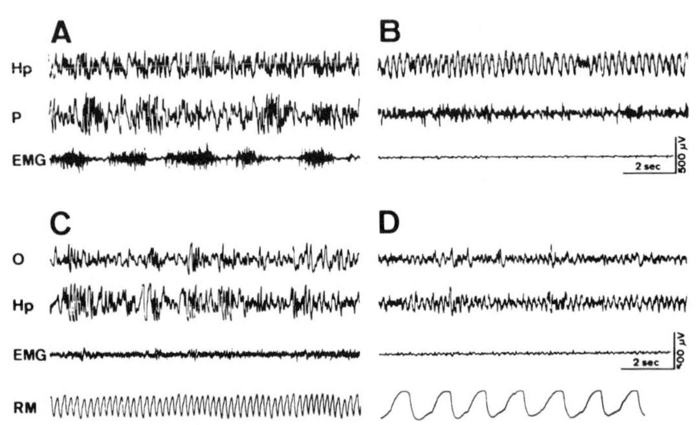

Figure 3. Recordings of EEG, EMG, and respiratory movements (RM) in a cat during sleep. The traces in *A* and *B* were obtained at a T_a of 6°C during SWS and PS, respectively. Bursts of shivering are evident in the neck muscle EMG during SWS, but not during PS. The traces in *C* and *D* were obtained at a T_a of 36.5°C during SWS and PS, respectively. Panting during SWS is shown by the high frequency of respiratory movements (255/min) in *C*. During the bout of PS shown in *D*, there is no panting, as evidenced by the slower, deeper respiratory movements. Abbreviations are as follows: Hp = hippocampus dorsalis; P = parietal; EMG = electromyogram of the neck muscles; O = occipital; RM = respiratory movements. (Reproduced from Parmeggiani and Rabini (69) with permission.)

transition from SWS to PS (Figure 3). These observations were confirmed in a more extensive study (68). Cessation of shivering during PS at a low T_a has subsequently been reported in armadillos (51, 70). Studies on men sleeping at neutral and high T_a values have revealed inhibition of sweating during PS (31, 32, 67, 71). The fact that PS is associated with inhibition of heat loss at high T_a values and inhibition of heat production at low T_a values suggests that the changes in brain temperature associated with this sleep stage are not regulated changes, but may represent a lack of regulation.

In light of the differences in thermoregulatory responses between SWS and PS, it is interesting to note the sleep-state-dependent changes in sympathetic output measured by Baust et al. (72). These researchers prepared cats for electromyogram (EMG), EEG, and electro-oculogram (EOG) recording, as well as for direct recording from the sympathetic renal nerve. They cite evidence indicating that changes in activity occur simultaneously in all sympathetic postganglionic branches. No significant differences in sympathetic output were observed during quiet wakefulness and SWS, but a dramatic reduction in sympathetic activity, interrupted only by phasic bursts, was seen during PS.

Changes in Hypothalamic Thermosensitivity During Sleep

The hypothesis that T_b is regulated at a lower level during SWS but not regulated during PS is strongly supported by experiments involving manipulations of T_{hy} while measuring thermoregulatory responses. Hypothalamic temperature is a major feedback signal to the regulator of T_b in mammals (73, 74). If the T_{hy} of an animal in a warm or neutral T_a is gradually lowered, a threshold T_{hy} will be reached, below which the rate of metabolic heat production is inversely proportional to T_{hy}. Similarly, if T_{hy} of an animal in a cold or neutral T_a is gradually increased, a threshold T_{hy} will be reached, above which the activity of heat loss effectors is directly proportional to T_{hy}. The symbol α is used to refer to the proportionality constants that relate the rates of thermoregulatory responses to T_{hy}. The characteristics of the central nervous regulator of T_b can be described by the following simple equation:

$$R_1 - R_0 = \alpha(T_{hy} - T_{set})$$

where R_0 is the basal rate of a thermoregulatory response when T_{hy} equals threshold (T_{set}) for that response. R_1 is the rate of that response when T_{hy} is above T_{set} for heat loss responses or below T_{set} for heat production/conservation responses. Other factors that influence the thermoregulatory system can be expressed as changes in T_{set} and/or α (19, 73, 74). In small mammals, T_{hy} is the major feedback signal to the regulator (75–78). Therefore, measurements of hypothalamic thermosensitivity during sleep stages, especially in small mammals, should provide an accurate description of changes in the thermoregulatory system associated with sleep.

Parmeggiani et al. (79) studied the responses of three cats to hypothalamic diathermic heating during sleep states. The cats were unrestrained and held at a T_a of 24–28°C to lower the T_{hy} threshold for the panting response. Results of experiments involving hypothalamic heating were reported for SWS and PS, but not for wakefulness. Marked thermal polypnea and even panting were elicited by hypothalamic heating during SWS; however, T_{hy} values which elicited polypnea during SWS had no effect on respiratory rate during PS. Some evidence was presented for a very modest increase in respiratory rate induced by strong hypothalamic heating. Parmeggiani et al. interpret their results to mean that the T_{hy} threshold for thermal polypnea increases during PS. But they state that "... the increase in preoptic threshold during [PS] may be considered as practically infinite." In light of their earlier work showing the cessation of shivering in a cold environment during PS, they further conclude that "... it is unlikely that the abolition of two functionally opposite thermoregulatory responses, panting and shivering, during [PS] may be the result of a simple resetting of the central thermostat." The results of Parmeggiani et al. (79) demonstrate that the central thermoregulatory system is functioning during SWS, but suggest that it is severely inhibited during PS. The questions to be asked next are 1) how do the characteristics of the CNS regulator of T_b during PS, SWS, and wakefulness compare, and 2) what is the nature of the apparent inhibition in the thermoregulatory system during PS?

In a study of hypothalamic thermosensitivity of kangaroo rats, Glotzbach and Heller (76) noted that metabolic heat production responses to hypothalamic cooling were drastically reduced during sleep. Moreover, the extent of the reduction depended on the qualitative stage of sleep as determined by behavioral criteria. Periods characterized by skeletal muscle atonia, frequent phasic movements of vibrissae and limbs, and irregularities in respiratory rate were believed to be PS, and no responses to hypothalamic cooling were observed during these periods. When animals were in a crouched posture with eyes closed, they were assumed to be in SWS. Responses to hypothalamic cooling were observed when the animals were in this condition, but they were lower than a similar cooling would elicit when the animal was obviously awake.

In subsequent experiments, Glotzbach and Heller (80) used electrophysiological criteria to define sleep states in kangaroo rats while manipulating T_{hy}. The EEG recordings obtained from these animals were similar to those reported for other rodents (81). With EEG and EMG recordings, it was possible to precisely identify responses to changes in T_{hy} with specific sleep states. The results of these experiments have confirmed the absence of thermoregulatory responses to hypothalamic cooling during PS, and, in addition, provide the first demonstration of thermoregulatory responses proportional to T_{hy} during SWS. During SWS, in comparison to wakefulness, the gain of the regulator is markedly reduced, and there is also a slight reduction in the T_{hy} threshold for the response (Table I and Figure 4).

These direct experiments on the characteristics of the thermoregulatory system during sleep support the suggestion of earlier studies that changes occur in the thermoregulatory system during sleep and that the nature of these changes is drastically different in SWS and PS. Proportional regulation of body temperature exists during SWS, but the decreased gain and threshold of the regulator

Table 1. Hypothalamic temperature thresholds and proportionality constants for the metabolic heat production response at T_a of 30°C during wakefulness, slow wave sleep, and paradoxical sleep in three kangaroo rats

Animal	Body weight (g)	$T_{set}A^a$ (°C)	$T_{set}SWS$ (°C)	$\alpha_A{}^b$	α_{SWS}	α_{PS}	α_A/α_{SWS}
MY2 (*D. ingens*)	130	36.1	35.8	−0.049	−0.021	+0.0021	2.4
n				23	20	24	
p^c				$p < 0.001$	$p < 0.02$	$p > 0.10$	
S4 (*D. ingens*)	125	35.7	35.2	−0.069	−0.042	−0.0004	1.6
n				19	14	22	
p^c				$p < 0.01$	$p < 0.01$	$p > 0.10$	
MY751 (*D. heermanni*)	65	36.3	36.1	−0.045	−0.026	+0.0010	1.8
n				14	19	17	
p^c				$p < 0.001$	$p < 0.01$	$p > 0.10$	

[a] T_{set}, hypothalamic temperature threshold; A, wakefulness, SWS, slow wave sleep; α, proportionality constant; PS, paradoxical sleep.

[b] α is expressed in cal g^{-1} min^{-1} $°C^{-1}$.

[c] Probability that α is not significantly different from zero.

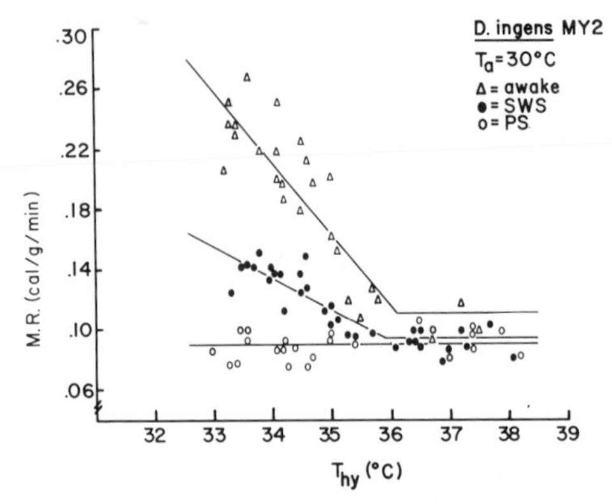

Figure 4. Responses in rate of metabolic heat production (*M.R.*) to short-term (2–10 min) manipulations of hypothalamic temperature in a kangaroo rat during wakefulness, SWS, and PS at an ambient temperature of 30°C. (Reproduced from Glotzbach and Heller (80) with permission. Copyright 1976 by the American Association for the Advancement of Science.)

result in regulation at a decreased level. During PS the thermoregulatory system appears to be inactivated. The alternative explanation for the lack of hypothalamic thermosensitivity during PS in our kangaroo rat experiments would be that the threshold T_{hy} for the metabolic heat production response was suppressed below the range of our T_{hy} manipulations. Parmeggiani et al. (79) showed no panting response to hypothalamic warming during PS in cats. This explanation would require the assumption that, simultaneously, during PS, the threshold T_{hy} values for heat production/conservation responses are severely depressed and the threshold T_{hy} values for heat loss are greatly elevated. All studies to date of conditions that alter T_{hy} threshold for thermoregulatory responses have shown that a stimulus or condition that induces a rise in the threshold for one class of responses induces rises in the thresholds for the other class as well. Similarly, a stimulus or condition that induces a rise in the threshold for one class of responses induces rises in the thresholds for the other class (33, 82). Therefore, it appears to be high unlikely that the T_{hy} thresholds for thermoregulatory responses are simply reset during PS; instead, we favor the interpretation that a severe inhibition, not compatible with the regulation of T_b, impinges upon the thermoregulatory system during PS.

The cessation of thermoregulatory responses such as shivering and panting during PS conceivably may be due to inhibition of skeletal muscle activity originating from the locus coeruleus and/or the pontine tegmentum during this sleep stage. Two observations suggest that this hypothesis is not adequate and that some degree of inhibition is occurring in the hypothalamus. First, Parmeggiani and Franzini (64) observed in the cat a decrease in firing rates of dorsal and anterior hypothalamic neurons during PS at both neutral and low environmental temperatures, in spite of the fact that increased firing rates are generally observed in other parts of the brain during PS (63, 83, 84). It must be noted, however, that a study of hypothalamic units in the rabbit did not produce similar results (63). Second, Hendricks et al. (85) have observed cats with lesions of the pontine tegmentum sleeping at low ambient temperatures. These lesions release the animals from the skeletal muscle atonia associated with PS (85). Even though skeletal muscle activity in these cats is not tonically inhibited during PS, shivering still ceases during PS.

In summary, changes in body temperatures and in the activities of thermoregulatory effector mechanisms occur during sleep. The nature of these changes is sleep-state-dependent. Changes in thermoregulatory responses associated with the transition from wakefulness to SWS indicate that T_b is regulated at a lower level during SWS. When an animal enters SWS from wakefulness in a neutral to warm environment, heat loss responses increase, and in a neutral to cold environment heat production/conservation responses decrease. During PS, active thermoregulatory responses always decrease in comparison to SWS or wakefulness so that the resulting change in T_b depends upon environmental temperature. Comparisons of thermoregulatory responses to manipulations of T_{hy} during different arousal states show that the characteristics of the central nervous regulator of T_b are reset during SWS so as to induce a decline in the regulated T_b, but the thermoregulatory system appears to be inactivated during PS.

EFFECTS OF HYPOTHALAMIC AND AMBIENT TEMPERATURES ON SLEEP STATES

The focus of attention in the preceding sections has been the influence of sleep states on the characteristics of the thermoregulatory system. In this section, studies on a complementary relationship, the effects of internal and external temperature on sleep, are reviewed. Anyone who has persevered through a dull lecture in a warm room will attest to the common experience that warmth, in the absence of arousing stimuli, promotes drowsiness and sleep. Experimental studies on animals lend credence to this observation.

Effect of Skin Temperature on EEG of Immobilized Cats

The effect of T_a on the electrocortical activity of cats immobilized with neuromuscular blocking agents was examined by Okuma et al. (87). After being immobilized with I.P. or I.M. injections of gallamine triethiodide (Flaxedil), succinylcholine cholride, or l-tubocurarine chloride, the cats were artificially respired and placed on a soft mat in a sound-attenuated chamber. Rectal temperature was monitored and maintained around 37°C. At a T_a of 20°C, the electrographic parameters were typical of arousal: low voltage, fast EEG from the neocortex, and rhythmic, slow activity (RSA) from the hippocampus. Hippocampal RSA became less evident and the neocortical EEG showed greater synchrony as the T_a was increased. As T_a approached and surpassed 30°C, the cortical EEG traces showed synchronized, high voltage, slow activity. The desynchronized, low voltage, fast (LVF) activity reappeared as T_a came close to or exceeded body temperature. A reversed progression of electrographic parameters was seen as T_a was slowly decreased to 20°C. Synchronization of the

EEG was also obtained by applying warm water bags to the abdomens of cats held at a cool T_a. When the warm bag was replaced with a cold one, the EEG immediately returned to a desynchronized pattern. The inferences to be drawn from these experiments are that peripheral temperatures outside of the thermoneutral zone promote arousal, whereas thermoneutral temperatures are permissive of and may facilitate sleep.

Influence of Ambient Temperature on Distribution of Sleep States

The first studies describing the influence of T_a upon normative sleep characteristics are those of Parmeggiani et al. (88, 89) on cats. The conventional electrographic criteria for defining arousal states were recorded in all these studies. In one series of experiments (89), the cats were maintained at seasonal T_a values ranging from 18–30°C and were acutely exposed to T_a values ranging from −10–37°C for 3-hr recording sessions. The maximal total sleep time (TST) occurred in a range of T_a values from 10–25°C; as T_a fell below 10°C or rose above 25°C, TST decreased. Parmeggiani et al. (88) define three sleep stages from their data: spindle sleep (SS), SWS, and PS. The percent occurrence of SS was the least affected by low and high T_a values. The decrease in TST was due, therefore, to parallel decreases in the amounts of SWS and PS at low and high T_a values. Sleep spindles are not a common component in the cortical EEG of some species, e.g. rodents. It is, therefore, of interest to look at the cat data of Parmeggiani et al. in terms of synchronized (SWS and SS) and desynchronized (PS) sleep. When spindle sleep is subsumed under SWS for the cat, the mean amount of SWS decreases less than the mean amount of PS at extreme T_a values. An important observation is that the ratio of SWS to PS increases as ambient temperature deviates farther and farther from thermoneutrality.

In another series of experiments (88), cats were exposed to low T_a values ranging from −15–10°C for 2–10 days while electrographic data were recorded continuously, or for 8–9 diurnal hours each day. Recordings were also made at a T_a of 22°C for several days before and after cold exposure. The results agree with those obtained during acute cold exposure. The relative amount of PS during cold exposure in comparison to the 22°C controls was positively correlated with the T_a. This decrease in PS at low T_a values resulted from a decrease in both the number of epochs and the average epoch length. The changes in the amounts of SS and SWS in comparison to the 22°C controls were less consistently related to the level of cold exposure. If SS is again subsumed under SWS, the relationship also emerges from this study that the SWS:PS ratio increases as T_a falls below thermoneutrality.

Similar results have been reported by Schmidek et al. (90) for the albino rat. For at least 2 months following weaning, groups of rats were adapted to T_a values of 30, 24, and 14°C, and then the frequency and duration of arousal states as defined by conventional electrographic criteria were measured at T_a values ranging from 18–36°C in 2° intervals. Total sleep time was clearly a function of T_a, but the relationship differed for the cold-adapted group (14°C) and the neutral and warm-adapted groups (24°C and 30°C). Maximal TST occurred at T_a values from 26–30°C for the cold-adapted group, but TST was greatest at 32°C for the neutral and warm-adapted rats. TST decreased progressively as T_a fell below or rose above these levels. Consistent changes in the bout duration, bout frequency, or total amount of SWS were not seen, but there was a trend in the neutral and warm-adapted animals for bout frequency to increase and bout duration to decrease at the lower ambient temperatures.

The changes in TST observed by Schmidek et al. (90) in rats at different ambient temperatures were largely due to changes in the total amount of PS. The maximal amount of PS was observed in the cold-adapted rats at ambient temperatures of 28–32°C and in the neutral and warm-adapted rats at 30–34°C, with the amount progressively decreasing at T_a values outside these ranges. These changes in total amount of PS were mostly the result of changes in the frequency of occurrence of individual bouts. There was a slight tendency for bout length to decrease at lower temperatures. The relationship between the SWS to PS ratio and T_a is the same for the rat as it was for the cat.

Valatx et al. (91) observed sleep patterns in rats chronically exposed to T_a values of 20, 25, 30, 34, and 36°C. Their results are in agreement with those of Schmidek et al. (90). Total sleep time was greatest at 34°C and decreased at lower T_a values. This decrease was mostly due to the decrease in the number and duration of bouts of PS.

These studies on cats and rats show a consistent relationship between thermoregulatory demands and relative distribution of sleep states. As T_a deviates farther from thermoneutrality, SWS accounts for an increasingly greater percentage of TST, and PS accounts for less. This relationship is especially interesting in the light of our findings that proportional regulation of body temprature exists during SWS, but not during PS (Figure 4).

Influence of Hypothalamic Temperature on Sleep

Hypothalamic temperature has also been shown to have an influence on sleep. Hemingway et al. (92) reported that local heating of the posterior hypothalamus of dogs induced drawsiness and sleep. Euler and Söderberg (33) observed the effects of heating the anterior hypothalamus on the cortical EEG of anesthetized cats and rabbits. Mild warming of the anterior hypothalamus in moderately anesthetized animals produced synchronization of the EEG in both cats and rabbits. Strong heating of the hypothalamus of the rabbits resulted in a desynchronized, "aroused" EEG. Diathermic warming of the anterior hypothalamus of unanesthetized, unrestrained cats and opossums induced relaxation and sleep (93, 94). Although these studies clearly indicate an influence of T_{hy} on sleep processes, the results have not been reported in terms of qualitative and quantitative changes in normative sleep patterns.

The influence of T_{hy} on the duration of individual PS epochs has been studied in cats by Parmeggiani et al. (95) at T_a values of 0 and 20°C. Onsets of PS were selected at random for hypothalamic heating of 1–2°C. Hypothalamic heating resulted in significant increases in length of PS epochs at both ambient temperatures.

A neurophysiological correlate of and possible mechanism for the sleep-inducing effect of hypothalamic warming has been described by De Armond and Fusco (96). Their experiments were based on the generally accepted assumption that arousal states are dependent upon the activity of the mesencephalic reticular formation (MRF) and its activation by afferent sensory inputs. They looked for the influence of radio frequency warming of the hypothalamus on evoked potentials and spontaneous firing rates of single units in the MRF of anesthetized and immobilized cats. Potentials in the MRF were evoked by electrical stimulation of the forepaw. In this study, hypothalamic warming decreased the amplitude of evoked potentials in the MRF in 63 of 80 experiments (79%). In 31% of the experiments, this reduction was greater than 10%, and in 48% the decrease was 10% or less. Nineteen MRF units with spontaneous firing rates were held long enough to test the influence of T_{hy} on their firing rates. Six units responded to hypothalamic warming by decreasing their firing rates. The responses of the other 13 were equivocal. Three of the units sensitive to hypothalamic warming were also driven by forepaw stimulation; hypothalamic warming decreased the driven activity of all three. The authors offer these results in support of the hypothesis that hypothalamic warming induces sleep by inhibiting the arousal-sustaining activity of the MRF.

Combined Influence of Hypothalamic and Ambient Temperatures on Distribution of Sleep States

Our current experiments on kangaroo rats (*Dipodomys*) (97) provide the only information of which we are aware on the separate and combined influences of hypothalamic and peripheral temperatures on the normative pattern of sleep and wakefulness. In these experiments, cortical EEG, neck muscle EMG, metabolic rate, T_{hy}, and movement are continuously recorded during the second half of the animal's daily inactive period. After an acclimation period of 1 hr, T_{hy} is either unmanipulated or clamped at some level for a 4-hr period. The experiments are conducted at a thermoneutral T_a and at two levels below thermoneutrality. The EEG recordings are scored in 1-min epochs. Typical sets of results are presented in Figures 5 and 6. The experiments in Figure 5 were conducted at a thermoneutral T_a of 30°C. The control (*A*) involved no manipulation of T_{hy}, and the figure shows the records of T_{hy}, metabolic rate, and arousal state. The total amount of SWS, PS, and wakefulness in minutes and the number of episodes are tabulated at the right of the figure. In experiment *B*, T_{hy} was clamped 1°C above normal. At the end of the 4-hr period, the clamp was released, and T_{hy} equilibrated with overall body temperature, which had been dropping during the experiment due to the effect of the elevated T_{hy}. The TST is markedly greater in the clamp experiment than in the control. This increase is largely due to an increase in the frequency of PS episodes. In contrast to the results of Parmeggiani et al. (95), who used pulse heatings of the hypothalamus during PS episodes, our long-term heatings of the hypothalamus did not have a dramatic influence on the average duration of PS episodes.

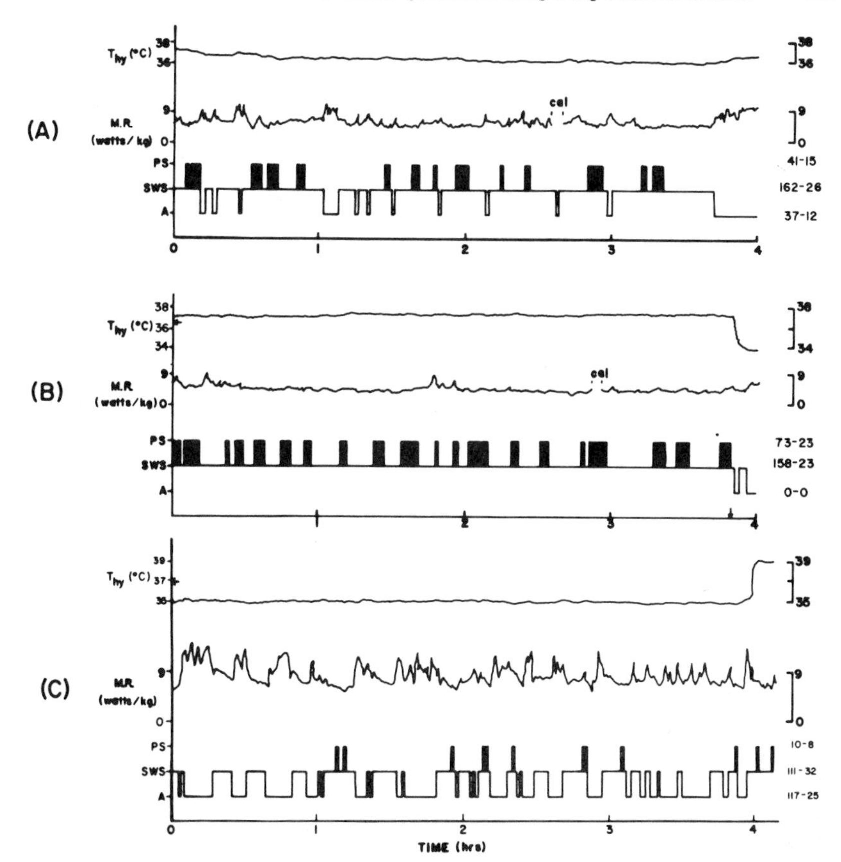

Figure 5. Four-hr recording of metabolic rate (*M.R.*) and arousal state in a kangaroo rat at an ambient temperature of 30°C with *A*, T_{hy} unmanipulated. *B*, T_{hy} warmed, and *C*, T_{hy} cooled. EEG was scored in 1-min epochs. Numbers to the right of the arousal state record represent number of minutes spent in each state during the 4-hr period and number of separate occurrences of that state. In *B* and *C*, the T_{hy} before warming or cooling is indicated by an *arrow*, and the T_b at the end of each experiment is shown by the level to which T_{hy} equilibrates when the hypothalamic thermal clamp is released.

A hypothalamic thermal clamp experiment and control performed at a T_a below thermoneutrality (20°C) are shown in Figure 6. A comparison of the control record (*B*) with the control at 30°C T_a reveals a decrease in TST at the lower T_a due to decreases in the amounts of both SWS and PS. In experiment *A*, hypothalamic temperature was elevated to and clamped at 39.5°C, slightly above the threshold for the metabolic heat production response. This clamp resulted in a greater imbalance between heat production and heat loss than was seen in the clamp experiment at T_a = 30°C, and T_b had fallen to 29°C by the end of the 4-hr period. Once again, hypothalamic warming produced a great increase in TST in spite of the profound hypothermia that developed. In contrast to the

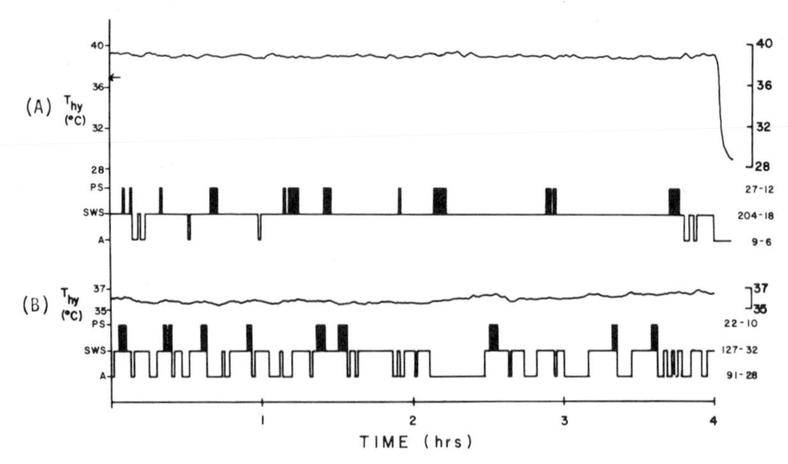

Figure 6. Four-hr recordings of metabolic rate and arousal state in a kangaroo rat at an ambient temperature of 20°C. In *A*, T_{hy} is clamped above normal (designed by *arrow* on *ordinate*) for the 4 hr. In *B*, T_{hy} is not manipulated. EEG was scored in 1-min epochs. Total minutes and number of bouts of each arousal state are shown at the right of each arousal state record.

experiment at T_a = 30°C, however, the increase in sleep time in this experiment is almost exclusively due to an increase in amount of SWS. At this T_a, hypothalamic warming had no influence on the frequency or duration of PS bouts. Again, these results do not appear to agree with those of Parmeggiani and colleagues on cats (95), which showed a marked increase in average duration of PS bouts when the hypothalamus was warmed, even at a T_a of 0°C. Because of the differences in experimental techniques (i.e., spot heating versus long-term heating of the hypothalamus), a resolution of this apparent discrepancy cannot be suggested until more work is done.

Experiments with mild hypothalamic cooling during the 4-hr recording period showed a consistent increase in wakefulness at the expense of both SWS and PS (Figure 5*C*).

The influence of peripheral temperature on the central nervous regulation of T_b must be kept in mind when interpreting these experiments. Cold peripheral stimuli raise the T_{hy} thresholds for thermoregulatory responses, and warm peripheral stimuli lower them (19, 33, 73, 74). Hypothalamic temperatures that deviate from these thresholds so as to generate driving signals for active thermoregulatory responses promote wakefulness. For example, hypothalamic cooling below the threshold for the metabolic heat production response increases wakefulness (Figure 5*C*). A low peripheral temperature raises the hypothalamic threshold for the metabolic heat production response above actual T_{hy}, which also increases wakefulness (Figure 6*B*). However, the kangaroo rat experiments indicate that peripheral thermal stimuli have an additional influence on sleep states apart from their roles in altering the relationships between hypothalamic temperature and thresholds for thermoregulatory responses. Information from peripheral temperature sensors indicating ambient conditions beyond the bounds of thermoneutrality has an inhibitory influence on PS regardless of T_{hy}. In summary, central signals indicating thermal error promote wakefulness, which is the most appropriate state for dealing with severe thermoregulatory problems. In the absence of central signals indicating thermal error, TST increases, but the distribution of sleep states is dependent upon peripheral temperature. If peripheral sensors indicate a potential thermal stress, the sleep state compatible with continuous thermoregulation (SWS) is favored over the sleep state during which the thermoregulatory system is inactivated (PS).

HIBERNATION AS EXTENSION OF SLEEP: ELECTROPHYSIOLOGICAL EVIDENCE

There are obvious behavioral similarities between hibernation and sleep, yet little work has been done to assess the possible physiological homologies between these two states of arousal. That hibernation is entered from a sleeping state or represents an extension of sleep or a "deeper" stage of sleep has been stated or implied by many authors (98–106). Mihailovic (104) has presented an excellent review of cortical and subcortical EEG activity in hibernation and hypothermia, pointing to the dynamic and coordinated changes in the brain of the hibernating animal during various stages of torpor. In this section, the electrophysiological evidence which lends itself to functional comparisons between hibernation and sleep are considered specifically.

Only a few studies have attempted to determine both the amount and type of sleep patterns persisting through various phases of the hibernating cycle. South and his colleagues (102) reported changes during hibernation in the electrophysiological recordings from marmots (*Marmota flaviventris*) implanted with chronic EEG and EMG needle electrodes. On the basis of cortical EEG and EMG/EKG records, South et al. found that in the initial phase of the entry into hibernation (as brain temperature (T_{br}) declined from euthermic levels, i.e., 37°C to 25°C), sleep patterns remained similar to the normothermic distribution of 80% SWS:20% PS seen in the marmot. Total sleep time during entry was not mentioned. In the remainder of the entrance and in deep hibernation, normative sleep stages could not be identified, although it is stated that during continuous deep hibernation at T_{br} values of 7–8°C the relative contribution of slow and fast activity in the EEG (presumably determined by the frequency spectral analysis) were about equal. On this basis, South et al. conclude that "... measurement of what might be analogous to SWS would indicate that this type of activity occupied no more than 50% of the period in deep hibernation and not less than 10%, the remainder being occupied either by PS or by very slight activity of indeterminant significance." In reaching this conclusion, the authors presumably used as a parameter for estimating PS the arrhythmia in the cardiac rate commonly seen in mammalian PS. The usefulness of cardiac irregularities in evaluating PS during hibernation is questionable, however, because in the ground squirrel (*Citellus lateralis*), cardiac arrhythmia is seen to occur not only during low voltage, fast (LVF) activity, but also during slow wave and spindle activity throughout the entrance (106) and, in the case of the hedgehog, even before the entrance begins (107).

Satinoff (103) studied the cortical EEG, EMG, and EOG of ground squirrels (*C. lateralis*) during entrance into hibernation. She discounts the hypothesis that hibernation is an extension of PS on the basis of data that indicate that during entry into hibernation there are few REMs, neck muscle atonia is not seen, and the characteristic EEG pattern is that of LVF activity with frequently intervening 12–20 Hz spindles. However, because the highest brain temperature Satinoff refers to during the entrance is 28°C, it is unknown whether her observations apply to T_{br} values in the initial part of the entry, between 37°C and 28°C.

Recently, Walker et al. (105, 106) obtained records of cortical and subcortical EEG, EOG, EMG, EKG, and T_{br} during entrance into hibernation in two species of ground squirrels, *C. lateralis* and *Citellus beldingi*. Sleep occupied 90% of the entrance from T_{br} values of 34°C to 25°C; brief periods of wakefulness occurred without disrupting the descent of T_{br}, but sustained wakefulness was associated with either a leveling off or an increase in T_{br}. In contrast with the data of South et al. (102), PS during this initial phase of entry was only 9% of TST (Figure 7). Because of the reduction in EEG amplitude below T_{br} values of 25°C, sleep states below this level could not be identified by conventional electrographic criteria.

A variety of sleep studies other than comparisons of sleep patterns within hibernation bouts have been conducted, but they tell us little about possible functional homologies between sleep and hibernation. Allison and Van Twyver (108) found that after arousal from hibernation in the echidna, total sleep time (TST) comprised 67% of total recording time (TRT), compared to 45% TST described before the animal entered hibernation. The authors suggest that this "rebound" may indicate possible sleep deprivation during hibernation, although they state that the stress of arousal cannot be overlooked in evaluating the significance of increased post-torpor sleep.

Van Twyver (81) recorded normative sleep patterns in five species of rodents at ambient temperatures of about 22°C. He found that in hibernators (*Mesocricetus auratus* and *Citellus tridecemlineatus*), sleep epochs were longer, the PS:TST ratio was larger, and TST was slightly higher than in nonhibernators (*Mus musculus, Rattus norvegicus,* and *Chinchilla laniger*). The significance of these differences is not clear. The time of year the data were collected was not mentioned; seasonal variation in sleep patterns very well may occur in hibernators and should be investigated.

In summary, electrographical studies of marmots and ground squirrels during the early phase of the entrance into hibernation when T_{br} is between 37°C and 25°C show that the animals are predominantly in slow wave sleep. Actual declines in T_{br} occur almost exclusively during sleep, with brief plateaus or rises in T_{br} being associated with epochs of wakefulness (105, 106). The hypothesis that hibernation is an extension of the thermoregulatory readjustments associated with SWS is tenable. However, the progressive changes in the EEG pattern

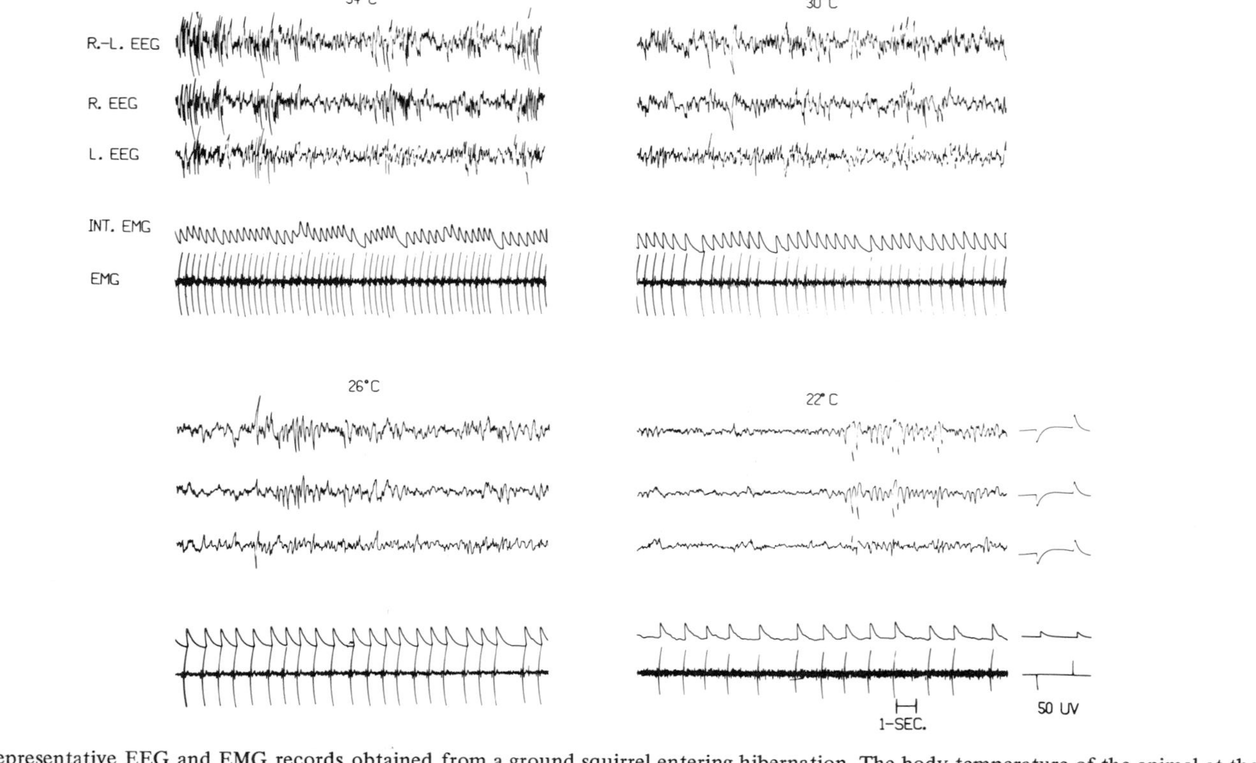

Figure 7. Representative EEG and EMG records obtained from a ground squirrel entering hibernation. The body temperature of the animal at the time of each recording is shown. (Reproduced from Walker et al. (106) with permission.)

as T_{br} declines below 25°C make analysis of sleep states questionable. It would be quite interesting to equilibrate animals in deep hibernation with higher and higher T_a values to compare the relative influences of T_{br} and possible active CNS mechanisms on the EEG. It would also be interesting to know whether the frequency and duration of sleep states change seasonally in hibernators.

THERMOREGULATION DURING HIBERNATION

The most striking manifestation of the suite of behavioral, physiological, cellular, and biochemical adaptations commonly called mammalian hibernation is the fall of the high, regulated mammalian body temperature to levels only slightly above ambient temperature, even when ambient temperature is as low as 0°C (98, 109, 110). Equally striking is the thermogenic ability of the hibernator. Unlike torpid ectotherms, the hibernating mammal can respond to a stimulus by mobilizing a tremendous metabolic effort which returns it to euthermia. Even in the absence of exogenous stimuli, the hibernator periodically arouses so that during the hibernation season the animal enters and arouses from many discrete bouts of hibernation or torpor which may vary in depth as well as length (111–113). Between bouts of torpor, the hibernator displays a normal, mammalian pattern of T_b regulation which may differ quantitatively, but not qualitatively, from that seen in nonhibernators (75, 114–116). In this section, our knowledge of thermoregulation in hibernators is reviewed in an effort to answer several questions. Does the entrance into torpor represent an inactivation of normal thermoregulatory mechanisms or does it represent a regulated decline in T_b? Is T_b regulated during a bout of hibernation or does it passively equilibrate with T_a? If T_b is regulated during deep hibernation, are the characteristics of regulation qualitatively the same as in euthermia? In other words, is the mammalian thermostat adjustable over the range of temperatures displayed by the hibernator?

Is T_b Regulated During Hibernation?

A large body of indirect information suggests that T_b is in some way regulated during hibernation, or at least it is defended against declines to dangerously low levels. As T_a approaches and falls below 0°C, many species have been reported to elevate metabolic heat production, but remain in hibernation with a larger gradient between T_b and T_a than would be expected if T_b passively equilibrated with T_a and metabolic rate were dependent only upon the tissue temperature (117–127). For example, hedgehogs maintained at a T_a of −5°C held T_b values at least 5°C higher than T_a for 2–3 days without fully arousing from torpor (124). Reite and Davis (123) exposed two species of bats (*Myotis lucifugus* and *Lasiurus borealis*) to ambient temperatures from 10°C to −5°C. Between T_a values of 10 and 7°C, the body temperatures of the dormant bats were only slightly above T_a, and the gradient between T_b and T_a remained rather constant. At T_a values below 4°C, however, the gradient between T_b and T_a progressively increased. Heart rate was also shown to increase as T_a fell below

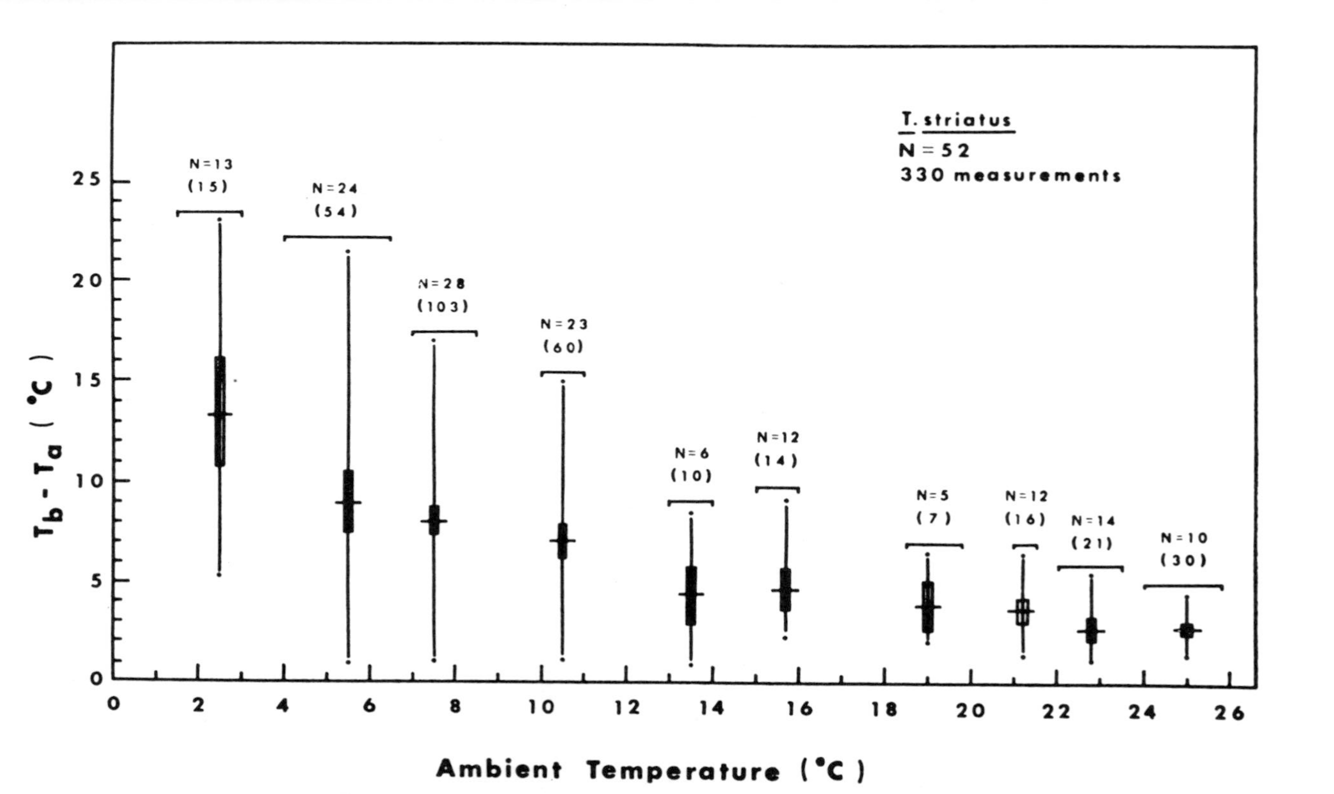

Figure 8. The differential between T_b and T_a in *T. striatus* hibernating at different ambient temperatures. *Vertical lines* represent the range, and *horizontal lines* the mean values. *Rectangular boxes* and $t \times$ standard error of the mean at $p = 0.05$. (Reproduced from Wang and Hudson (125) with permission.)

4°C. In their study of hibernating eastern chipmunks (*Tamias striatus*), Wang and Hudson (125) found that the animals usually maintained T_b between 10 and 24°C over a T_a range of 1.5–9.5°C. The gradient between T_b and T_a increased as T_a decreased (Figure 8).

In an extensive study of hibernation in the California ground squirrel, Strumwasser (111, 128–130) showed that at the beginning of the hibernation season animals exposed to T_a values of 5.5–8.0°C entered and aroused from torpor on a daily basis before they remained in torpor for more than a day. During each of these successive short bouts of torpor, the animal allowed T_b to fall lower than it had during the previous bout. The minimal T_b reached during each of these "test drops" was termed a "critical temperature." The fact that these critical temperatures encompassed the range of body temperatures from levels occurring during sleep to the lowest levels of deep hibernation, and the fact that the critical temperature could be maintained for periods long enough to require regulation led Strumwasser to the conclusion that the California ground squirrel ". . . can probably maintain and regulate its temperature anywhere between 6 and 39°C" (130). Animals in deep hibernation, with a constant brain temperature several degrees above T_a, exhibited oscillations in skin temperatures, indicating that vasomotor activity was modulating the rate of heat loss from the animal; therefore, the rate of heat loss was not passively following the rate of heat production (129). Strumwasser presents one 33-hr record (Figure 9) which shows a very interesting relationship between heart rate and T_{br} during hibernation. At the beginning of this record, T_{br} falls; it then plateaus for 10 hr, rises slightly to a new plateau which is maintained for 6 hr, and finally begins an irregular but progressive rise at the end of the record. Heart rate was lowest while T_{br} was falling; it then increased and remained at that higher level during

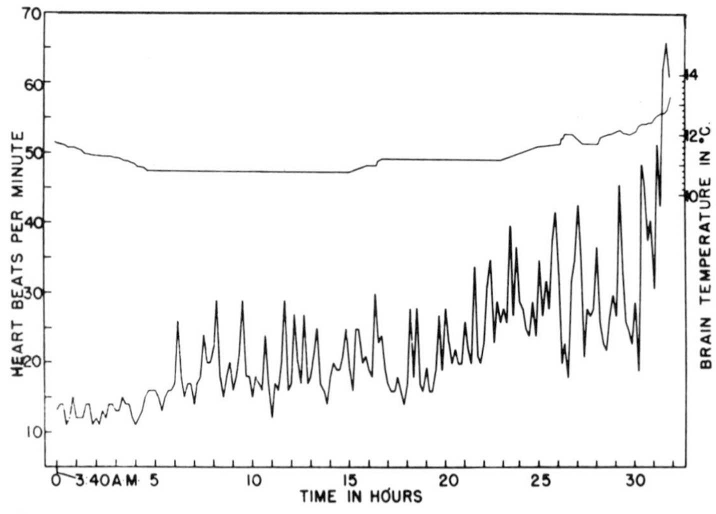

Figure 9. Heart rate and brain temperature during the lowest portion of a short bout of torpor in the ground squirrel. (Reproduced from Strumwasser (129) with permission.)

the first plateau of T_{br}. The rise in T_{br} from the first to the second plateau was accompanied by another increase in heart rate. At the end of the second plateau, both heart rate and T_{br} began an irregular, progressive rise. Although heart rate is not a thermoregulatory response, it is controlled by sympathetic nervous activity, as are thermoregulatory responses. These results indicate that falls in T_{br} are associated with low levels of sympathetic activity and increases in T_{br} are associated with increases in sympathetic activity, and the level at which T_{br} is maintained during hibernation is related to the level of sympathetic activity. All of Strumwasser's results convincingly support the hypothesis that T_{br} is regulated during hibernation.

In addition to the hibernators which undergo seasonal torpor, there are many species of small mammals that frequently undergo daily episodes of lowered T_b as an adaptation for energy conservation. Some of these species have been observed to maintain large differentials between T_b and T_a during torpor, indicating that T_b is regulated during torpor. The hispid pocket mouse (*Perognathus hispidus*) allowed its T_b to fall within 0.2–2.0°C of T_a when T_a was equal to or greater than 14°C. However, T_b–T_a during torpor increased as T_a fell below 14°C (131). Similarly, in a study of five species of *Peromyscus*, Morhardt (132) showed a large range of T_b–T_a differentials during torpor at t_a values ranging from 0–25°C. Only one animal permitted its T_b to fall within 1°C of T_a.

The phenomena of seasonal and daily torpor are not found as commonly in avian as in mammalian species; however, daily torpor has been observed in many species of hummingbirds. Measurements in three species during torpor have shown active metabolic resistance to declines in T_b below certain levels. Above these lower limits, T_b passively equilibrates with T_a, and below these lower limits, T_b remains constant or rises slightly as T_a drops to lower levels (133, 134).

Thermosensitivity During Hibernation

In light of the developing story of the mammalian regulator of body temperature, it seemed unlikely to Hammel (101) at the time of the Third International Hibernation Symposium that the metabolic defenses of hibernators against extremely low T_b values could be controlled by the same system responsible for the regulation of T_b during euthermia. Hammel and collaborators (135) investigated this question by doing simultaneous direct and indirect calorimetry on hibernating ground squirrels (*C. lateralis*). They allowed the hibernating animals to come into equilibrium with T_a values between 2°C and 13°C and used T_{hy} as an indication of the temperature of the well circulated core of the animal. The results showed that the rate of metabolic heat production of the hibernating ground squirrel was directly proportional to T_{hy} with a Q_{10} of 2.4 over this range of temperatures. These results indicated that the rates of heat production in these animals were dependent upon only the tissue temperatures and were in no way modified by CNS control. In this same study, however, it was observed that animals could not be kept in hibernation at T_a values below 2.3–2.8°C. Lower T_a values induced thermogenic responses that were suppressible by elevating T_a. The authors interpreted these bursts of increased heat production as being partial or incipient arousal responses and not proportional thermoregulatory responses. They predicted that "heating the hypothalamus to 5.0°C while cooling the air to 1.5°C, for example, would not inhibit the arousal response. Conversely, cooling the hypothalamus to 1.5°C while maintaining the T_a at or above 3.0°C would not activate arousal."

This prediction was tested by Heller and Hammel (136), who used chronically implanted, water-perfused thermodes to manipulate the T_{hy} of hibernating golden-mantled ground squirrels while continuously measuring the rate of metabolic heat production. Experiments were performed on animals hibernating at equilibrium T_{hy} values ranging from 2–9°C. Manipulations of T_{hy} ranging from 0.4–15°C were shown to have no influence on rates of metabolic heat production. These results seemed to verify the prediction mentioned above. However, in nine experiments in which animals were hibernating at equilibrium T_{hy} values of 3.0–6.6°C, hypothalamic coolings ranging from 0.6–2.5°C elicited what appeared to be arousal responses. In one of these cases, the cooling was simply terminated after the arousal was initiated, and in three cases T_{hy} was heated to 2.8, 6.5, or 7.0°C; in all of these experiments, the arousals were complete. In contrast, after the other five arousals were initiated, T_{hy} was heated to between 11.5°C and 12.5°C, and they were all reversed. These results demonstrated that hypothalamic cooling can initiate an arousal. The reversals of these arousals by hypothalamic heating indicated the existence of a rising T_{hy} threshold for the metabolic heat production response during arousal. Another experiment demonstrated that as long as T_{hy} was held above the alarm level by thermode perfusion, T_a and extrahypothalamic T_b could be lowered below the observed alarm levels, and even below 0°C, without eliciting an increase in metabolic heat production. Clearly, hypothalamic thermosensitivity existed during deep hibernation.

Similar conclusions were also reached by Lyman and O'Brien (137, 138), who used an external spoon-shaped thermode to heat or cool the heads of hibernating ground squirrels (*C. lateralis* and *C. tridecemlineatus*) and golden hamsters (*M. auratus*). They showed that subzero T_a values did not elicit increases in heart rate or respiratory rate so long as temperature of the head was held above 4°C by means of the external thermode. Lowering head temperature below 3°C, however, resulted in increases in heart rate and respiratory rate leading in some cases to arousal.

The generality that thermogenic or arousal responses or both are mediated in the hibernating animal by thermosensitivity of the hypothalamus or some other location in the head was contradicted by experiments on dormice (*Glis glis*) (138). This species is particularly sensitive to drops in T_a during hibernation whether or not head temperature was held constant with an external thermode. Unlike the ground squirrel species and the hamster, the dormouse hibernates on its back with its feet pointing up. Lyman and O'Brien devised a clever means of manipulating the temperature of just the hind feet of their subjects and discovered that cooling the hind feet alone resulted in increased heart rate, respiratory rate, and muscle action potentials, which usually led to full arousals. There can be no doubt that in dormice peripheral temperature receptors are functional during deep hibernation even though this does not seem to be true of *C. lateralis, C. tridecemlineatus,* and *M. auratus.* At present no information exists to permit conclusions regarding central thermosensitivity in hibernating dormice.

Studies on marmots have also revealed hypothalamic thermosensitivity during deep hibernation (116, 139–141). Increases in heart rate, respiratory rate, muscle action potentials, and EEG activity were induced in hibernating marmots by cooling the hypothalamus. Although these parameters in themselves are not necessarily indicative of significant thermogenic responses, postcooling T_{hy} was elevated in comparison to precooling T_{hy}. In several cases, hypothalamic cooling resulted in full arousals from hibernation. The hibernating marmot was also shown to respond to hypothalamic warming with decreases in heart rate and EMG/EKG activity. Peripheral thermosensitivity is indicated in the hibernating marmot by the rapidity with which its heart rate responds to changes in T_a (139).

The studies reviewed above demonstrate that hibernators have the means of detecting alterations in deep T_b and, in at least some species, changes in skin temperature also. However, these results cast no light on the nature of the system which couples temperature detection to the thermogenic responses. Are these thermal stimuli triggering a simple arousal response which is analogous to an on-off controller, or does the hibernating mammal have proportional control over thermoregulatory responses when in deep hibernation, as it has when euthermic? If proportional regulation of T_b occurs during hibernation, is it mediated by the same regulating system that is operative during euthermia?

The Nature of Regulation During Hibernation

The possibility of on-off control of thermogenesis was suggested by observations of hibernating ground squirrels which maintained fairly large gradients between T_b and T_a by cyclic bursts of heat production (126, 136). It appeared that whenever T_{hy} fell below a certain level a thermogenic burst was triggered which resulted in a rise in T_{hy}, thereby diminishing or removing the stimulus for the elevated thermogenesis. When T_{hy} was held above the triggering level by means of chronically implanted thermodes straddling the hypothalamus, the bursts of O_2 consumption were eliminated and metabolic rate remained at minimal levels. The cyclical bursts resumed when hypothalamic warming was terminated. An alternative hypothesis to explain the cyclical bursts of thermogenesis is that a hypothalamic threshold temperature (T_{set}) for the metabolic heat production response exists during hibernation and, if T_{hy} falls below this threshold, increases in rate of metabolic heat production are induced which are proportional to ($T_{set} - T_{hy}$). The cyclical nature of the response pattern could be explained by the spatial separation between the sites of temperature reception (the hypothalamus) and the major sites of heat production (brown fat deposits). When T_{hy} falls below threshold, sympathetic stimulation activates brown fat metabolism, which remains elevated until enough of the heat produced is conducted to the hypothalamus to raise T_{hy} above T_{set}.

Proportional regulation of T_b in hibernating golden-mantled ground squirrels was demonstrated by Heller and Colliver (126). In these experiments, the rate of metabolic heat production was measured continuously as the hypothalamus was heated and cooled. In comparison to the experiments of Heller and Hammel (136), the changes of T_{hy} were in smaller steps. Clear hypothalamic thresholds for the metabolic heat production response were determined, and when T_{hy} was cooled below those thresholds, the increase in the rate of metabolic heat production was proportional to the difference between threshold and T_{hy} ($T_{set} - T_{hy}$) (Figure 10). When T_{hy} was cooled too low, arousal responses ensued, as reported by Heller and Hammel (136). Proportional thermoregulatory responses and arousal responses were clearly distinguishable. If the T_{hy} threshold was higher after the response than before, the response was called a partial arousal, but if T_{set} were the same following the response as before, the response was considered a proportional thermoregulatory one.

The demonstration of proportional regulation of T_b during hibernation still left open the question of whether the same populations of hypothalamic neurons were responsible for regulation both in euthermia and in hibernation. The *a priori* rationale against this possibility was presented by Hammel (101) and is illustrated in Figure 11. The proportionality constant for the metabolic heat production response in the dog is on the order of 1 watt kg^{-1} $°C^{-1}$. If the hypothalamic thermosensitivity of the hibernator were similar to the dog's and if a biologically reasonable Q_{10} of 2.5 were applied to this value over the range of

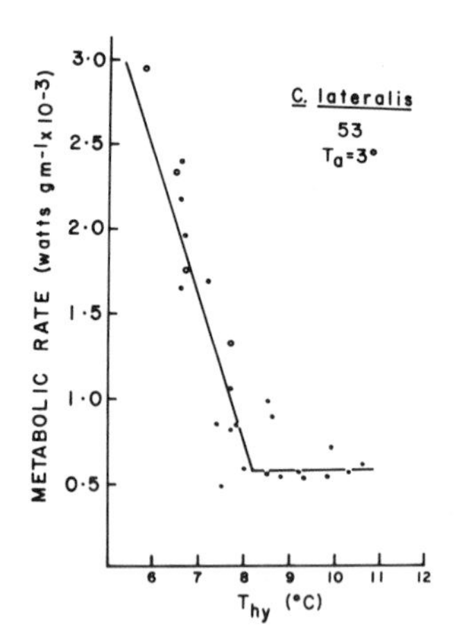

Figure 10. Responses in rate of metabolic heat production induced by manipulations of T_{hy} in a ground squirrel. The animal was maintaining T_b considerably above T_a by bursts of thermogenesis. *Open circles* represent rate of metabolic heat production following naturally occurring declines in T_{hy}, and *solid points* represent levels of metabolic heat production during manipulations of T_{hy}. (Reproduced from Heller and Colliver (126) with permission.)

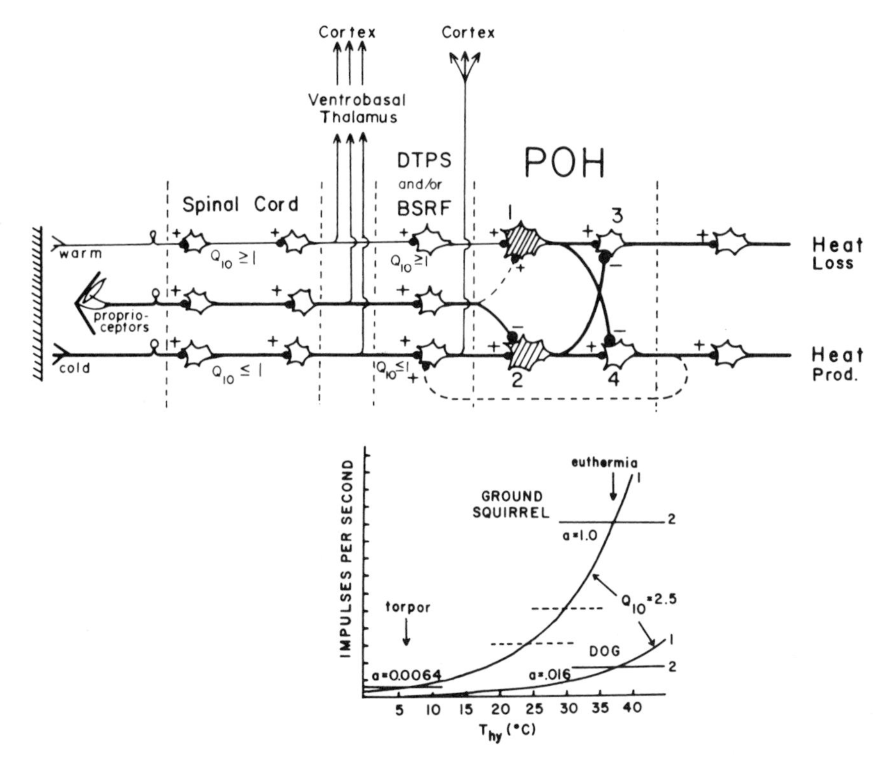

Figure 11. Hammel's neuronal model of the regulator of T_b in mammals and predictions based on this model for changing characteristics of the regulator if T_b could be regulated at any level from 0–40°C. Heat loss and heat production represent target organs for affecting rates of heat loss from the body and heat production by thermogenic tissue. Cell populations are represented by single neurons; four basic types are indicated in the POH. Neurons 1 and 2 differ in their degree of temperature dependence, and they both facilitate and inhibit other neurons such as 3 and 4, which have no spontaneous activity or temperature dependence. Neurons 1 and 2 receive excitatory and inhibitory inputs from such sources as peripheral temperature receptors, proprioceptors, temperature-sensitive interneurons in the spinal cord, brainstem reticular formation, etc. The *graph* at the bottom shows the hypothetical firing rates of neurons 1 and 2 as a function of POH temperature for the dog and for the ground squirrel. In both animals, when the firing rates of the two populations are equal, neurons 3 and 4 are both inhibited more than facilitated, and there is no net driving force for heat loss or heat production. If the POH is cooled, neuron 4 will begin to fire, and if the POH is heated, neuron 3 will begin to fire. The threshold temperature (T_{set}) for a thermoregulatory response depends, therefore, on the relationships between neurons 1 and 2, and this can be changed by the inputs to the regulator mentioned above. For example, the entrance into hibernation may be achieved by decreased input from the BSRF to population 2. As the activity of population 2 decreases (*dashed lines* on ground squirrel curves), T_{set} would decrease. In both the curves for the ground squirrel and the curves for the dog, the Q_{10} of population 1 is 2.5, but the curves for the ground squirrel are shifted to the left. At any T_{set}, the proportionality constant for thermoregulatory responses (α) is a function of the difference in slopes of the firing rate versus temperature curves of neurons 1 and 2; hence, at any T_{set} the regulator of the ground squirrel would produce a greater α than the regulator of the dog. This may be an adaptation for thermoregulation at the low T_b values of hibernation. The numbers labeled *a* represent α values in cal g^{-1} min^{-1} $°C^{-1}$ which have been measured experimentally. (Reproduced from Heller and Colliver (126) with permission.)

T_{set} values from euthermia to deep hibernation (roughly 30°C), the result would be an inadequate driving force to explain the thermoregulatory responses of the hibernator. This theoretical impediment to the concept of a unitary regulator operating both in euthermia and hibernation was eliminated by the discovery of extremely high hypothalamic thermosensitivity of the euthermic golden-mantled ground squirrel (75). The proportionality constant relating rate of metabolic heat production to T_{hy} was almost an order of magnitude greater in the ground squirrel than in the dog. Comparison of the proportionality constants for the ground squirrel in euthermia and hibernation yielded a Q_{10} of 2.5, which is a reasonable value if the same regulatory system is operating over the wide range of temperatures experienced by the hibernator.

Is it reasonable to propose a population of mammalian neurons with continuous firing rates positively correlated with temperature over a 30°C range? Wünnenberg et al. (142) have recently reported an abundance of warm-sensitive neurons in the preoptic anterior hypothalamus (POH) of anesthetized golden hamsters which had firing rates of 30–70 impulses/s at a T_{hy} of 38°C. As the tissue was cooled with an array of silver needle thermodes, these units decreased their activities with a Q_{10} of about 3 and were continuously active down to temperatures of 10°C (Figure 12). Such units could be involved in a thermoregulatory system capable of operating over the range of temperatures experienced

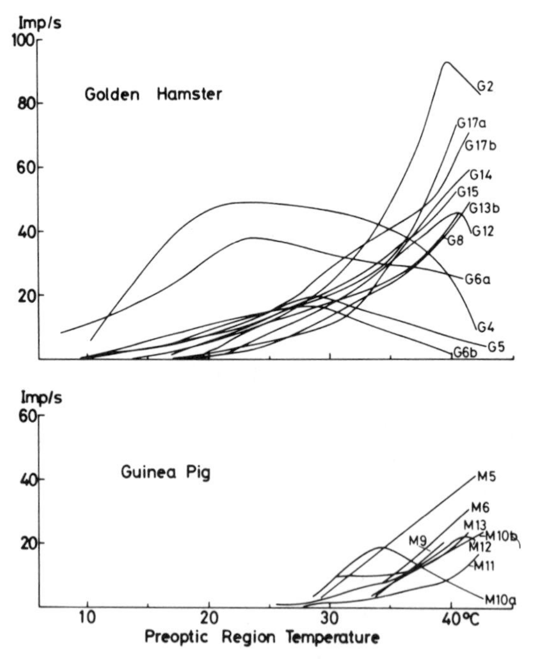

Figure 12. Firing rate as a function of temperature recorded extracellularly from units in the preoptic region of a hibernator, the golden hamster, and a nonhibernator, the guinea pig. (Reproduced from Wünnenberg, Merker, Speulda (142) with permission.)

by hibernators. Units with similar properties were not found by Wünnenberg et al. in a nonhibernator, the guinea pig. In another comparative study of POH unit characteristics in a hibernator and a nonhibernator, significant differences were not observed. Boulant and Bignall (143, 144) recorded firing rates of POH neurons in ground squirrels and rats while manipulating the temperature of the units from 30–40°C with a water-perfused thermode. In general, the firing rates recorded at any T_{hy} from warm-sensitive units in the ground squirrel were lower than those reported for the hamster. The slopes of the firing rates of these units versus temperature were also lower than the values reported for the hamster.

Proportional regulation of T_b in the hibernating marmot has now been demonstrated by Florant and Heller (145). Each marmot received a pair of water-perfused thermodes chronically implanted astraddle the hypothalamus. The rate of metabolic heat production was measured as a function of T_{hy} during euthermia and hibernation in three animals. The proportionality constants during euthermia averaged 1.1 watts kg^{-1} °C^{-1}, and during hibernation they averaged 0.08 watts kg^{-1} °C^{-1}. In contrast to the observations of South et al. (141), Florant and Heller used larger displacements of T_{hy} and were able to locate a T_{hy} threshold for the metabolic heat production response and elicit proportional responses to coolings below threshold at all times during the hibernation bout. Of great interest is the fact that the threshold progressively decreased during most of the bout. For example, in Figure 13 the *right curve* presents data obtained on the first 2 days of a bout and the *left curve* gives data obtained during the following 2 days of the bout. A progressive increase in the threshold began a day or so before a spontaneous arousal, and the threshold was seen to rise above actual T_b just before the active arousal was initiated. This progressive change in the relationship between T_{hy} and T_{set} during a bout of hibernation may be the mechanism that generates periodic arousals and also may be the mechanism underlying the increasing irritability of the hibernator toward the end of a bout as observed by Twente and Twente (146) and by Beckman and Stanton (147).

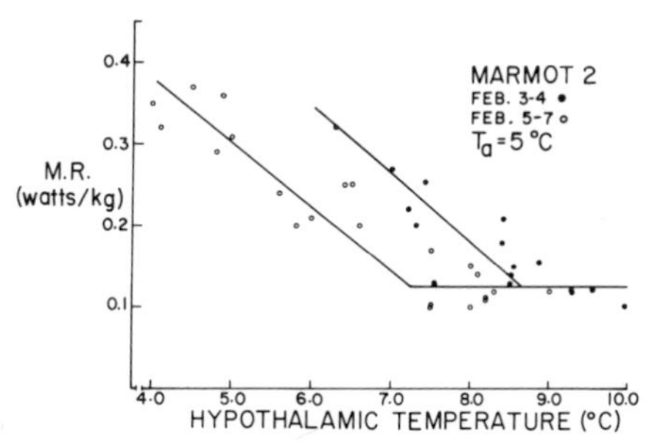

Figure 13. Responses of a marmot to manipulations of hypothalamic temperature at two different times within one bout of hibernation showing a clear change in the threshold T_{hy} for the metabolic heat production response. (Reproduced from Florant and Heller (145) with permission.)

The concept of a unitary hypothalamic regulator of T_b continuously operating between normal mammalian T_b values and the levels of deep hibernation gains additional support from studies of the entrance into hibernation. It has been frequently observed that the fall of T_b during entrance into hibernation or shallow torpor does not always describe a passive cooling curve (74, 113, 125, 127, 128, 131, 148). Transient bursts of shivering and increases in oxygen consumption, heart rate, and respiratory rate suggest braking mechanisms slowing or even halting the entry into hibernation. Strumwasser (128, 130) described continuous recordings of T_{br} from the California ground squirrel (*Citellus beechyi*) entering hibernation as consisting of a series of plateaus and small step drops. Simultaneous recordings of skin temperature and EMG showed that vasodilation occurred just prior to step-downs in brain temperature. Vasoconstriction and shivering coincided with the beginning of a plateau. These observations suggested that T_{br} was definitely regulated during entrance, but at successively lower levels. It is possible, then, that the entry into hibernation is due to a downward resetting of the body temperature regulator. If so, the rate of entry into hibernation would be limited by the rate at which the threshold for the metabolic heat production response falls. As long as the threshold T_{hy} for the metabolic heat production response falls faster than T_{hy}, the cooling curve of the animal would appear passive, but whenever T_{hy} falls below threshold, a thermogenic response would be elicited which would interrupt or "brake" the passive cooling curve and slow the entrance. If a rapid fall in T_{hy} or a sudden rise in threshold generated a large error signal ($T_{set}-T_{hy}$), an arousal response could be induced which would reverse the entrance and return the animal to the euthermic condition. It is conceivable that the "test drops" first described by Strumwasser (111) and subsequently seen in other species at the beginning of the hibernation season are entirely homologous with the arousal responses from deep hibernation induced by cooling T_{hy}.

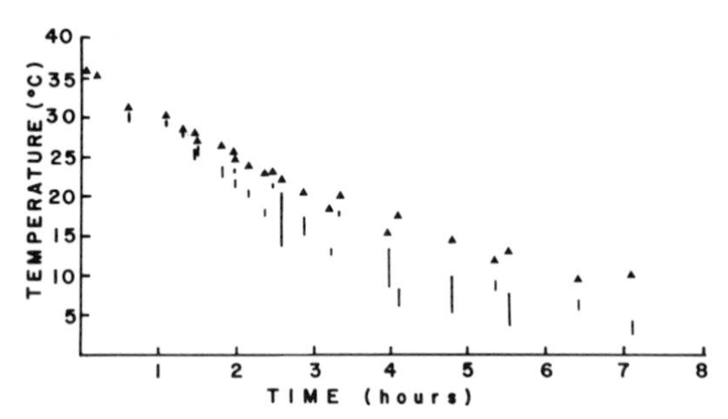

Figure 14. Data from four separate experiments in which the hypothalamus was heated and cooled while a ground squirrel was entering hibernation. *Triangles* represent actual T_b, and *vertical lines* show the range of T_{hy} values within which the threshold T_{hy} for the metabolic heat production response was at different times during the entrances. Each range is delimited by the highest T_{hy} that elicited a response and the lowest T_{hy} that did not elicit a response, as determined by manipulations of T_{hy} over a 15–60-min period. The longer periods of manipulation were required at the lower temperatures.

Experiments in which T_{hy} was periodically manipulated while golden-mantled ground squirrels entered hibernation demonstrate the continuity of the thermoregulatory system from euthermia to deep hibernation (149). At all points during the entrance, it was possible to elicit a thermogenic response with hypothalamic cooling without initiating a partial arousal. Because the threshold for the thermogenic response was continuously dropping, it was impossible to pinpoint it precisely at any one time; however, over a 15–20 min period, it could be bracketed by finding the lowest T_{hy} that did not induce a response and the highest T_{hy} that did induce a response. An example of the results obtained is presented in Figure 14, which is a composite of data obtained on one animal during four separate entries into hibernation. Similar data have been obtained on five other individual *C. lateralis*. These results demonstrate that the entrance into hibernation is not a turning off of the T_b regulator, but a gradual resetting of the regulator to progressively lower levels.

CONCLUSIONS

The conclusions drawn from the literature reviewed can be summarized as follows:

1. Alterations in the thermoregulatory system occur during sleep, and the nature of these alterations is sleep state-dependent.
2. There is proportional regulation of T_b during SWS, but the T_{hy} thresholds and the proportionality constants relating response rates to T_{hy} are lower than in the wakeful state.
3. The thermoregulatory system is inactivated during PS.
4. Animals enter hibernation predominantly through SWS.
5. The entrance into hibernation involves a progressive lowering of the T_{hy} threshold for the metabolic heat production response.
6. There is proportional regulation of body temperature as well as a low temperature alarm mechanism during hibernation.

These conclusions support the hypothesis that SWS and hibernation are physiologically homologous. The phenomenological similarities between SWS and hibernation are evidenced by electrographic studies. Ground squirrels and marmots are predominantly in SWS, defined electrographically, as they begin to enter hibernation. However, the EEG of the hibernator gradually deviates from euthermic patterns as T_b falls, and below a T_b of about 25°C arousal states cannot be identified by standard electrophysiological criteria. Functional similarities between SWS and hibernation are seen in thermoregulatory studies. Both states involve a lowering of T_{hy} thresholds and the proportionality constants for thermoregulatory responses.

It seems logical to propose that SWS, shallow torpor, and hibernation may be homologous thermoregulatory adaptations. However, hibernation is a polyphyletic phenomenon (150, 151), and at present comparative knowledge is meager. Mechanisms other than those associated with SWS cause changes in the thermoregulatory system, and it is not inconceivable that torpor in some species could have evolved as an extension of one of these other mechanisms.

The apparent homology between SWS and hibernation in the species studied tempts us to engage in speculation about the evolution of these adaptations. The fact that the regulator of T_b is reset downward during SWS lends credence to the hypothesis that a primitive and perhaps primary function of SWS is energy conservation (108, 152–154). Slow wave sleep is a general characteristic of endotherms, but its existence in ectotherms is unclear and debatable (152, 154–156). An important selective pressure in the evolution of SWS could have been the high energetic costs associated with endothermy. An adaptation to lower the regulated T_b during periods of obligate inactivity could have resulted in significant energetic savings in early mammals. SWS as seen in modern mammals enables a lowering of T_b during inactivity without an abandonment of proportional regulatory capabilities. The thermoregulatory adjustments associated with SWS could have been preadaptations for species exposed to dangerously long periods of food unavailability, which may have been seasonal in animals adapting to the increasing climatic fluctuations of the Cenozoic period or daily in animals evolving to fill the ecological niches presently occupied by the smallest endotherms. The strong selective pressures for increased energy conservation in these species would have favored lower and lower body temperatures during SWS, along with the necessary subsidiary adaptations now seen in animals that undergo shallow torpor and hibernation.

ACKNOWLEDGMENTS

We appreciate the permission of many of our colleagues to republish illustrations from their papers. The following persons were most kind to take the time to read the manuscript critically and offer useful suggestions: Ralph J. Berger, Gregory L. Florant, Barbara D. Snapp, Roderick Van Buskirk, and James M. Walker.

REFERENCES

1. Aschoff, J. (1955). Der Tagesgang der Körpertemperatur beim Menschen. Klin. Wochenschr. 33:545.
2. Aschoff, J. (1965). Circadian rhythms in man. Science 148:1427.
3. Aschoff, J. (1970). Circadian rhythm of activity and of body temperature. *In* Physiological and Behavioral Temperature Regulation, pp. 905–920. J. D. Hardy, A. P. Gagge, and J. A. J. Stolwijk (eds.), Charles C Thomas, Springfield.
4. Chossat, C. (1843). Recherches expérimentales sur l'inanition. Des effets de l'inanition sur la chaleur animale. Ann. Sci. Naturelles 2 Serie 20:293.
5. Jürgensen, T. (1873). Die Körperwärme des gesunden Menschen. Leipzig.
6. Cumming, M. C., and Morrison, S. D. (1960). The total metabolism of rats during fasting and refeeding. J. Physiol. (Lond.) 154:219.
7. Chew, R. M., Lindberg, R. G., and Hayden, P. (1965). Circadian rhythm of metabolic rate in pocket mice. J. Mammal. 46:477.

8. Kleitman, N. (1923). Studies on the physiology of sleep. I. The effects of prolonged sleeplessness on man. Am. J. Physiol. 66:67.
9. Kleitman, N. (1963). Sleep and Wakefulness, Ed. 2, p. 552. University of Chicago Press, Chicago.
10. Kreider, M. B. (1961). Effects of sleep deprivation on body temperature. Fed. Proc. 20:214.
11. Timbal, J., Colin, J., Boutelier, C., and Guieu, J. D. (1972). Bilan thermique en ambiance controlée pendent 24 heures. Pfluegers Arch. 335:97.
12. Johansson, J. E. (1898). Ueber die Tagesschwankungen des Stoffwechsels und der Koerpertemperatur in nuechternem Zustaende und vollstaendiger Muskelruhe. Scand. Arch. Physiol. 8:85.
13. Heusner, A. (1957). Mise en évidence d'une variation nycthémérale de la calorification indépendante di cycle de l'activité chez le rat. C. R. Soc. Biol. (Paris) 150:1246.
14. Heusner, A. (1959). Variation nycthémérale de la température centrale chez le rat adapté à la neutralité thermique. C. R. Soc. Biol. (Paris) 153:1258.
15. Morrison, P. (1962). Modification of body temperature by activity in Brazilian hummingbirds. Condor 64:315.
16. Aschöff, J., Gerecke, U., and Wever, R. (1967). Phasenbeziehungen zwischen den circadianen perioden der Aktivität und der Kerntemperatur beim Menschen. Pfluegers Arch. 295:173.
17. Lund, R. (1974). Circadian periodicity of physiological and psychological variables in seven blind subjects with and without time cues. Ph.D. thesis, Technische Universität München.
18. Wever, R. (1975). The circadian multi-oscillator system of man. Int. J. Chronobiol. 3:19.
19. Hammel, H. T., Jackson, D. C., Stolwijk, J. A. J., Hardy, J. D., and Strømme, S. B. (1963). Temperature regulation by hypothalamic proportional control with an adjustable set point. J. Appl. Physiol. 18:1146.
20. Fredericq, L. (1901). Le courbe diurne de la température des centres nerveux sudoripares fonctionnant sous l'influence de la chaleur. Arch. Biol. 17:557.
21. Fox, R. H., Crockford, G. W., Hampton, I. F. G., and MacGibbon, R. (1967). A thermoregulatory function test using controlled hyperthermia. J. Appl. Physiol. 23:267.
22. Smith, R. E. (1969). Circadian variations in human thermoregulatory responses. J. Appl. Physiol. 26:554.
23. Hildebrandt, G. (1974). Circadian variations of thermoregulatory responses in man. *In* Chronobiology, Ed. 6, pp. 234–240. L. E. Scheving, F. Halberg, and J. E. Pauly (eds.), Igaku Shoin Ltd., Tokyo.
24. Timbal, J., Colin, J., and Boutelier, C. (1975). Circadian variations in the sweating mechanism. J. Appl. Physiol. 39:226.
25. Wenger, C. B., Roberts, M. F., Stolwijk, J. A. J. and Nadel, E. R. (1976). Nocturnal lowering of thresholds for sweating and vasolidation. J. Appl. Physiol. 41:15.
26. Geschickter, E. H., Andrews, P. A., and Bullard, R. W. (1966). Nocturnal body temperature regulations in man: a rationale for sweating in sleep. J. Appl. Physiol. 21:623.
27. Day, R. (1941). Regulation of body temperature during sleep. Am. J. Dis. Child. 61:734.

28. Kirk, E. (1931). Untersuchungen über den Einfluss des normalen Schlafes auf die Temperatur der Füsse. Scand. Arch. Physiol. 61:71.
29. Kreider, M. B., Buskirk, E. R., and Bass, D. E. (1958). Oxygen consumption and body temperatures during the night. J. Appl. Physiol. 12:361.
30. Kreider, M. B., and Iampietro, P. F. (1959). Oxygen consumption and body temperature during sleep in cold environments. J. Appl. Physiol. 14:765.
31. Satoh, T., Ogawa, T., and Takaji, K. (1965). Sweating during daytime sleep. Jpn. J. Physiol. 15:523.
32. Takagi, K. (1970). Sweating during sleep. *In* Physiological and Behavioral Temperature Regulation, pp. 669–675. J. D. Hardy, A. P. Gagge, and J. A. J. Stolwijk (eds.), Charles C Thomas, Springfield.
33. Euler, C. V., and Söderberg, U. (1957). The influence of hypothalamic thermoreceptive structures on the electroencephalogram and gamma motor activity. Electroencephalogr. Clin. Neurophysiol. 9:391.
34. Adams, T. (1963). Hypothalamic temperature in the cat during feeding and sleep. Science 139:609.
35. Abrams, R., and Hammel, H. T. (1964). Hypothalamic temperature in unanesthetized albino rats during feeding and sleeping. Am. J. Physiol. 206:641.
36. Aserinsky, E., and Kleitman, N. (1955). Two types of ocular motility occurring in sleep. J. Appl. Physiol. 8:1.
37. Dement, W. C., and Kleitman, N. (1957). The relation of eye movements during sleep to dream activity: an objective method for the study of dreaming. J. Exp. Psychol. 53:339.
38. Jouvet, M. (1967). Neurophysiology of the states of sleep. Physiol. Rev. 47:117.
39. Serota, H. M. (1939). Temperature changes in the cortex and hypothalamus during sleep. J. Neurophysiol. 2:42.
40. Delgado, J. M. R., and Hanai, T. (1966). Intracerebral temperatures in freely moving cats. Am. J. Physiol. 211:755.
41. Baker, M. A., and Hayward, J. N. (1967). Autonomic basis for the rise in brain temperature during paradoxical sleep. Science 157:1586.
42. Hull, C. D., Buchwald, N. A., Dubrovsky, B., and Garcia, J. (1965). Brain temperature and arousal. Exp. Neurol. 12:238.
43. Kundt, H. W., Brück, K., and Hensel, H. (1957). Hypothalamustemperature und Hautdurchblutung der nichtnarkotisierten Katze. Pfluegers Arch. 264:97.
44. Satoh, J. (1968). Brain temperature of the cat during sleep. Arch. Ital. Biol. 106:73.
45. Hayward, J. N., and Baker, M. A. (1969). A comparative study of the role of the cerebral arterial blood in the regulation of brain temperature in five mammals. Brain Res. 16:417.
46. Parmeggiani, P. L., Agnati, L. F., Zamboni, G., and Cianci, T. (1975). Hypothalamic temperature during the sleep cycle at different ambient temperatures. Electroencephalogr. Clin. Neurophysiol. 38:589.
47. Kovalzon, V. M. (1973). Brain temperature variations during natural sleep and arousal in white rats. Physiol. Behav. 10:667.
48. Kawamura, H., Whitmoyer, D. I., and Sawyer, C. H. (1966). Temperature changes in the rabbit brain during paradoxical sleep. Electroencephalogr. Clin. Neurophysiol. 21:469.
49. Hamilton, C. L. (1963). Hypothalamic temperature records of a monkey. Proc. Soc. Exp. Biol. (N. Y.) 112:55.

50. Fusco, M. M. (1963). Temperature pattern throughout the hypothalamus in the resting dog. *In* Temperature and Control in Science and Industry, pp. 585–587. J. D. Hardy (ed.), Reinhold, New York.
51. Van Twyver, H., and Allison, T. (1974). Sleep in the armadillo *Dasypus novemcinctus* at moderate and low ambient temperatures. Brain Behav. Evol. 9:107.
52. Van Twyver, H., and Allison, T. (1970). Sleep in the opossum *Didelphis marsupialis.* Electroencephalogr. Clin. Neurophysiol. 29:181.
53. Rechtschaffen, A., Cornwell, P., and Zimmerman, W. (1965). Brain temperature variations with paradoxical sleep in the cat. Association of Psychophysiological Study of Sleep, Washington, D.C.
54. Hayward, J. N. (1968). Brain temperature regulation during sleep and arousal in the dog. Exp. Neurol. 21:201.
55. Kawamura, H., and Sawyer, C. H. (1965). Temperature elevation in the rabbit brain during paradoxical sleep. Science 150:912.
56. Allison, T., and Van Twyver, H. (1970). Sleep in the moles, *Scalopus aquaticus* and *Condylura cristata.* Exp. Neurol. 27:564.
57. Reite, M. L., and Pegram, G. V. (1968). Cortical temperature during paradoxical sleep in the monkey. Electroencephalogr. Clin. Neurophysiol. 26:36.
58. Taylor, C. R. (1966). The vascularity and possible thermoregulatory function of the horns in goats. Physiol. Zool. 39:127.
59. Parmeggiani, P. L., Franzini, C., Lenzi, P., and Cianci, T. (1971). Inguinal subcutaneous temperature changes in cats sleeping at different environmental temperatures. Brain Res. 33:397.
60. Baust, W. (1966). Local blood flow in different regions of the brainstem during natural sleep and arousal. Electroencephalogr. Clin. Neurophysiol. 22:365.
61. Reivich, M., Isaacs, G., Evarts, E., and Kety, S. (1968). The effect of slow wave sleep and REM sleep on regional cerebral blood flow in cats. J. Neurochem. 15:301.
62. Shapiro, C. M., and Rosendorff, C. (1975). Local hypothalamic blood flow during sleep. Electroencephalogr. Clin. Neurophysiol. 39:365.
63. Findlay, A. L. R., and Hayward, J. N. (1969). Spontaneous activity of single neurons in the hypothalamus of rabbits during sleep and waking. J. Physiol. (Lond.) 201:237.
64. Parmeggiani, P. L., and Franzini, C. (1971). Changes in the activity of hypothalamic units during sleep at different environmental temperatures. Brain Res. 29:347.
65. Euler, C. V., and Söderberg, U. (1958). Co-ordinated change in temperature thresholds for thermoregulatory reflexes. Acta Physiol. Scand. 42:112.,
66. Allison, T., Van Twyver, H., and Goff, W. R. (1972). Electrophysiological studies of the echidna, *Tachyglossus aculeatus.* I. Waking and sleep. Arch. Ital. Biol. 110:145.
67. Ogawa, T., Satoh, T., and Takagi, K. (1967). Sweating during night sleep. Jpn. J. Physiol. 17:135.
68. Parmeggiani, P. L., and Sabattini, L. (1972). Electromyographic aspects of postural, respiratory and thermoregulatory mechanisms in sleeping cats. Electroencephalogr. Clin. Neurophysiol. 33:1.
69. Parmeggiani, P. L., and Rabini, C. (1967). Shivering and panting during sleep. Brain Res. 6:789.
70. Affanni, J. M., Lisogorsky, E., and Scaravilli, A. M. (1972). Sleep in the

giant South American armadillo *Priodontes giganteus* (Edentata, Mammalia). Experientia 28:1046.

71. Shapiro, C. M., Moore, A. T., Mitchell, D., and Yodaiken, M. L. (1974). How well does man thermoregulate during sleep? Experientia 30:1279.
72. Baust, W., Weidinger, H., and Kirchner, F. (1968). Sympathetic activity during natural sleep and arousal. Arch. Ital. Biol. 106:379.
73. Hammel, H. T. (1968). Regulation of internal body temperature. Ann. Rev. Physiol. 30:641.
74. Hammel, H. T., Heller, H. C., and Sharp, F. R. (1973). Probing the rostral brainstem of anesthetized, unanesthetized, and exercising dogs and of hibernating and euthermic ground squirrels. Fed. Proc. 32:1588.
75. Heller, H. C., Colliver, G. W., and Anand, P. (1974). CNS regulation of body temperature in euthermic hibernators. Am. J. Physiol. 227:576.
76. Glotzbach, S. F., and Heller, H. C. (1975). CNS regulation of metabolic rate in the kangaroo rat *Dipodomys ingens.* Am. J. Physiol. 228:1180.
77. Heller, H. C., Henderson, J. A., and Glotzbach, S. F. (1975). The relationship between POH thermosensitivity and body size in mammals. Biometeorology 6:18.
78. Heller, H. C., and Henderson, J. A. (1976). Hypothalamic thermosensitivity and regulation of heat storage behavior in a day-active desert rodent, *Ammospermophilus nelsoni.* J. Comp. Physiol. (B) 108:255.
79. Parmeggiani, P. L., Franzini, C., Lenzi, P., and Zamboni, G. (1973). Threshold of respiratory responses to preoptic heating during sleep in freely moving cats. Brain Res. 52:189.
80. Glotzbach, S. R., and Heller, H. C. (1976). Central nervous regulation of body temperature during sleep. Science 194:537.
81. Van Twyver, H. (1969). Sleep patterns of five rodent species. Physiol. Behav. 4:901.
82. Hellstrøme, B., and Hammel, H. T. (1967). Some characteristics of temperature regulation in the unanesthetized dog. Am. J. Physiol. 213:547.
83. Evarts, E. V. (1967). Unit activity in sleep and wakefulness. *In* The Neurosciences, p. 545. G. C. Quarton, T. Melnechuk, and F. O. Schmitt (eds.), Rockefeller University Press, New York.
84. Manohar, S., Noda, H., and Adey, W. R. (1972). Behavior of mesencephalic reticular neurons in sleep and wakefulness. Exp. Neurol. 34:140.
85. Hendricks, J. C., Bowker, R. M., and Morrison, A. R. (1976). Functional characteristics of cats with pontine lesions during sleep and wakefulness: further studies. Sleep Res. 5:23.
86. Henley, K., and Morrison, A. R. (1974). A re-evaluation of the effects of lesions of the pontine tegmentum and locus coeruleus on phenomena of paradoxical sleep in the cat. Acta Neurobiol. Exp. 34:215.
87. Okuma, T., Fujimori, M., and Hayashi, A. (1965). The effect of environmental temperature on the electrocortical activity of cats immobilized by neuromuscular blocking agents. Electroencephalogr. Clin. Neurophysiol. 18:392.
88. Parmeggiani, P. L., Rabini, C., and Cattalani, M. (1969). Sleep phases at low environmental temperature. Arch. Sci. Biol. 53:277.
89. Parmeggiani, P. L., and Rabini, C. (1970). Sleep and environmental temperature. Arch. Ital. Biol. 108:369.
90. Schmidek, W. R., Hoshino, K., Schmidek, M., and Timo-Iaria, C. (1972). Influence of environmental temperature on the sleep-wakefulness cycle in the rat. Physiol. Behav. 8:363.

91. Valatx, J. L., Roussel, B., and Curé, M. (1973). Sommeil et température cérébrale du rat au cours l'exposition chronique en ambiance chaude. Brain Res. 55:107.
92. Hemingway, A., Rasmussen, T., Wickoff, H., and Rasmussen, A. T. (1940). Effects of heating hypothalamus of dogs by diathermy. J. Neurophysiol. 3:329.
93. Roberts, W. W., Bergquist, E. H., and Robinson, T. C. L. (1969). Thermoregulatory grooming and sleep-like relaxation induced by local warming of preoptic area and anterior hypothalamus in opossum. J. Comp. Physiol. Psychol. 67:182.
94. Roberts, W. W., and Robinson, T. C. L. (1969). Relaxation and sleep induced by warming of preoptic region and anterior hypothalamus in cats. Exp. Neurol. 25:282.
95. Parmeggiani, P. L., Zamboni, G., Cianci, T., Agnati, L. F., and Ricci, C. (1974). Influence of anterior hypothalamic heating on the duration of fast wave sleep episodes. Electroencephalogr. Clin. Neurophysiol. 36:465.
96. De Armond, S. J., and Fusco, M. M. (1971). The effect of preoptic warming on the arousal system of the mesencephalic reticular formation. Exp. Neurol. 33:653.
97. Glotzbach, S. F., and Heller, H. C. (1976). Effects of hypothalamic and ambient temperatures on sleep states in the kangaroo rat. Fed. Proc. 35:559.
98. Lyman, C. P., and Chatfield, P. O. (1955). Physiology of hibernation in mammals. Physiol. Rev. 35:403.
99. Suomalainen, P. (1961). Hibernation and sleep. *In* The Nature of Sleep, pp. 307–316. G. E. W. Wolstenholme and M. O'Connor (eds.), J. A. Churchill, Ltd., London.
100. Twente, J. W., and Twente, J. A. (1965). Regulation of hibernating periods by temperature. Proc. Natl. Acad. Sci. USA 54:1058.
101. Hammel, H. T. (1967). Temperature regulation and hibernation. *In* Mammalian Hibernation III, pp. 86–96. K. C. Fisher, A. R. Dawe, C. P. Lyman, E. Schönbaum, and F. E. South (eds.), Oliver and Boyd Ltd., London.
102. South, F. E., Breazile, J. E., Dellman, H. D., and Epperly, A. D. (1969). Sleep, hibernation, and hypothermia in the yellow-bellied marmot (*M. flaviventris*). *In* Depressed Metabolism, pp. 277–312. X. J. Musacchia and J. F. Saunders (eds.), Elsevier, New York.
103. Satinoff, E. (1970). Hibernation and the central nervous system. *In* Progress in Physiological Psychology, Vol. 3, pp. 201–236. E. Stellar and J. M. Sprague, (eds.), Academic Press, New York.
104. Mihailovic, L. T. (1972). Cortical and subcortical electrical activity in hibernation and hypothermia: a comparative analysis of the two states. *In* Hibernation–Hypothermia: Perspectives and Challenges, pp. 487–534. F. E. South, J. P. Hannon, J. R. Willis, E. T. Pengelly, and N. R. Alpert (eds.), Elsevier, London.
105. Walker, J. M., Glotzbach, S. F., Berger, R. J., and Heller, H. C. (1975). Sleep and hibernation. I. Electrophysiological observations. Sleep Res. 4:67.
106. Walker, J. M., Glotzbach, S. F., Berger, R. J., and Heller, H. C. (1977). Sleep and hibernation in ground squirrels (*Citellus*): electrophysiological observations. Amer. J. Physiol. In press.
107. Toutain, P. L., and Ruckebusch, Y. (1975). Arousal as a cyclic phe-

nomenon during sleep and hibernation in the hedgehog (*Erinaceus europeanus*). Experientia 31:312.

108. Allison, T., and Van Twyver, H. (1972). Electrophysiological studies of the echidna, *Tachyglossus aculeatus*. II. Dormancy and hibernation. Arch. Ital. Biol. 110:185.
109. Kayser, C. (1961). The Physiology of Natural Hibernation. Pergamon Press, Oxford.
110. Mrosovsky, N. (1971). Hibernation and the Hypothalamus. Appleton-Century-Crofts, New York.
111. Strumwasser, F. (1959). Factors in the pattern, timing and predictability of hibernation in the squirrel, *Citellus beecheyi*. Am. J. Physiol. 196:8.
112. Twente, J. W., and Twente, J. A. (1967). Seasonal variation in the hibernating behavior of *Citellus lateralis*. *In* Mammalian Hibernation III, pp. 47–63. K. C. Fisher, A. R. Dawe, C. P. Lyman, E. Schönbaum, and F. E. South (eds.), Elsevier, New York.
113. Wang, L. C. H. (1973). Radiotelemetric study of hibernation under natural and laboratory conditions. Am. J. Physiol. 224:673.
114. Williams, B. A., and Heath, J. E. (1970). Responses to preoptic heating and cooling in a hibernator *Citellus tridecemlineatus*. Am. J. Physiol. 218:1654.
115. Williams, B. A., and Heath, J. E. (1971). Thermoregulatory responses of a hibernator to preoptic and environmental temperatures. Am. J. Physiol. 221:1134.
116. Mills, S. H., Miller, V. M., and South, F. E. (1974). Thermoregulatory responses of a hibernator to hypothalamic and ambient temperature. Cryobiology 11:465.
117. Wyss, O. A. M. (1932). Winterschlaf und Wärmehaushalt, untersucht am Siebenschläfer (*Myoxus glis*). Pfluegers Arch. 229:599.
118. Lyman, C. P. (1948). The oxygen consumption and temperature regulation of hibernating hamsters. J. Exp. Zool. 109:55.
119. Chao, I., and Yeh, C. J. (1950). Hibernation of the hedgehog. I. Body temperature regulation. Chin. J. Physiol. 17:343.
120. Kayser, C. (1953). L'hibernation des mammifères. L'Année Biologique, 3me serie 29:109.
121. Pengelley, E. T. (1964). Responses of a new hibernator (*Citellus variegatus*) to controlled environments. Nature 203:892.
122. Kristoffersson, R., and Soivio, A. (1964). Hibernation in the hedgehog (*Erinaceus europaeus* L.) Ann. Acad. Sci. Fenn. (Biol.) 82:3.
123. Reite, O. B., and Davis, W. H. (1966). Thermoregulation in bats exposed to low ambient temperatures. Proc. Soc. Exp. Biol. Med. 121:1212.
124. Soivio, A., Tähti, H., and Kristoffersson, R. (1968). Studies on the periodicity of hibernation in the hedgehogg (*Erinaceus europaeus* L.). III. Hibernation in a constant ambient temperature of -5°C. Ann. Zool. 5:224.
125. Wang, L. C. H., and Hudson, J. W. (1971). Temperature regulation in normothermic and hibernating eastern chipmunks, *Tamaias striatus*. Comp. Biochem. Physiol. 38:59.
126. Heller, H. C., and Colliver, G. W. (1974). CNS regulation of body temperature during hibernation. Am. J. Physiol. 227:583.
127. Pivorun, E. B. (1976). A biotelemetry study of the thermoregulatory patterns of *Tamias striatus* and *Eutamias minimus* during hibernation. Comp. Biochem. Physiol. 53A:265.

128. Strumwasser, F. (1959). Thermoregulatory, brain and behavioral mechanisms during entrance into hibernation in the squirrel, *Citellus beecheyi.* Am. J. Physiol. 196:15.
129. Strumwasser, F. (1959). Regulatory mechanisms, brain activity and behavior during deep hibernation in the squirrel, *Citellus beecheyi.* Am. J. Physiol. 196:23.
130. Strumwasser, F. (1960). Some physiological principles governing hibernation in *Citellus beecheyi.* Bull. Mus. Comp. Zool. Harvard 124:285.
131. Wang, L. C. H., and Hudson, J. W. (1970). Some physiological aspects of temperature regulation in the normothermic and torpid hispid pocket mouse, *Perognathus hispidus.* Comp. Biochem. Physiol. 32:275.
132. Morhardt, J. E. (1970). Body temperatures of white-footed mice (*Peromyscus* sp.) during daily torpor. Comp. Biochem. Physiol. 33:423.
133. Hainsworth, F. R., and Wolf, L. L. (1970). Regulation of oxygen consumption and body temperature during torpor in a hummingbird, *Eulampis jugularis.* Science 168:368.
134. Wolf, L. L., and Hainsworth, F. R. (1972). Environmental influence on regulated body temperature in torpid hummingbirds. Comp. Biochem. Physiol. 41A:167.
135. Hammel, H. T., Dawson, T. V., Abrams, R. M., and Andersen, H. T. (1968). Total calorimetric measurements on *Citellus lateralis* in hibernation. Physiol. Zool. 41:341.
136. Heller, H. C., and Hammel, H. T. (1972). CNS control of body temperature during hibernation. Comp. Biochem. Physiol. 41A:349.
137. Lyman, C. P., and O'Brien, R. C. (1972). Sensitivity to low temperature in hibernating rodents. Am. J. Physiol. 222:864.
138. Lyman, C. P., and O'Brien, R. C. (1974). A comparison of temperature regulation in hibernating rodents. Am. J. Physiol. 227:218.
139. Hartner, W. C., South, F. E., Jacobs, H. K., and Luecke, R. H. (1971). Preoptic thermal stimulation and temperature regulation in the marmot (*M. flaviventris*). Cryobiology 8:312.
140. Mills, S. H., and South, F. E. (1972). Central regulation of temperature in hibernation and normothermia. Cryobiology 9:393.
141. South, F. E., Hartner, W. C., and Luecke, R. H. (1975). Responses to preoptic temperature manipulation in the awake and hibernating marmot. Am. J. Physiol. 229:150.
142. Wünnenberg, W., Merker, G., and Speulda, E. (1976). Thermosensitivity of preoptic neurones in a hibernator (golden hamster) and a non-hibernator (guinea pig). Pfluegers Arch. 363:119.
143. Boulant, J. A., and Bignall, K. E. (1973). Changes in thermosensitive characteristics of hypothalamic units over time. Am. J. Physiol. 225:331.
144. Boulant, J. A., and Bignall, K. E. (1973). Hypothalamic neuronal responses to peripheral and deep body temperatures. Am. J. Physiol. 225:1371.
145. Florant, G., and Heller, H. C. Central nervous regulation of body temperature in euthermic and hibernating marmots. Am. J. Physiol., in press.
146. Twente, J. W., and Twente, J. A. (1968). Progressive irritability of hibernating *Citellus lateralis.* Comp. Biochem. Physiol. 25:467.
147. Beckman, A. L., and Stanton, T. L. (1976). Changes in CNS responsiveness during hibernation. Am. J. Physiol. 231:810.
148. Lyman, C. P. (1958). Oxygen consumption, body temperature and heart rate of woodchucks entering hibernation. Am. J. Physiol. 194:83.
149. Heller, H. C., Colliver, G., and Beard, J. Thermoregulation during entrance into hibernation. Pfluegers Arch., in press.

150. Cade, T. (1964). The evolution of torpidity in rodents. Ann. Acad. Sci. Fenn. (Biol.) 71:77.
151. Bligh, J. (1973). Temperature Regulation in Mammals and Other Vertebrates. North-Holland, Amsterdam.
152. Walker, J. M., and Berger, R. J. (1973). A polygraphic study of the tortoise (*Testudo denticulata*): absence of electrophysiological signs of sleep. Brain Behav. Evol. 8:453.
153. Berger, R. J. (1975). Bioenergetic functions of sleep and activity rhythms and their possible relevance to aging. Fed. Proc. 34:97.
154. Walker, J. M., and Berger, R. J. Slow-wave sleep: an energy conservation process. In preparation.
155. Flanigan, W. F., Jr., Knight, C. P., Hartse, K. M., and Rechtschaffen, A. (1974). Sleep and wakefulness in chelonian reptiles. I. The box turtle, *Terrapene carolina.* Arch. Ital. Biol. 112:227.
156. Flanigan, W. F., Jr. (1974). Sleep and wakefulness in chelonian reptiles. II. The red-footed tortoise, *Geochelone carbonaria.* Arch. Ital. Biol. 112:253.

International Review of Physiology
Environmental Physiology II, Volume 15
Edited by David Robertshaw
 University Park Press Baltimore

6
Role of the Adrenal Medulla in Thermoregulation

D. ROBERTSHAW
Indiana University School of Medicine,
Bloomington, Indiana

METHODS OF STUDY 192
  Chemical Assay 192
  Surgical Intervention 192
  Pharmacological Blocking Agents 194

EFFECTS OF COLD EXPOSURE 194
  Nonshivering Thermogenesis 197
  Shivering Thermogenesis 200
  Basal Metabolism 202
  Heat Conservation Mechanisms 202
  Conclusions 203

EFFECTS OF HEAT EXPOSURE 204
  Panting 206
  Sweating 207
    Apocrine Sweat Glands 207
    Eccrine Sweat Glands 208
    Sweating in Disease States 210

CONCLUSIONS 212

The autonomic nervous system is classically divided into two parts, the parasympathetic and sympathetic. These two divisions are concerned with the vegetative functions of the body, and both are continuously active in making the adjustments necessary for the maintenance of homeostasis. It is usually considered, however, that under conditions of rest parasympathetic activity predominates, whereas under stressful situations the sympathetic nervous system is the major controlling influence. The sympathetic division of the autonomic nervous system has an extension or additional component, the adrenal medulla. It is the function of this gland with which we are concerned in this review. This endocrine gland is apparently entirely under neural control and until a few years ago was considered to be the only endocrine gland controlled in such a manner. It, therefore, appeared to hold a somewhat unique position until, with the development in recent years of our understanding of the role of hypothalamic hormones in the control of the anterior pituitary, a whole new discipline developed, namely neuroendocrinology.

There is continuous activity in the nerve supplying the adrenal medulla which results in a constant "basal" secretion of adrenal medulla hormones, i.e. epinephrine and norepinephrine, into the circulation. Under conditions which might be construed as "life-threatening," adrenal medullary secretion is increased due to increased activity in the nerve supply to the gland. The proportion of epinephrine to norepinephrine secreted by the gland varies between animal species and with the stimulus. Thus, for example, insulin hypoglycemia cause only the release of epinephrine, whereas hypotension results in a mixed secretion. It has been suggested, therefore, that there are cells in the gland which secrete specific hormones and have a discrete nerve supply of their own with independent control within the central nervous system. The hormones being released into the blood are carried, of course, to all tissues in the body, and unless there is a barrier between the blood supply and a particular tissue the hormones have access to any receptor site that a particular tissue may contain. In this way, a receptor could be stimulated either by transmitters released from a nearby sympathetic nerve terminal or by blood-borne catecholamines. Thus, the hormones can be considered as supplementing the transmitter released by sympathetic nerve endings and magnifying the effects of sympathetic stimulation. However, it is difficult to envisage blood-borne concentrations of hormone having any major additive effect on transmitter released from sympathetic stimulation because the concentration achieved at the receptor site by neural stimulation will be many times greater than that which can be achieved by blood-borne stimulation. Generally speaking, therefore, the effects of adrenal medullary hormones are different from those produced by neural stimulation. The adrenal medullary hormones, particularly epinephrine, have profound metabolic effects and cause increases in circulating free fatty acids and glucose. Obviously, therefore, these effects will provide substrate for any increased activity in tissues whose metabolism requires enhancement. As a broad generalization, the effects of epinephrine are mediated through β receptors and those of norepinephrine by α receptors. β Receptors probably, therefore, have no innervation, although this is still an open question. Epinephrine, in addition, will stimulate α receptors, but for the reasons discussed above it may have relatively small effects under physiological conditions when compared to the effects of norepinephrine released from nerve endings.

If heat or cold exposure can be considered potentially life-threatening, then it might be expected that the adrenal medulla would be stimulated under these circumstances. If so, the way in which adrenal medullary secretion, particularly an outpouring of epinephrine, assists the animal in counteracting these types of stress must be considered. Figure 1 diagrammatically illustrates the problem. Although it is known that there may be direct pathways from the cerebral cortex and other parts of the brain to the adrenal medulla, stimulation of areas of the hypothalamus can evoke adrenal medullary secretion. It might be assumed, therefore, that stimulation of the adrenal medulla in response to heat or cold exposure would act via the hypothalamus because it is also the site of thermoregulatory control. A hypothetical center within the hypothalamus controlling adrenal medullary secretion is, therefore, proposed, and connections between the center and that controlling temperature regulation are suggested by the *arrow* (Figure 1). If either heat or cold exposure results in increased adrenal medullary secretion, then a direct connection between these centers would be suggested. Likewise, stimulation of the adrenal medullary center or areas of the hypothalamus concerned with autonomic function may elicit thermoregulatory responses; the *broken arrow* (Figure 1) suggests such a possibility. An example of a thermoregulatory response to a nonthermal stimulus is the sweating response to fright, the so-called "cold sweat." The release of epinephrine from the adrenal medulla is shown on the diagram and is linked to the heat loss mechanisms of vasodilation, panting, or sweating and the heat conservation

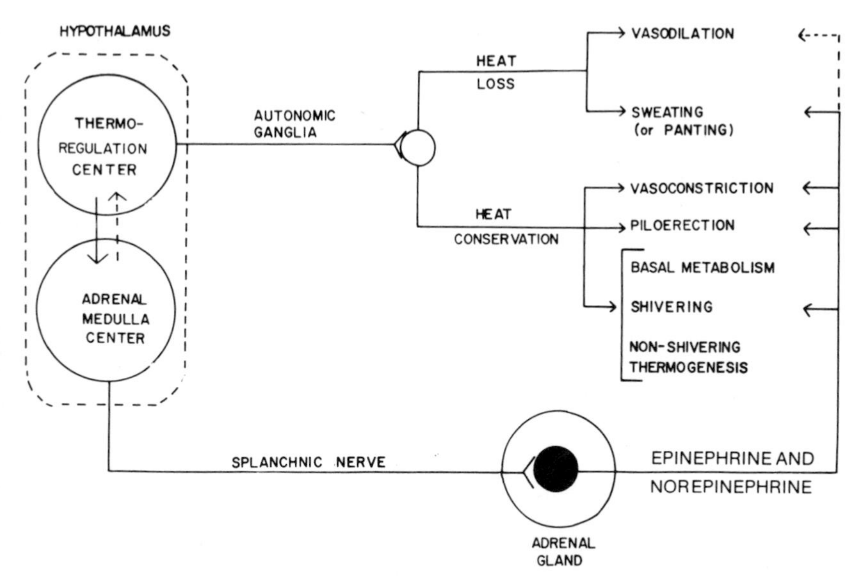

Figure 1. Diagrammatic representation of the control of thermoregulation by the sympathetic nervous system. Control is by neural pathways except where the output of the adrenal medulla gains access to the effector system via the blood stream. The *arrows* connected by interrupted lines indicate that there is no evidence, at present, for such a relationship.

mechanisms of hypermetabolism, vasoconstriction, and piloerection. The *broken lines* indicate areas that have not yet been investigated, and *continuous lines* represent parts of the thermoregulatory effector system in which epinephrine may have a significant physiological role.

METHODS OF STUDY

An assessment of adrenal medullary function can be made in three ways: 1) chemical assay of the secretion; 2) surgical intervention with or without removal of the adrenal medulla; and 3) pharmacological agents which block either the release of hormones or the receptors on which they act.

Chemical Assay

Progress in the study of adrenal medullary function has been greatly hampered by the lack of sensitive methods for the measurement of catecholamines in plasma. It is generally considered that circulating epinephrine is of adrenal medullary origin, whereas norepinephrine originates both from the adrenal medulla and from sympathetic nerve endings. Norepinephrine levels cannot be said to actually reflect either adrenal medullary function or sympathetic nervous activity because most of the transmitter released from nerve endings is taken up again by the nerve; that appearing in the circulation only represents a small portion of the transmitter which has diffused unchanged from the synaptic region. The concentration of circulating catecholamines has proved, generally speaking, too low for chemical assay except by using large quantities of blood. However, chemical assay by fluorometric techniques has been refined recently to such an extent that estimations can now be made on 1-ml samples of plasma (1) and radiometric techniques have been developed for the separate estimation of epinephrine and norepinephrine on much smaller plasma samples (2). Already significant advances are being made in the understanding of the physiology of the autonomic nervous system. Free catecholamines appear in urine in a higher concentration than in plasma and can be more readily measured than plasma catecholamines; however, they only represent a very small proportion, possibly less than 1%, of the secreted material, the rest being metabolized. Nevertheless, valuable information has been obtained from measurements on the urinary excretion of free catecholamines, although they give no indication of the minute-by-minute changes in rate of secretion.

Surgical Intervention

Adrenalectomy not only removes the medulla but also, of course, the adrenal cortex. Replacement therapy of adrenocortical hormones theoretically restores that deficiency, leaving a functionally demedullated animal. Such a preparation suffers from its inability to produce normal changes in adrenal cortical function, and many situations which elicit increases in adrenal medullary discharge also affect adrenocortical activity. It is difficult, therefore, to positively identify an adrenal-medullary function in adrenalectomized animals. Adrenal demedullation has also been attempted, but such a preparation requires rigorous postmortem examination of the integrity of the cortex and the effectiveness of the demedullation.

Because adrenal medullary secretion is under neural control, section of the nerves supplying the adrenal medulla suppresses both the resting secretion and prevents the release of hormone under stressful situations. Vogt (3) has shown that, although denervation reduces normal resting secretion, there continues to be a small output of hormone. This may represent either a continuous nonneural synthesis and release or incomplete denervation. The nerve supply to the adrenal gland shows considerable species variation, but in all species the major innervation is from the greater splanchnic nerve. Contributions from other nerves vary, originating mainly from the first two lumbar sympathetic ganglia and sometimes described as the lesser and least splanchnic nerves. The innervation is illustrated in Figure 2. Complete denervation can usually be achieved by a combination of section of the greater splanchnic nerve and removal of the first two lumbar sympathetic ganglia. The lesser and least splanchnic nerves are usually too small to identify, and an effective denervation requires removal of the sympathetic ganglia. The effectiveness of denervation can be deduced from chemical estimations of urinary catecholamines. Complete denervation should result in the disappearance of urinary epinephrine, with little or no effect on norepinephrine excretion. Table 1 shows the effect of section of the greater splanchnic nerve on the urinary excretion of epinephrine in the stump-tail macaque monkey (*Macaca speciosa*). The urinary excretion of epinephrine is reduced to 20% of that of the intact animal. The residual excretion is further reduced to unmeasurable levels by lumbar sympathectomy. This suggests that approximately 80% of the innervation is derived from the greater splanchnic nerve, the remainder from contributions originating in the first two lumbar ganglia. The major disadvantage of this technique is that total adrenal medullary denervation denervates other structures as well, for example, the liver and spleen obtain their innervation from the greater splanchnic nerve, as does a portion of

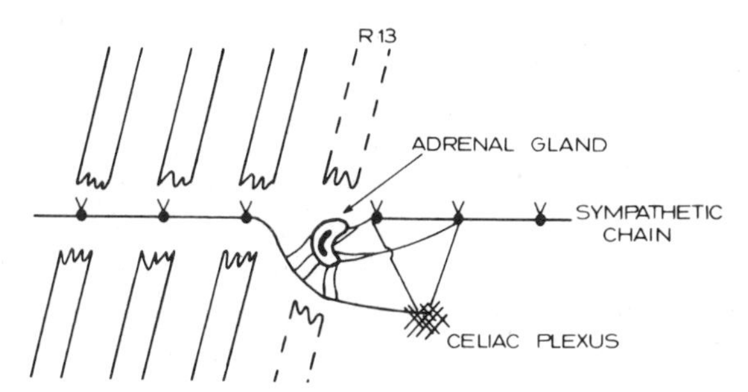

Figure 2. Diagram of the nerve supply to the adrenal gland. The outline of the ribs is indicated; in many species the 13th rib (R13) needs to be resected for ready access to the greater splanchnic nerve. The diagram shows an interruption in the sympathetic chain and fusion of the greater splanchnic nerve with the sympathetic chain. This is not universal, and in primates the greater splanchnic nerve separates from the sympathetic chain within the thorax and a continuity of the thoracic and lumbar parts of the chain is usually present.

Table 1. Effect of splanchnicotomy and splanchnicotomy plus lumbar sympathectomy on the urine concentration[a] of epinephrine and norepinephrine

	Epinephrine	Norepinephrine
Control	6.6	8.5
Splanchnicotomy	1.1	6.9
Splanchnicotomy plus lumbar sympathectomy	0.1	9.6

[a]In ng/ml. Limit of assay < 0.5 ng/ml.

the gut. Lumbar sympathectomy also denervates some of the vasculature to the hind limbs and results in cutaneous vasodilation. Interpretation of results, therefore, should take into account these possible effects.

Pharmacological Blocking Agents

Pharmacological inactivation of the adrenal medulla can best be achieved by the use of ganglionic blocking agents which block transmission within the adrenal medullary synapse. The major disadvantage with this technique is that transmission is blocked in all other autonomic ganglia. It, therefore, is difficult to identify the specific role of the adrenal medulla under these circumstances. Likewise, the use of specific receptor-blocking drugs does not separate out effects specifically due to epinephrine, because epinephrine has both α and β receptor-stimulating actions. Furthermore, norepinephrine occasionally appears to have β receptor-stimulating properties; for example, it is almost as potent as epinephrine in its inotropic effect on the heart, an organ which from pharmacological evidence appears to possess only β receptors. This has led to a further subdivision of β receptors into β_1 and β_2 subtypes, the receptors of the heart belonging to the β_1 variety and β_2 receptors being found typically in the bronchial musculature (4).

EFFECTS OF COLD EXPOSURE

Because cutaneous vasoconstriction is one of the responses to cold exposure and is mediated by stimulation of the sympathetic nerves supplying the blood vessels of the skin, it might be expected that there would be an increased excretion of urinary norepinephrine as a result of cold exposure. This, in fact, occurs, but is also accompanied by an increase in epinephrine excretion (5). Activation of heat conservation mechanisms by localized cooling of the hypothalamus also results in an increased excretion of urinary epinephrine (6). Figure 3 shows the effect on the urinary excretion of epinephrine and norepinephrine in rats exposed to 22°C (thermoneutrality) and 3°C. It will be seen that during continued cold exposure for a period of 36 days there is a decline in urinary catecholamine excretion. Thus, as cold acclimation progresses, the response diminishes; but

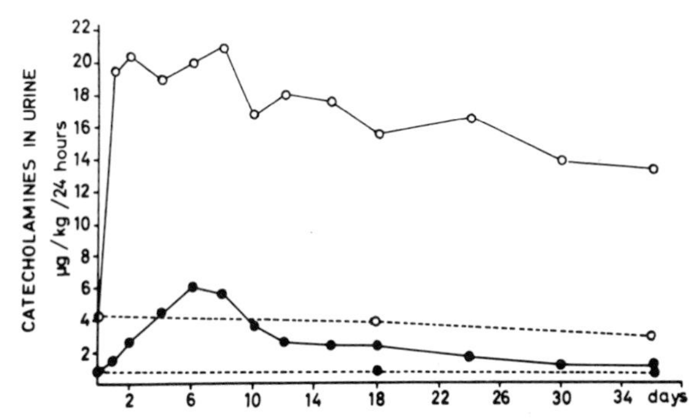

Figure 3. Urinary excretion of epinephrine (•) and norepinephrine (○) in rats at 3°C (*solid line*) and 22°C (*dashed line*). Each *point* represents the mean of six individual rats. (Reproduced from Leduc (5) by courtesy of Acta Physiol. Scand.)

even after several months of cold exposure, epinephrine excretion is still 50–100% higher than that of the control group maintained at a thermoneutral temperature (5). There is also a direct relationship between the level of catecholamine excretion and the intensity of cold stimulus (7). Repetitive hypothalamic cooling causes a reduction in the magnitude of the catecholamine response, epinephrine excretion being much less affected by repetitive cooling than norepinephrine excretion (6). Thus, localized central cooling can, to some extent, mimic acclimation to cold exposure. It might be argued that central cooling more closely simulates hypothermia than it does cold exposure, and a pronounced adrenal medullary response to this stimulus might be indicative of the type of response that is expected when heat conservation mechanisms prove to be inadequate. There appears to be, therefore, a definite functional link between the thermoregulatory center of the hypothalamus and that controlling adrenal medullary function. Adrenal medullary secretion might be looked upon as part of the normal response associated with activation of heat conservation mechanisms.

The role of the adrenal medulla in adaptation to cold and its mechanism of action in response to cold exposure have recently been thoroughly reviewed by Himms-Hagen (8). This review quite rightly points out the difficulty in evaluating the role of the adrenal medulla in both the response to cold exposure and acclimation in that animals functionally deprived of the adrenal medulla appear to be able to compensate for the deficiency by utilizing other parts of the sympathetic nervous system.

The physiological responses to cold exposure which allow an animal to maintain homeothermy under cold conditions can be classified into two parts: 1) a reduction of heat loss which is brought about by peripheral cutaneous vasoconstriction and piloerection and 2) an increase in metabolism through shivering, and other means of increased thermogenesis, broadly classified as nonshivering thermogenesis.

Animals which do not have a functional adrenal medulla (most experiments have been performed on rats) cannot maintain a constant body temperature when exposed to cold, a situation which can be rectified by injections of epinephrine (8). There is, however, a difference between young and old rats, young animals being much more dependent upon the integrity of the adrenal medulla than older animals. The intolerance to cold exposure of animals without an adrenal medulla and the cause of the hypothermia have never been fully identified. Most studies have been directed toward the thermogenic response to cold exposure on the assumption that demedullated rats are unable to show an adequate metabolic adjustment. Most of these studies have been carried out in the rat; it must be recognized that there are wide species differences and that adrenal demedullation or denervation do not necessarily result in cold-induced hypothermia in other species.

Hypothermia will occur when heat loss exceeds heat production. The hypothermia of adrenal demedullated animals may be due to defective heat production mechanisms or inadequate heat conservation. As air temperature falls and the gradient between body temperature and air temperature increases, heat loss will also increase. In order to maintain a constant body temperature, heat production must increase in proportion to the increase in heat loss, until a maximal heat production is achieved. If heat loss continues to increase, then body temperature falls. This maximal heat production is known as summit metabolism. Summit metabolism has three components: 1) basal metabolism, which is the metabolism of a post-absorptive resting animal; 2) shivering, which is a largely uncontrolled rapid contraction of certain skeletal muscles and is controlled, at least in part, by the somatic nervous system; and 3) nonshivering thermogenesis, which is an increased heat production without the contraction of skeletal muscle and is under the control of the autonomic nervous system. A reduction in any of these three components reduces summit metabolism. The contributions of each of these components to summit metabolism vary between species, particularly with respect to the relative proportions of shivering and nonshivering thermogenesis. For example, the newborn of many species depend almost entirely upon nonshivering thermogenesis, as do many hibernating species. In contrast, other species show either no nonshivering thermogenesis or an increase in nonshivering thermogenesis only on cold acclimation. This is particularly true in the case of the rat.

The only study which has definitely indicated that the maximal heat production of cold-exposed rats is reduced as a result of demedullation is that of Ring (9). He showed that adrenal demedullation caused a fall in maximal heat production from 56.1 to 49.5 cal/m^2·hr. Calculations from the data of Manara et al. (10) show the heat production of demedullated rats exposed to 4°C to be 60.4 cal/m^2·hr as compared to 62.5 cal/m^2·hr in intact animals. These latter results may not represent the maximal catabolic response to cold exposure because measurements were not made of body temperature. However, in other experiments from the same laboratory, it was shown that exposure to 4°C for 1 hr resulted in hypothermia, the hypothermia being more severe in the adrenal-

Table 2. Separation of various components of summit metabolism of lamb

	Birth		One month	
	(O_2/kg·hr)	percentage of metabolism	(O_2/kg·hr)	percentage of metabolism
Basal	0.8	23	0.5	18
Shivering	1.6	46	2.1	78
Nonshivering	1.1	31	0.1	4
Total	3.5	100	2.7	100

From Alexander and Williams (12).

demedullated animals (11). One may assume, therefore, that summit metabolism had been achieved. The fall in body temperature would, however, cause a fall in heat production by virtue of the direct relationship between metabolic rate and body temperature.

Drugs have been used to partition the hypermetabolism of cold exposure. For example, autonomic-blocking drugs and muscle relaxants separate nonshivering and shivering thermogenesis, respectively. A reduction in summit metabolism after treatment with curariform drugs is assumed to be due to abolition of shivering thermogenesis; the fall in metabolism after sympathetic blockade is attributed to a loss of nonshivering thermogenesis. The residual metabolism after the combined effects of both muscular paralysis and sympathetic blockade should be equivalent to the basal metabolism. Alexander and Williams (12) have used this approach with lambs; the data in Table 2 are taken from their experiments. Their results show that the proportion of the nonshivering component is greater in the newly born lamb than at 1 month of age and that basal metabolism also declines with age.

Nonshivering Thermogenesis

The role of the adrenal medulla in nonshivering thermogenesis has been elegantly demonstrated by Cottle and Carlson (13); their results are shown in Figure 4. Curarized rats, which are unable to shiver, showed differing responses to cold exposure, depending upon their state of acclimation. Cold-acclimated animals increased their metabolism, whereas warm-acclimated animals became hypothermic and their metabolism decreased. The increase in metabolism shown by curarized, cold-acclimated individuals is considered, by definition, to be nonshivering thermogenesis. It was reduced in adrenal-demedullated animals, thereby demonstrating a role for circulating epinephrine in contributing to nonshivering thermogenesis. In the noncurarized animal, i.e. one which is able to shiver, the deficiency in nonshivering thermogenesis brought about by the loss of adrenal medullary secretion may be counteracted by an increase in the intensity of shivering at levels of cold exposure that do not elicit summit metabolism.

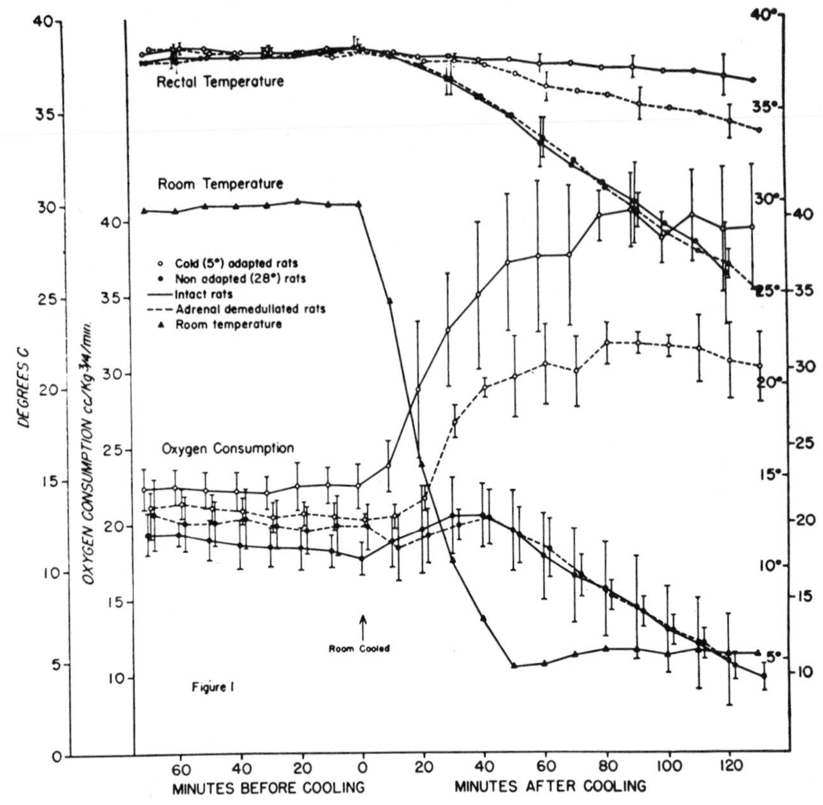

Figure 4. Metabolic response of curarized rats to low ambient temperature. Mean oxygen consumption and rectal temperature of each group along with ambient temperature are plotted on the *ordinate*; before and after cooling on the *abscissa*. *Vertical lines* indicate standard deviations. Effect of adrenal demedullation on warm-acclimated and cold-acclimated rats is shown. (Reproduced from Cottle and Carlson (13) by courtesy of Proc. Soc. Exp. Biol. Med.)

There will be, however, a reduced summit metabolism, and hypothermia will develop at higher temperatures than in intact animals.

The nature of nonshivering thermogenesis has been the subject of much discussion. Webster (14), in the previous volume in this series, concluded that nonshivering thermogenesis is entirely due to the presence of brown fat and that an absence of brown fat indicates that nonshivering thermogenesis does not exist. These conclusions seem well supported in most cases by anatomical and physiological studies that show that nonshivering thermogenesis is predominant in newborn and hibernating animals and that as the brown fat diminishes with age so nonshivering thermogenesis declines and is replaced by shivering thermogenesis. Likewise, cold acclimation in certain species can induce the formation of brown fat tissue; this parallels an increase in nonshivering thermogenesis. Furthermore, the surgical removal of brown fat from cold-acclimated rats reduces the adaptive process (15), although this requires several days to become effective. It is the fact that the partial loss of acclimation after surgery is not immediate which provides some evidence for the hypothesis that thermogenesis can occur in tissues other than brown fat itself and yet be dependent upon the integrity of brown fat depots.

In early studies, brown fat appeared to be located predominantly in a subcutaneous position between the scapulae and, in some species, around the kidney. If not located in these sites, it was stated that brown fat did not exist. However, Alexander et al. (16) have emphasized that electron microscopy must be used in order to determine the presence of brown fat anatomically. They have detected it in the newborn calf, an animal which was assumed from light microscopy not to possess brown fat (17). They have demonstrated a widespread distribution in the calf, and it may, therefore, be much more ubiquitous in other species than was previously supposed. This does not, however, prevent a search for nonshivering thermogenesis in tissues other than brown fat. One example of extremely active thermogenesis is the large increase in body temperature that occurs during arousal from hibernation. The presence of well defined brown fat areas in hibernating animals has suggested that this is the major site of heat production for arousal. However, Hayward and Ball (18) have shown that removal of the interscapular brown fat pad in the big brown bat (*Eptesicus fuscus*) does not prevent arousal from hibernation and that in vitro measurements of the oxygen consumption of brown fat stimulated by either epinephrine or norepinephrine could only account for 5–7% of the oxygen consumption observed during arousal. The brown fat of the cold-acclimated rat has been calculated to provide 6–13% of the heat of nonshivering thermogenesis (19). On the other hand, nonshivering thermogenesis of the newborn rabbit appears to reside almost totally in the interscapular fat pad, as its removal virtually abolishes the metabolic response to cold exposure (20). In some species, therefore, there may be other tissues which can increase their metabolism in response to cold exposure. The delay in the loss of nonshivering thermogenesis following surgical removal of brown fat (15) has led to the theory that tissues other than brown fat are responsible for nonshivering thermogenesis but are, at least in part, dependent upon a substance released from brown fat whose presence persists for a few days after brown fat extirpation. The tissue involved has been identified as skeletal muscle, but neither the chemical composition nor the mechanism of action of the postulated hormone released from brown fat is understood (19).

Removal of the adrenal gland in hibernating ground squirrels fails to prevent arousal (C. P. Lyman, personal communication). This finding suggests that the adrenal medulla is not essential for arousal from hibernation, although its presence may enhance the rate of increase of body temperature and that, unless there are other major sites of brown fat within the animal, another source of heat must be required. Furthermore, R. Elizondo (personal communication) has shown that the thermogenesis associated with fever in rhesus monkeys (*Macaca mulatta*) does not appear to be caused by shivering and is not affected by adrenal medullary denervation. Thus, the increase in metabolism is due to nonshivering thermogenesis, although, because the rhesus monkey is not considered to have brown fat, another tissue which is not controlled by circulating catecholamines must be responsible.

Nonshivering thermogenesis can be stimulated in two ways: 1) the nerve supply to brown fat releases a transmitter, norepinephrine, which increases both blood flow and heat production of the brown fat tissue itself (circulating epinephrine may supplement the neural stimulation) and 2) in addition, some unidentified substance is possibly released from the brown fat into the bloodstream and is carried to muscle tissue, which as a result of either sympathetic neural stimulation or circulating epinephrine can increase its heat production (19). Because Cottle and Carlson (13) have shown that adrenal medullary secretion is essential for the maximal nonshivering thermogenic response to cold exposure, it would be interesting to know whether epinephrine is acting on brown fat or some other tissue. The only other tissue so far implicated in nonshivering thermogenesis is the liver, which appears to be capable of a small increase in thermogenesis. This possibly reflects the increased energy demands for gluconeogenesis and ureagenesis (21).

In conclusion, it appears that the adrenal medulla, at least in the cold-acclimated rat, contributes to nonshivering thermogenesis, although its site of action, which is likely to be either brown fat or skeletal muscle, remains to be determined. The presence of the adrenal medulla does not seem to be essential for the increase in heat production associated with either arousal from hibernation or fever.

Shivering Thermogenesis

A role for the adrenal medulla in shivering thermogenesis is speculative and derives from four sources of information.

1. It is known that infused epinephrine can induce tremor. It might be expected, therefore, that elevated levels of epinephrine during cold exposure may potentiate shivering thermogenesis by augmentation of the magnitude of the tremor. This has never been critically studied. Shivering in splanchnicotomized animals appears to be normal, but the magnitude of shivering thermogenesis during summit metabolism both before and after adrenal medullary perturbation has never been examined. Likewise, the surgically sympathectomized cats of Cannon et al. (22) were able to shiver, and this certainly suggested that there was no role for either the peripheral sympathetic nervous system or the adrenal medulla. Adrenal-demedullated rats show visible signs of shivering, although they are less capable of maintaining body temperature than intact rats (11).
2. Ganglionic-blocking agents and other autonomic-blocking drugs have been shown to reduce overt shivering (11). This evidence has been used to conclude that the autonomic nervous system is essential for shivering, although shivering is considered to be solely a function of the somatic nervous system. The inhibition of shivering produced by ganglionic-blocking drugs can be restored by catechola-

mines (11). In the experiments of Andersson et al. (6), localized cooling of the hypothalamus resulted in shivering which could be blocked by the ganglionic-blocking agent chlorisondamine and restored by catecholamines, norepinephrine being more potent than epinephrine. Thus, the surgical and chemical methods of sympathectomy produce differing results. The action of ganglionic-blocking drugs may be related to changes in hypothalamic blood flow that these drugs induce. Andersson (6) noted that, following treatment with the ganglionic-blocking agent, there was apparently an increased hypothalamic blood flow because it was necessary to increase the rate of cold water perfusion of the thermode in order to obtain the same degree of brain cooling. He also noted that the opposite occurred during norepinephrine and, to a lesser extent, epinephrine infusion. This seems an unlikely explanation because it is well known that cerebral blood flow assists in the dissipation of heat from the brain (23); an increased blood flow would, therefore, tend to lower hypothalamic temperature and thereby potentiate shivering.

3. A relationship between baroreceptor activity and shivering has been shown by Mott (24). Baroreceptor stimulation enhances shivering, and, if an increase in circulating epinephrine is sufficient to cause an increase in blood pressure, enhanced baroreceptor activity may potentiate the shivering response. A loss of baroreceptor stimulation may provide the explanation for the loss of shivering following treatment with ganglionic-blocking drugs. In contrast, chemoreceptor stimulation causes an inhibition of shivering and a fall in body temperature. Epinephrine and norepinephrine both directly stimulate the chemoreceptors, and it is considered that hormones released from the adrenal medulla during physiological situations may be sufficient to increase the concentration of circulating catecholamines to a level that will increase chemoreceptor stimulation (25). Thus, an enhancement of shivering through baroreceptor stimulation may be offset to some extent by the inhibition resulting from chemoreceptor stimulation.

4. Both shivering and nonshivering thermogenesis require substrate for the increase in metabolic rate. These substrates are glucose and free fatty acids (FFA) and derive either from intracellular stores of glycogen and triglyceride within the active tissue or from the blood. Blood levels of both these substrates rise during cold exposure (26) and presumably have a sparing action on tissue stores. The mobilization of FFA seems to be a consequence of the activation of triglyceride lipase in white adipose tissue by norepinephrine released from adrenergic nerve endings (26). On the other hand, the increase in blood glucose levels during cold exposure is abolished in both adrenal-demedullated (26) and splanchnicotomized animals (27), which suggests a definite role for epinephrine in glucose mobilization. The mobilization of glucose by epinephrine is complex and represents an increased glycogenolysis and gluconeogenesis, accompanied by a decreased insulin secretion. Failure to mobilize adequate glucose substrate may in part be responsible for the decreased survival time of demedullated rats in the cold (9, 11); the fall in rectal temperature parallels the onset of hypoglycemia

(26). If the inability to mobilize glucose is responsible for the decreased survival time of demedullated rats, the hypoglycemia should be accompanied by a fall in summit metabolism. Unfortunately, this has never been measured.

Basal Metabolism

It is generally thought that the level of basal metabolism is controlled by thyroid hormone secretion and is not a function of the adrenal medulla. Furthermore, basal metabolism is not affected by short-term cold exposure, and there is doubt that it is altered in cold acclimation (14). However, D. Robertshaw, C. R. Taylor, and V. Rowntree (unpublished observations) have recently observed that the basal metabolism of splanchnicotomized stump-tailed macaque monkeys (*M. speciosa*) is reduced by approximately 25%. These observations have subsequently been confirmed in the Rhesus monkey (*M. mulatta*). The basal metabolism can be restored to presurgical levels by the intravenous administration of either epinephrine or norepinephrine, epinephrine being more effective. Although the basal metabolism was reduced, the response to cold exposure was unaffected and animals were able to maintain body temperature. However, because basal metabolism is a component of summit metabolism, it can be predicted that summit metabolism would be reduced by an amount equivalent to the reduction in basal metabolism. If basal metabolism is reduced as a result of loss of adrenal medullary function in species other than monkeys, this might explain the reduction in summit metabolism obtained by Ring in the rat (9). He noted a reduction in summit metabolism equivalent to 20–25% of basal metabolism. The mechanism of this reduction in basal metabolism is unknown, but it does suggest that basal metabolic rate may represent a synergistic effect between circulating catecholamines and thyroid hormones and is not simply a manifestation of thyroid hormone status alone.

Heat Conservation Mechanisms

The only evidence to suggest that the adrenal medullary secretion may be important in heat conservation mechanisms as opposed to increased metabolism arises from the studies of Manara et al. (10) and Maickel et al. (11). The animals used in their experiments were rats, and it is assumed that they were not cold-acclimated. Maickel et al. (11) noted that adrenal-demedullated rats showed hypothermia after 1-hr exposure to 4°C, and that this progressed so that their survival time was considerably less than that of control animals. On the other hand, Manara et al. (10) showed that the heat production of adrenal-demedullated rats was not significantly different from that of control animals under the same conditions. The conclusion, therefore, must be made that the hypothermia of demedullated animals is not due to a diminished heat production but rather to an increased heat loss. In fact, by calculating the change in heat content, it can be shown that heat loss of the adrenal-demedullated animal is some 4 times that of the intact animals. This conclusion has never been fully tested, and one can only speculate upon the possible mechanism. The heat conservation mechanisms are a) piloerection, which increases insulation by extending the thickness of the layer of air entrapped by fur, and b) cutaneous vasoconstriction, which lowers skin temperature and thereby reduces convective heat loss. Possible means by which the adrenal medulla may influence heat conservation are as follows:

1. Enhancement of the piloerection, which is known to be mediated by adrenergic nerve fibers (28). Intradermal injections of epinephrine in concentrations up to 10^{-3} do not produce any observable piloerection (29); it would seem, therefore, that circulating epinephrine has little or no effect upon the magnitude of the piloerector response to cold exposure.
2. Potentiation of the vasoconstrictor effect of the sympathetic stimulation of cutaneous blood vessels. It is known that intravascular infusions of epinephrine cause cutaneous vasoconstriction; it may be, therefore, that blood-borne epinephrine has a role in the response of the skin blood vessels to cold exposure. However, the work of Celander (30) clearly indicated that at maximal rates of adrenal medullary secretion the effects of circulating catecholamines on cutaneous blood flow would be trivial in comparison to those evoked by sympathetic nerve fibers. However, Celander did not demonstrate a possible summation of neural and humoral influences. It may be, therefore, that maximal cutaneous vasoconstriction is dependent upon both neural and humoral components, and that loss of one or the other results in incomplete vasoconstriction. Some findings that may support such an hypothesis arise from the studies of Graham and Keatinge (31). They found that the inner layers of the muscular coat of the carotid artery are much more sensitive to stimulation by catecholamines than are the outer layers. The innervation of the smooth muscle of the artery is found in the adventitial layers, which means that the layers of muscle closest to the nerve supply will be exposed to high concentrations of transmitter substance. As the transmitter diffuses inward toward the lumen of the vessel, it will be diluted by interstitial fluid and the inner layers will thereby be subjected to a much lower concentration. This may be the simple explanation of Graham and Keatinge's findings. On the other hand, the sensitive inner layers may respond to circulating catecholamines which would, of course, be present at a much lower concentration than is found in the immediate vicinity of sympathetic nerve endings. This is an intriguing possibility and has yet to be examined.

Conclusions

In conclusion, it would appear that secretion from the adrenal medulla represents part of the physiological response to cold exposure. The role of increased circulating catecholamines in maintaining a normal body temperature during cold exposure has still been incompletely determined. Evidence is presented that suggests that adrenal medullary secretion may affect a) basal metabolism, b) shivering thermogenesis, c) nonshivering thermogenesis, and d) cutaneous vasoconstriction.

It may be that one or all of these factors are implicated in the role that adrenal medullary secretion, particularly epinephrine, plays in thermal homeostasis during cold exposure.

EFFECTS OF HEAT EXPOSURE

Arguing teleologically, it is hard to envisage properties that epinephrine may possess that would assist in the dissipation of a heat load. Thus, stimulation of heat production and cutaneous vasoconstriction represents effects of epinephrine which would be contrary to the requirements of an animal exposed to high temperatures. The suggestion by Evans and Smith (32) that the sweat glands of the horse are controlled by circulating epinephrine raises the possibility that, in this species at least, heat dissipation may be related to an increase in adrenal medullary secretion. Hartman and Hartman (33) suggested an increased adrenal medullary secretion during heat exposure of anesthetized dogs. Similarly, Symbas et al. (34) demonstrated increased levels of circulating total catecholamines during hyperthermia. Robertshaw and Whittow (35) examined the effects of heat exposure on circulating plasma catecholamines in the ox and demonstrated that exposure to a hot, dry environment which did not cause hyperthermia had no effect on plasma catecholamine levels. However, if the humidity were raised to almost saturation, so that evaporative cooling could not occur, body temperature increased. When it reached levels approximately 1.5–2°C above normal, plasma epinephrine levels were elevated (Figure 5). In some instances, the raised levels of plasma epinephrine were accompanied by increases in plasma norepinephrine; in others, there was no change in plasma norepinephrine concentration. In experiments carried out on splanchnicotomized animals, hyperthermia had no effect on plasma norepinephrine levels, and plasma epinephrine levels were so low that they were beyond the limits of the method of assay (Figure 6). It would seem, therefore, that any increase in plasma norepinephrine levels in certain individuals was due to an increased secretion of norepinephrine from the adrenal medulla. These findings also demonstrate that the stimulus for adrenal medullary secretion was via the splanchnic nerve and was not a direct effect of elevated body temperature on the rate of secretion of the adrenal medullary cells. Further experiments demonstrated that the stimulus for secretion did not appear to be due to a direct effect of high temperature on the thermoregulatory center, because localized heating of the hypothalamus to temperatures similar to those achieved by heat exposure (up to 41.5°C activated evaporative heat loss mechanisms, i.e., panting and sweating), but had no effect on plasma catecholamine levels. These findings contrasted with those of Proppe and Gale (36), who determined urinary catecholamine excretion during hypothalamic heating in the baboon and noted a fall in both epinephrine and norepinephrine excretion. Their findings could have been complicated by psychic stimulation of their animals before hypothalamic heating which would elevate adrenal medullary secretion. Andersson et al. (6) have shown that hypothalamic heating in a cold environment prevents the normal increase in adrenal medullary excretion. Thus, ele-

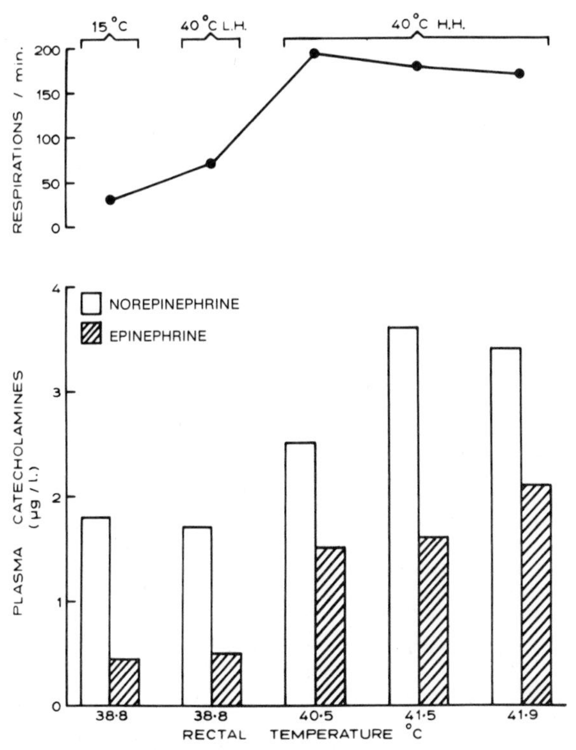

Figure 5. Plasma epinephrine and norepinephrine concentrations and respiration rate measured at various rectal temperatures. The *upper figures* refer to the environmental conditions. L. H., low humidity; H. H., high humidity. (Reproduced from Robertshaw and Whittow (35) by courtesy of J. Physiol. (Lond.).)

vated adrenal medullary secretion induced by other means can be inhibited by hypothalamic heating; it may be that the baboons of Proppe and Gale which were restrained in a primate chair were subjected to sufficient psychic stimulation to elevate their adrenal medullary secretion. The conclusion drawn from these experiments is that the response to hyperthermia is not due to stimulation of the adrenal medullary center as a result of a direct connection from the hypothalamic thermoregulatory center and must be due to either a direct effect of temperature on the adrenal medullary center or to nonthermal changes that occur during hyperthermia. For example, in panting animals such as the ox, hyperthermia is associated with respiratory alkalosis, plasma pH levels approaching a value of 8 (37). Tamura (38) has shown that when blood pH is above 7.76 there is an increased epinephrine secretion which does not occur in splanchnicotomized animals. Thus, the hyperthermic stimulation of the adrenal medulla in a panting animal may be the result of respiratory alkalosis acting on some center within the central nervous system.

It is difficult to envisage a mechanism whereby circulating epinephrine released during hyperthermia may assist in heat dissipation. The sympatho-

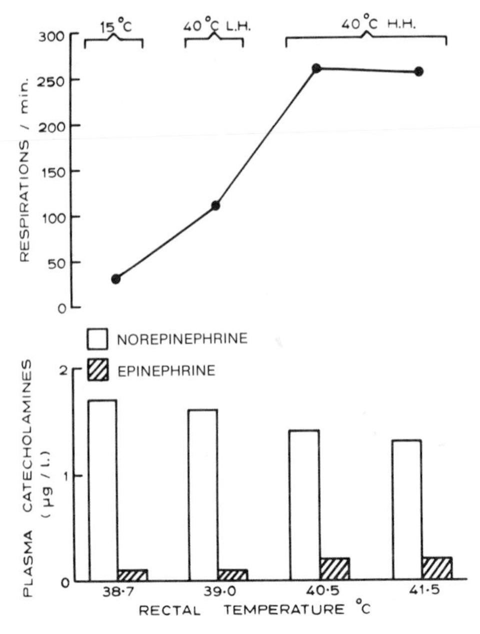

Figure 6. Plasma epinephrine and norepinephrine concentrations measured at various rectal temperatures in a splanchnicotomized animal. L. H., low humidity; H. H., high humidity. The epinephrine concentrations measured at rectal temperatures of 38.7°C and 39°C are not true values because they represent values which are less than the sensitivity of the assay. (Reproduced from Robertshaw and Whittow (35) by courtesy of J. Physiol. (Lond.).)

adrenal response may be related more to the cardiovascular adjustments necessary during hyperthermia than to enhancement of heat loss. Whittow (39) has demonstrated a decrease in peripheral resistance during hyperthermia, and circulating catecholamines may assist in the increase in cardiac output necessary to maintain blood pressure.

The evaporative heat loss mechanisms commonly found in mammals are panting and sweating.

Panting

There seems to be little evidence to suggest that the normal resting secretion of the adrenal medulla plays any part in the panting of a heat-exposed animal, although this fact has never been critically tested and panting in a splanchnicotomized animal appears to be normal (29). However, during hyperthermia panting changes from the typical rapid, shallow form to a slower, deeper respiration. The rapid, shallow panting is largely confined to movements of the diaphragm, but the intercostal muscles become involved with the slower, deeper form of respiration (40). The fact that the increase in circulating catecholamines is coincident with the change in respiration, coupled with the observation of Joels (25) that an elevation of either circulating epinephrine or norepinephrine will stimulate the chemoreceptors, suggests a possible causal relationship. However, because splanchnicotomized animals are able to show the change in pattern of respiration during hyperthermia and because the change in respiration from rapid and shallow to slow and deep can be induced by a combination of heat exposure and localized heating of the hypothalamus without alteration in circulating catecholamines (35), it would appear that elevated circulating catecholamines are not responsible for or contribute little to the alteration in panting associated with hyperthermia.

Sweating

Because it has been shown that heat exposure without hyperthermia does not cause stimulation of the adrenal medulla, it might seem an unnecessary exercise to speculate upon its role in the control of sweating. However, other situations occur in which the adrenal gland is stimulated either physiologically or pathologically, and it is under these circumstances that an effect on the sweat glands might be demonstrated. Furthermore, there are two types of sweat glands, apocrine and eccrine, and they show considerable species variation in their relative distribution on the body surface. The apocrine glands are associated with hair follicles; therefore, in species with a poor hair covering (e.g., man), their distribution is confined to the hair-covered areas of the skin (e.g., the axilla). The skin of man is, therefore, covered with the eccrine type of gland, and it is this gland which has a thermoregulatory function. In animals with a hairy skin, which involves most other species, the glands of the general body surface are apocrine, eccrine glands being confined to the glabrous areas, such as the footpads. In some species (e.g., horse and ox), the apocrine glands have a thermoregulatory function; in others, such as man and dog, apocrine sweat glands have, as yet, an undetermined function. The difference between apocrine and eccrine glands is not only anatomical, but physiological, apocrine glands being adrenergic and eccrine glands cholinergic (41).

Apocrine Sweat Glands Because the apocrine glands are adrenergic and contain adrenergic receptors, it is possible that the circulating catecholamines may be important in their function. As previously indicated, this was first suggested by Evans et al. (32), who examined sweat gland function of the horse. They postulated that, because the glands are much more responsive to epinephrine than to norepinephrine and because they were unable to demonstrate a response to cutaneous sympathetic nerve stimulation, sweating in this species must be entirely under the control of circulating catecholamines. However, since Robertshaw and Whittow (35) had demonstrated that heat exposure does not increase circulating catecholamines in the ox, which also possesses adrenergic apocrine sweat glands, then either the horse possesses a different type of sweat gland mechanism or sweating is controlled by local changes in cutaneous blood flow which thereby delivers varying amounts of the hormone to the secretory cells of the gland. The possibility that sweating in the ox is controlled entirely by a humoral substance has been investigated by the simple procedure of sectioning the nerves to the skin and examining the activity of the sweat glands of the denervated area to heat exposure. If the glands are controlled by circulating epinephrine, denervation would have no effect on sweat gland function. Findlay and Robertshaw (29) showed that denervation of the skin inhibited sweating from the denervated area, thereby demonstrating that an intact nerve supply was essential. Furthermore, because adrenal medullary catecholamines did not increase during heat exposure and because sweating appeared to be normal in splanchnicotomized animals, it was deduced that circulating catecholamines were either of little significance or played only a small role in the sweat gland response to heat exposure. The question remains, however, as to whether or not the increase in circulating epinephrine levels during hyperthermia may summate with the effect produced by neural stimulation. The sweat gland response to heat exposure attains maximal rates of secretion before hyperthermia reaches a level sufficiently high to elevate circulating catecholamines; intravenous infusion of epinephrine at the time of maximal stimulation fails to increase sweat rate any further (D. Robertshaw, unpublished observations). Under these circumstances, therefore, circulating epinephrine would be unlikely to have any stimulant action on the already maximally stimulated sweat glands. These results in the ox on the control of apocrine sweating led Robertshaw and Taylor (42) to re-examine in more detail the control of sweating in the horse. They showed that section of the nerves to the skin abolished heat-induced sweating and concluded that, as in the ox, the sweat gland response to heat exposure was not mediated by the adrenal medulla. They then examined the control of sweating during exercise, a situation which combines both a heat load and adrenal medullary stimulation. The heat of the increased metabolism of exercise imposes a load on the body which, like an external heat load, must be either dissipated or stored. They found that in the horse epinephrine released during exercise could stimulate denervated glands and may, therefore, supplement the sweating due to neural stimulation. The proportion of the sweat gland response to exercise that is due to neural stimulation and that which is due to humoral stimulation could not be determined because denervated glands demonstrated denervation supersensitivity. The partitioning of sweating into its neural and humoral components could be attempted, however, in the ox, given the problems involved in inducing exercise in animals of the bovine species, because the sweat glands of this species do not show denervation supersensitivity (29). The exact role, therefore, of circulating epinephrine in sweating from apocrine sweat glands is undetermined as yet.

Eccrine Sweat Glands As stated above, eccrine sweat glands on the general body surface of man and primates are cholinergic in nature; they respond to the local administration of acetylcholine and cholimimetic drugs and can be inhibited by atropine. Early observations that the sweat glands would also respond to the localized administration of catecholamines (43) suggested that either the stimulus is nonspecific and has no physiological significance or that the glands have a dual innervation, i.e., both cholinergic and adrenergic. On the other hand, they may possess receptors which are not innervated and are specifically stimulated by blood-borne catecholamines. Allen and Roddie (44) have shown that the intravenous infusion of both epinephrine and isoprenaline increases sweat production and that this still occurs after cholinergic blockade by atropine.

Sweating is a marked feature in certain pathological conditions associated with elevated plasma catecholamines. Thus, in pheochromocytomas, sweating is observed during paroxysms of hypertension, and in insulin hypoglycemia, a situation which induces increased epinephrine secretion, generalized sweating is a common feature. Prout and Wardell (45) have shown that sweating in patients with pheochromocytoma can be blocked by the local application of hyoscine, a substance with an atropine-like action. Likewise, the sweating of insulin hypoglycemia would appear to be different from that of isoprenaline or epinephrine infusion because insulin-induced sweating can be blocked by atropine (46). Furthermore, sweating is known to occur during insulin hypoglycemia in adrenalectomized subjects (47), thus ruling out the adrenal medulla as the source of the stimulus. Thus, it has been suggested that the effect of catecholamines on cholinergic eccrine glands has no physiological or pathological significance and that the action of sympathomimetic compounds on sweating simply represents a phylogenetic vestige of an earlier form of sudomotor transmission (48). This sort of explanation, however, is a convenient way of dismissing a problem that has no obvious solution, in the same way that the human appendix is blithely labeled as a redudant organ which can be readily dispensed with. Experiments have shown that the receptors at which both acetylcholine and epinephrine act are quite distinct, blockade of one receptor having no effect on the responses to the other agonist (49). Furthermore, there are specific adrenergic receptors present, which have been variously identified as α (49, 50) and β receptors (44). Their location has not been identified, i.e., they may be either pre- or postsynaptic. The presence of an adrenergic receptor system would suggest a specific role for catecholamines, whether they be locally released by adrenergic sudomotor nerves or conveyed in the circulation. Even if the adrenergic influence on cholinergic sweating is considered to be of vestigial significance, there may be situations in which the sweat glands can be physiologically stimulated by catecholamines. Sonnenschein et al. (49) have shown that epinephrine has a potentiating effect on acetylcholine-induced sweating, and Kennard (51) has suggested that epinephrine can reduce the threshold of sweat glands so that they are able to respond more readily to neural stimulation. Kuno (52) suggested that epinephrine released during exercise may have a stimulant effect on sweating and thereby assist in the dissipation of the additional heat load. These findings, together with the fact that circulating epinephrine has been shown to play a role in the stimulation of apocrine glands during exercise, led Robertshaw et al. (53) to test this hypothesis by using the stump-tailed macaque (*M. speciosa*) as a model for man. In this species, sweating is, as in man, under cholinergic control and is the main mechanism of evaporative heat loss. The workers showed that sweating of resting animals produced by heat exposure reached a maximal level in an environment of 40°C; at temperatures in excess of this, hyperthermia developed. However, when animals were exercised on a treadmill, there was a

50% increase over the maximal sweat rate achieved at rest. If these animals were then splanchnicotomized, this increment in sweating was abolished but could be restored by epinephrine infusion (Figure 7). They concluded, therefore, that circulating catecholamines released from the adrenal medulla were able to enhance sweating resulting from neural stimulation and thereby assist in the dissipation of the heat load of exercise.

In summary, circulating catecholamines appear to potentiate sweating in species with apocrine, adrenergic thermoregulatory sweat glands, as well as those with cholinergic eccrine glands. This is of particular significance in assisting the dissipation of the heat of exercise and can be considered part of the general adjustments to exercise that are brought about by circulating catecholamines.

Sweating in Disease States As indicated, the sweating noted in pheochromocytoma appears not to be due to the direct effect of catecholamines on the sweat glands, but is the result of neural stimulation. Although the central nervous system is protected from the effects of circulating catecholamines by

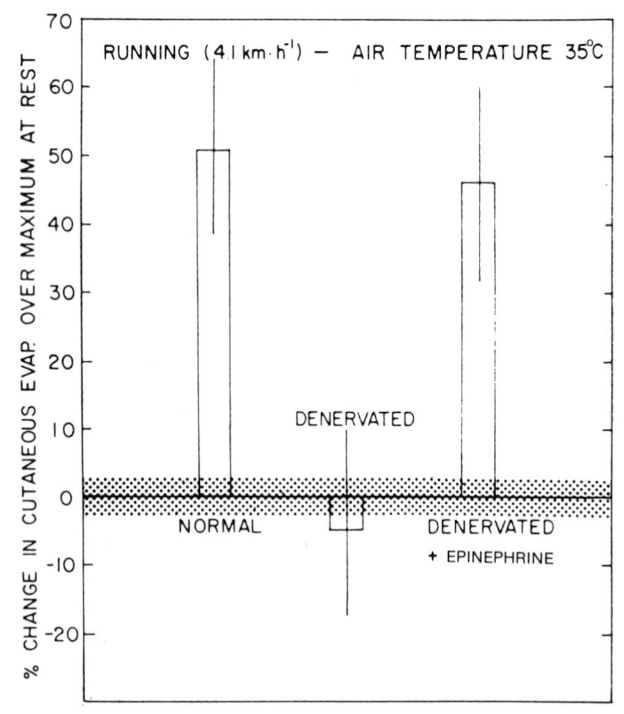

Figure 7. Enhancement of maximal sweating rates during exercise, its abolition by splanchnicotomized animals. Stump-tailed macaques (*M. speciosa*) ran on a treadmill at 4.1 km/hr at an air temperature of 35°C. Sweating rate while running is compared to the maximal sweat rate observed in the heat and is represented as zero on the graph (± twice the standard error as shown by the *hatched area*). *Vertical bars* represent the mean ± twice the standard error of 6–14 measurements under each condition. (Reproduced from Robertshaw, Taylor, and Mazzia (53) by courtesy of Am. J. Physiol.)

the blood-brain barrier, there must be means by which circulating catecholamines can affect central sudomotor mechanisms. At high concentrations of catecholamines, the barrier to epinephrine can be overcome (54), although it is not clear whether the concentrations of circulating epinephrine found during paroxysms observed in pheochromocytoma reach levels which might pass through the blood-brain barrier. The barrier is quantitatively similar for both epinephrine and norepinephrine, and there appears to be no difference in the sweat response from either epinephrine- or norepinephrine-secreting tumors, the degree of sweating being better related to the magnitude of the hypertension (W. S. Peart, personal communication). This is further evidence to support the contention that the sweating observed in cases of pheochromocytoma is of central and not peripheral origin. It has been shown that the main portion of the brain that is permeable to circulating catecholamines, other than the pituitary and pineal gland, which are not protected by the blood-brain barrier, is the hypothalamus (54). The uptake in the hypothalamus may be confined to the median eminence, which is, in reality, part of the pituitary and therefore devoid of any barrier (55). If the concentration of plasma catecholamines, whether they be epinephrine or norepinephrine, is sufficient to penetrate the blood-brain barrier and gain access to synapses within the hypothalamus, they are likely to stimulate adrenergic receptors within the brain and evoke responses which are normally mediated centrally by catecholamines. Evidence has been produced that the central control of thermoregulation is mediated at some point by an adrenergic synapse. In primates, the local administration of catecholamines into the hypothalamus is known to activate heat loss mechanisms (56). These mechanisms would include sweating and cutaneous vasodilation. Vasodilation, however, does not occur in pheochromocytoma. In fact, sweating is associated with a vasoconstricted skin, the vasoconstriction probably resulting from a direct action of circulating catecholamines, particularly epinephrine on the cutaneous vasculature.

The most likely cause of sweating would be retention of heat caused by cutaneous vasoconstriction coupled with an increase in metabolism brought about by the thermogenic action of catecholamines. Evidence against this theory of the etiology of sweating is that there is no increase in rectal temperature associated with the sweating and that sweating is stimulated within a few seconds of the onset of a hypertensive paroxysm. Furthermore, intravenous infusions of epinephrine or norepinephrine do not produce sweating (57). These arguments are not entirely convincing. Rectal temperature is a poor index of deep body temperature under transient conditions and only responds slowly. A subject in an environment close to the upper end of the thermoneutral zone might show sweating in response to a sudden reduction in convective heat loss brought about by cutaneous vasoconstriction. The thermoregulatory system of man is controlled with such precision that sweating would be rapid in onset, before any rise in rectal temperature could be detected, and, as indicated below, would apparently be "profuse." Likewise, the argument that intravenous infusions of catecholamines do not elicit sweating may not be valid because the infusion rate may not be comparable to the rate of secretion from tumor tissue, any attempt to increase the dose still further being obviously hazardous. The nature of sweating during hypertensive paroxysms of pheochromocytoma remains unresolved, but may be caused by substances other than catecholamines released from the tumor which gain access to the hypothalamic thermoregulatory centers and activate the sweat gland mechanism.

The sweating associated with insulin hypoglycemia and hypotension, whether caused by hemorrhage or syncope, both potent stimuli for the secretion of catecholamines, may be different from that observed in pheochromocytoma. Vogt (58) has argued that the maximal adrenal secretion evoked by both these stimuli is insufficient to raise plasma catecholamine levels to a level sufficient to penetrate the blood-brain barrier. Furthermore, because sweating is known to occur during insulin hypoglycemia of adrenalectomized subjects (47), circulating catecholamines of adrenomedullary origin play little or no part in the sweat gland response to hypoglycemia. One is led to the conclusion, therefore, that the sweat gland response to hypoglycemia, hemorrhage, or hypotension is part of a general sympathetic stimulation which includes sweating but is not caused by elevated levels of circulating catecholamines. As in pheochromocytoma, sweating occurs from a vasoconstricted skin caused either by neural or humoral stimulation—the well known "cold-sweat" response. With a low skin blood flow, there can only be minimal evaporation, much of the sweat accumulating on the skin surface. This gives the impression that sweating is "profuse." In fact it may be no greater than exposure to moderate heat when, with a high skin blood flow, evaporation may equal sweat secretion and result in little or no surface accumulation. It is difficult to understand the beneficial effects of hypoglycemic or hypotensive sweating. On the other hand, the benefit of sweating to prepare for "fight or flight" is readily appreciated. It may be that the sympathetic nervous system has not evolved sufficiently to discriminate between various stimuli and match the response accordingly. Some responses may, therefore, seem inappropriate.

CONCLUSIONS

The case is presented that an increased adrenal medullary secretion represents part of the normal response to cold exposure and that this, in some as yet undefined way, promotes either heat production or heat conservation. In contrast, there appears to be no adrenal medullary component to the physiological response to heat exposure unless the heat exposure is sufficiently intense to produce hyperthermia. If, however, the adrenal medulla is stimulated by exercise, then the released secretion appears to be able to assist the heat loss mechanisms, particularly sweating, in dissipating the heat of exercise. The sweating produced by pathological situations in which there are increased levels of circulating catecholamines does not appear to be due to a direct action of the catecholamines on the sweat glands.

REFERENCES

1. Campuzano, H. C., Wilkerson, J. E., and Horvath, S. M. (1975). Fluorometric analysis of epinephrine and norepinephrine. Anal. Biochem. 64: 578.
2. Passon, P. G., and Peuler, J. D. (1973). A simplified radiometric assay for plasma norepinephrine and epinephrine. Anal. Biochem. 51:618.
3. Vogt, M. (1952). The secretion of the denervated adrenal medulla of the cat. Br. J. Pharmacol. 7:325.
4. Lands, A. M., Arnold, A., McAuliff, J. P., Luduena, F. P., and Brown, T. G. (1967). Differentiation of receptor systems activated by sympathomimetic amines. Nature 214:597.
5. Leduc, J. (1961). Catecholamine production and release in exposure and acclimation to cold. Acta Physiol. Scand. (Suppl.) 183:5.
6. Andersson, B., Gale, C. C., Hökfelt, B., and Ohga, A. (1963). Relation of the preoptic temperature to the function of the sympathico-adrenomedullary system and the adrenal cortex. Acta Physiol. Scand. 61:182.
7. Webster, A. J. F., Heitman, J. H., Hays, F. L., and Olynyk, G. P. (1969). Catecholamines and cold thermogenesis in sheep. Can. J. Physiol. Pharmacol. 47:719.
8. Himms-Hagen, J. (1975). Role of the adrenal medulla in adaptation to cold. *In* Handbook of Physiology, Endocrinology, Vol. VI, Adrenal Gland, pp. 637–665. H. Blaschko, G. Sayers, and A. D. Smith (eds.), American Physiological Society, Washington, D.C.
9. Ring, G. C. (1942). The importance of the thyroid in maintaining an adequate production of heat during exposure to cold. Am. J. Physiol. 137:582.
10. Manara, L., Costa, E., Stern, D. N., and Maickel, R. P. (1965). Effect of chemical sympathectomy on oxygen consumption by the cold-exposed rat. Int. J. Neuropharmacol. 4:301.
11. Maickel, R. P., Matussek, N., Stern, D. N., and Brodie, B. B. (1967). The sympathetic nervous system as a homeostatic mechanism. 1. Absolute need for sympathetic nervous function in body temperature maintenance of cold-exposed rats. J. Pharmacol. Exp. Ther. 157:103.
12. Alexander, G., and Williams, D. (1968). Shivering and non-shivering thermogenesis during summit metabolism in young lambs. J. Physiol. (Lond.) 198:251.
13. Cottle, W. H., and Carlson, L. D. (1956). Regulation of heat production in cold-adapted rats. Proc. Soc. Exp. Biol. Med. 92:845.
14. Webster, A. J. F. (1974). Physiological effects of cold exposure. *In* MTP International Review of Science, Physiology Series 1, Vol. 7, Environmental Physiology, pp. 33–69. D. Robertshaw (ed.), Butterworths, London.
15. Himms-Hagen, J. (1969). The role of brown adipose tissue in the calorigenic effect of adrenaline and noradrenaline in cold-acclimated rats. J. Physiol. (Lond.) 205:393.
16. Alexander, G., Barnett, J. W., and Gemmell, R. T. (1975). Brown adipose tissue in the newborn calf (*Bos taurus*). J. Physiol. (Lond.) 244:223.
17. Jenkinson, D. McE., Noble, R. C., and Thompson, G. E. (1968). Adipose tissue and heat production in the newborn ox (*Bos taurus*). J. Physiol. (Lond.) 195:639.
18. Hayward, J. S., and Ball, E. G. (1966). Quantitative aspects of brown adipose tissue thermogenesis during arousal from hibernation. Biol. Bull 131:94.

19. Himms-Hagen, J. (1976). Cellular thermogenesis. Annu. Rev. Physiol. 38:315.
20. Hull, D., and Segall, M. M. (1965). The contribution of brown adipose tissue to heat production in the newborn rabbit. J. Physiol. (Lond.) 181:449.
21. Stoner, H. B. (1973). The role of the liver in non-shivering thermogenesis in the rat. J. Physiol. (Lond.) 232:285.
22. Cannon, W. B., Newton, H. F., Bright, E. M., Menken, V., and Moore, R. M. (1929). Some aspects of the physiology of animals surviving complete exclusion of sympathetic nerve impulses. Am. J. Physiol. 89:84.
23. Hayward, J. N., and Baker, M. A. (1969). A comparative study of the role of the cerebral arterial blood in the regulation of brain temperature in five mammals. Brain Res. 16:417.
24. Mott, J. C. (1963). The effects of baroreceptor and chemoreceptor stimulation on shivering. J. Physiol. (Lond.) 166:563.
25. Joels, N. (1975). Reflex respiratory effects of circulating catecholamines. *In* Handbook of Physiology, Endocrinology, Adrenal Gland, Vol. VI, pp. 491–505. H. Blaschko, G. Sayers, and A. D. Smith (eds.), American Physiological Society, Washington, D.C.
26. Gilgen, A., Maickel, R. P., Nikodijevic, O., and Brodie, B. B. (1962). Essential role of catecholamines in the mobilization of free fatty acids and glucose after exposure to cold. Life Sci. 1:709.
27. Andersson, B., Ekman, L., Hökfelt, B., Jobin, M., Olsson, D., and Robertshaw, D. (1967). Studies on the importance of the thyroid and the sympathetic nervous system in the defence to cold of the goat. Acta Physiol. Scand. 69:111.
28. Thompson, G. E., Robertshaw, D., and Findlay, J. D. (1969). Adrenergic innervation of the arrectores pilorum muscles of the ox (*Bos taurus*). Can. J. Physiol. Pharmacol. 47:309.
29. Findlay, J. D., and Robertshaw, D. (1965). The role of the sympatho-adrenal system in the control of sweating in the ox (*Bos taurus*). J. Physiol. (Lond.) 179:258.
30. Celander, O. (1954). The range of control exercised by the sympathico-adrenal system. Acta Physiol. Scand. (Suppl.) 116:1.
31. Graham, J. M., and Keatinge, W. R. (1972). Differences in sensitivity to vasoconstrictor drugs within the wall of the sheep carotid artery. J. Physiol. (Lond.) 221:447.
32. Evans, C. L., and Smith, D. F. G. (1956). Sweating responses in the horse. Proc. R. Soc. Lond. (Biol.) 145:61.
33. Hartman, F. A., and Hartman, W. B. (1923). Influence of temperature changes on the secretion of epinephrine. Am. J. Physiol. 65:612.
34. Symbas, P. N., Jellinek, M., Cooper, T., and Hanlon, C. (1964). Effect of hyperthermia on plasma catecholamines and histamine. Arch. Int. Pharmacodyn. Ther. 150:132.
35. Robertshaw, D., and Whittow, G. C. (1966). The effect of hyperthermia and localized heating of the anterior hypothalamus on the sympatho-adrenal system of the ox (*Bos taurus*). J. Physiol. (Lond.) 187:351.
36. Proppe, D. W., and Gale, C. C. (1970). Endocrine thermoregulatory responses to local hypothalamic warming in unanesthetized baboons. Am. J. Physiol. 219:202.
37. Hales, J. R. S., and Findlay, J. D. (1968). Respiration of the ox: normal values and the effects of exposure to hot environments. Respir. Physiol. 4:333.
38. Tamura, K. (1968). Effect of metabolic alkalosis on adrenal medullary secretion in the dog. Tohoku J. Exp. Med. 95:403.

39. Whittow, G. C. (1965). The effect of hyperthermia on the systemic and pulmonary circulation of the ox (*Bos taurus*). Q. J. Exp. Physiol. 50:300.
40. Hales, J. R. S. (1973). Effects of heat stress on blood flow in respiratory and non-respiratory muscles in the sheep. Pfluegers Arch. 345:123.
41. Robertshaw, D. (1975). Catecholamines and control of sweat glands. *In* Handbook of Physiology, Endocrinology, Adrenal Glands, Vol. VI, pp. 591–603. H. Blaschko, G. Sayers, and A. D. Smith (eds.), American Physiological Society, Washington, D.C.
42. Robertshaw, D., and Taylor, C. R. (1969). Sweat gland function of the donkey (*Equus asinus*). J. Physiol. (Lond.) 205:79.
43. Wada, M. (1950). Sudorific action of adrenaline on the human sweat glands and determination of their excitability. Science 111:376.
44. Allen, J. A., and Roddie, I. C. (1972). The role of circulating catecholamines in sweat production in man. J. Physiol. (Lond.) 227:1972.
45. Prout, B. J., and Wardell, W. M. (1969). Sweating and peripheral blood flow in patients with phaeochromocytoma. Clin. Sci. 36:109.
46. Chalmers, T. M., and Keele, C. A. (1952). The nervous and chemical control of sweating. Br. J. Dermatol. 64:43.
47. Ginsburg, J., and Paton, A. (1956). Effects in man of insulin hypoglycemia after adrenalectomy. J. Physiol. (Lond.) 133:59P.
48. Kuno, Y. (1956). Human Perspiration. Thomas, Springfield, Illinois.
49. Sonnenschein, R. R., Kobrin, H., Janowitz, H. D., and Grossman, M. I. (1951). Stimulation and inhibition of human sweat glands by intradermal sympathomimetic agents. J. Appl. Physiol. 3:573.
50. Chalmers, T. M., and Keele, C. A. (1951). Physiological significance of the sweat response to adrenaline in man. J. Physiol. (Lond.) 114:510.
51. Kennard, D. W. (1963). The nervous regulation of the sweating apparatus of the human skin, and emotive sweating in thermal sweating areas. J. Physiol. (Lond.) 165:457.
52. Kuno, Y. (1965). The mechanism of human sweat secretion. Proc. Int. Congr. Physiol. Sci. XXIII, Tokyo 5:8.
53. Robertshaw, D., Taylor, C. R., and Mazzia, L. M. (1973). Sweating in primates: role of secretion of the adrenal medulla during exercise. Am. J. Physiol. 224:678.
54. Whitby, L. G., Axelrod, J., and Weil-Malherbe, H. (1961). The fate of H^3-norepinephrine in animals. J. Pharmacol. 132:193.
55. Loizou, L. A. (1970). Uptake of monoamines into central neurones and the blood-brain barrier of the infant rat. Br. J. Pharmacol. 40:800.
56. Myers, R. D., and Yaksh, T. L. (1969). Control of body temperature in the unanesthetised monkey by cholinergic and aminergic systems in the hypothalamus. J. Physiol. (Lond.) 202:483.
57. Barcroft, M., and Swan, H. J. C. (1953). Sympathetic Control of Human Blood Vessels, p. 165. Arnold, London.
58. Vogt, M. (1975). Influence of circulating catecholamines on the central nervous system. *In* Handbook of Physiology, Endocrinology, Adrenal Gland, Vol. VI, pp. 667–668. H. Blaschko, G. Sayers, and A. D. Smith (eds.), American Physiological Society, Washington, D.C.

International Review of Physiology
Environmental Physiology II, Volume 15
Edited by David Robertshaw

7 Physiological Responses and Adaptations to High Altitude

S. LAHIRI
University of Pennsylvania,
Philadephia, Pennsylvania

REGULATION OF PULMONARY VENTILATION 219
Initiation of Chemoreflexes 219
Hypoxic Threshold for Ventilation 220
Peripheral Versus Central Drive 220
Cerebral Acidosis 221
Central Sensitization 221
Residual Ventilation 222
Role of Carbon Dioxide 222
Acid-Base Homeostasis 223
Role of Cerebrospinal Fluid 223
Role of Cerebral Blood Flow and Metabolism 223
Hypoxic Sensitivity 224
Euoxic Ventilatory Drive in High Altitude Natives 224
Effect of Lifelong Hypoxia at Sea Level 226
Environmental Versus Genetic Effect 226
Loss of Hypoxic Drive 227
Reversibility of Blunted Hypoxic Drive 227
Animal Model 228
Role of Neonatal Hypoxia 229
Effect of Erythrocythemia 231
Carotid Body and Chronic Hypoxia 231
Hypoxic Chemoreflex Versus Central Depression 231
Summary 232

This research was supported in part by Grant HL-08805 from the United States Public Health Service.

The abbreviations are: P_{IO_2}, partial pressure of O_2 in inspired air saturated at body temperature; P_{aO_2}, partial pressure of O_2 in arterial blood; $\dot{V}_I$, inspiratory air flow (ventilation) in liter min^{-1}; P_B, barometric pressure; P_{aCO_2}, partial pressure of CO_2 in arterial blood; P_{vO_2}, partial pressure of O_2 in venous blood; $P_{50,74}$, P_{O_2} (mmHg) at 50% saturation with O_2 at pH 7.4 at a specified temperature; $\dot{V}_{O_2}$, O_2 uptake in liters min^{-1} at standard temperature, pressure, dry; K_m, substrate concentration or pressure at half-maximal rate; BTPS, gas volume at body temperature, pressure, saturated with water vapor.

HYPOXIC EXERCISE 232

TEMPERATURE REGULATION: INTERACTION BETWEEN HYPOXIA AND COLD 233

ACUTE MOUNTAIN SICKNESS 233

MOUNTAIN DISEASES AND MOUNTAIN MEDICINE 235

SLEEP AND HYPOXIC RESPONSE 235

O_2 AND CO_2 TRANSPORT BY LUNGS AND BLOOD 236

HIGH ALTITUDE ANIMALS 238

PHYSICAL PERFORMANCE CAPACITY 239

CELLULAR AND SUBCELLULAR ADAPTATION 240

GENERAL CONCLUSION 242

At sea level, the O_2 stores of the human body can support life for only about 5 min at the resting metabolic rate when the O_2 supply is terminated (1). At high altitude, where inspired oxygen tension is lower, the O_2 stores are also decreased. Because resting metabolic rate does not decrease at altitude, the O_2 stores will last even less time than at sea level. This continuing threat of oxygen shortage is further compounded by the narrow pressure difference for O_2 between the air source and the sink in the mitochondrial material of the cells in which it is consumed. The body responds to this threat by making short-term and long-term adjustments, not only to ensure an adequate supply of O_2 to the tissues, but also to use O_2 at a normal rate. It must be stated at once that higher oxygen pressure is toxic for mammals (2, 3), and there is no known adaptation to it. Indeed, there is the notion that oxygen pressure at sea level may even be harmful, and nitrogen pressure may cause mild narcosis (4–7). The origin of life and the evolution of the earth's atmosphere seem to provide one line of evidence in support of this idea. Because oxygen is absent from volcanic effluents, it is believed that at the time of the origin of life the earth's atmosphere was oxygen poor. As molecular oxygen concentration was increased from the dissociation of water, complex cellular organisms evolved first in sea and then on land. The organisms which needed little or no oxygen were inhibited by its preponderance. It is well known that oxygen above the pressure at sea level proved harmful and decreased life span.

Emphasis in the previous review (8) was on adaptation in lifelong hypoxia versus chronic hypoxia of shorter duration. Interest has continued to grow in the area and has focused on the developmental effect of hypoxia, particularly in pulmonary function, O_2 transport, and physical performance. The idea is emerging that many of the physiological characteristics observed in the adults native to high altitude are acquired during a prolonged period of chronic hypoxia. The sojourners seem to acquire all the processes, including a normal reproductive ability, if exposed to hypoxia at a suitable age for a suitable length of time. In the area of cerebrospinal fluid (CSF) pH regulation and control of ventilation during acclimatization to high altitude, there is less certainty now than there was a few years ago.

This review is limited to homeothermic mammals. Successful survival of homeotherms depends upon homeostasis. Because homeostasis reflects integrative physiological control, which is the key to adaptative process, the theme of this review will be the control system. Much of the physiological control in mammals is accomplished through nervous and endocrine systems. At birth, these systems are not fully functional, and hence adaptation is not fully developed. It develops in man, for example, after birth, whereupon it remains fully functional until old age. With aging, speed of homeostasis declines, and failure of homeostasis results in disease and death. Thus, the ability to maintain homeostasis under hypoxic conditions is most certainly dependent upon the age of exposure to hypoxia. Although systematic data on many crucial aspects of the subject are not available, a meaningful trend can be traced and working hypotheses formed.

Although this chapter concerns hypoxia of high altitude, it is worth noting that environmental conditions other than P_{IO_2} change as one moves from sea level to high altitude. For example, a decrease in ambient temperature and humidity may assume great physiological significance, particularly in influencing physical performance. It is important to remember that subjects adapted to hypoxia of high altitude are also adapted to these other environmental conditions. The problem of hypoxia will be considered from these points of view: successive oxygen tensions in the transport system between atmospheric air and the mitochondrial system (ambient air→lungs→arterial blood→capillary blood→cells→mitochondria) and the control of reaction of oxygen at the sites in mitochondria.

This review is a personal assessment of this trend. Some areas, such as endocrine responses to acute and chronic hypoxia, have not been included, mainly because the little information that is available has been reviewed recently (9). During the last few years, several review articles on various aspects of hypoxic adaptation have appeared (9–11).

REGULATION OF PULMONARY VENTILATION

Initiation of Chemoreflexes

The ventilatory effect of acute hypoxia is mediated through the peripheral chemoreceptors within the carotid and aortic bodies in mammals. Complete denervation of all these structures abolishes the prompt ventilatory response to acute hypoxia in man and animals. Both whole nerve and single fiber preparations of the carotid sinus nerve of the cat have shown that chemoreceptor activity increases progressively as P_{aO_2} is decreased from a very high to a low value (12–15). There is no evidence to date that the peripheral chemoreceptor discharge for a given decrease in P_{aO_2} is diminished by chronic hypoxia. Åstrand (16) was the first to demonstrate that the carotid chemoreceptor afferent activity of the cat was normal after several days of chronic hypoxia at 4,000 m, and Hornbein and Severinghaus (17) reported that the hypoxic response of the carotid chemoreceptor afferent of cats born and raised at 4,000 m was not less than that of cats living at sea level.

Hypoxic Threshold for Ventilation

The ventilatory response to hypoxia seems to show a threshold value. In acute hypoxia, the ventilatory response obtained from steady state methods is barely measurable at a P_{aO_2} above 60 mmHg, yet a few days of chronic hypoxia of the same degree increases ventilation considerably. The question of what initiates this change during chronic hypoxia has not been answered satisfactorily. That mild hypoxia (P_{aO_2} ~ 60 mmHg) is a ventilatory drive can be shown by various techniques, e.g., by the transient "O_2 test" (18), by a combination of muscular exercise and hypoxia (19), and by hypoxia on a background of hypercapnia (20); these conclusions concerning ventilation also correspond to the continuous relation between chemoreceptor activity and P_{aO_2}. Consequently, the observed apparent ventilatory threshold for hypoxia does not correspond to a threshold for the peripheral chemoreceptors; there is a continuous increase of chemoreceptor activity as P_{aO_2} decreases below the sea level value. The question then remains: what prevents the increase in ventilation during mild acute hypoxia despite an increase in the activity of peripheral chemoreceptors? A plausible hypothesis is that chemoreceptor activity is too feeble in steady state mild acute hypoxia to overcome its secondary effects on ventilation. With the prolongation of the same hypoxia, central "sensitization" occurs which in turn increases the effect of chemoreceptor activity on ventilation.

Peripheral Versus Central Drive

The concomitant respiratory alkalosis during acute hypoxia limits the stimulatory effect of hypoxia on ventilation. Before the discovery of medullary chemosensitivity to CO_2-H^+, it was generally believed that arterial blood pH, after an initial rise, returned toward normal through renal compensation, and that this restored blood [H^+] stimulated ventilation until the full effect of hypoxia was reached. As the phenomenon of medullary chemosensitivity was unraveled (21), Kellogg (22) pointed out convincingly that ventilatory acclimatization preceded the return of arterial blood pH to normal, and Severinghaus et al. (23) subsequently reported that CSF pH returned to normal ahead of arterial blood pH during high altitude acclimatization to 3,880 m. They argued that since the medullary chemosensitive areas were readily influenced by the CSF pH the ventilation increase in the second phase was related to this prompt restoration of CSF pH, thereby substantiating Kellogg's original hypothesis. In a preliminary study, Severinghaus (24) extended these observations on acclimatization to human subjects who were primed with metabolic alkalosis and acidosis. He reported a prompt decrease in CSF [HCO_3^-] at high altitude, regardless of plasma [HCO_3^-]; plasma [HCO_3^-] was of little importance in regulating CSF [HCO_3^-]. More recently, Dempsey and Forster and their colleagues (25, 26) refuted the foregoing claims of Severinghaus and asserted that a significant alkalinity in CSF continued parallel to arterial blood in sojourners at 3,100 and 4,300 m as ventilation increased and P_{CO_2} fell. They concluded that a change in CSF [HCO_3^-] was critically dependent upon a change in plasma [HCO_3^-]. Thus, we have gone a full circle without a clearer understanding of the ventilatory acclimatization to chronic hypoxia. According to the current trend of thought on acid-base regulation of CSF (27), a decrease in CSF [HCO_3^-] during acclimatization to high altitude would not be expected if plasma [HCO_3^-] were maintained at sea level value by the administration of $NaHCO_3$. This line of evidence exists only in a preliminary report by Severinghaus (24). However, because the peripheral chemoreflex response seems to remain unchanged during early acclimatization, it may be concluded that the process of ventilatory acclimatization is a central phenomenon.

It is difficult to pinpoint the reason for the difference between the two results of the two groups (24–26). The data provided by Dempsey and his colleagues are more numerous than those of Severinghaus and his colleagues. The question still hinges on the crucial question of acid-base regulation of its own environment by the central nervous system itself, particularly under hypoxic condition.

Cerebral Acidosis

Under conditions of severe hypoxia, cerebral acidosis may occur, presumably because of an augmented anaerobic metabolism. This acidosis component has been demonstrated by exposing animals in which the arterial chemoreceptors have been surgically denervated to hypoxia (28). However, if this cerebral acidosis is gradually compensated in this preparation by an increase of medullary [HCO_3^-], then ventilatory drive (and consequently ventilation) may be expected to decrease with time, unlike the normal process of ventilatory response to chronic hypoxia. Observations of the author on the time course in the bilaterally carotid chemoreceptor-denervated goats exposed to 5,000 m for several weeks showed that both arterial blood and CSF acidity were no greater than at sea level despite a small decrease of [HCO_3^-] (29).

Central Sensitization

Having failed to identify CSF or arterial acid-base as a stimulus to ventilatory acclimatization, several investigators revived the concept of central sensitization originally proposed by Nielsen (30). According to this idea, a central arousal develops slowly during chronic hypoxia and also disappears slowly during deacclimatization. The phenomenon presumably also plays a role in the initiation of acclimatization and maintenance of breathing in newborns. But the evidence is only circumstantial.

It is relevant to mention at this point that animals and man after denervation of either carotid or of both aortic and carotid bodies at sea level usually increase their arterial P_{CO_2} and $[HCO_3^-]$ in blood because of a loss of ventilatory drive from the chemoreceptors (8). Our own data on goats and sheep showed gradual increases in P_{CO_2} and $[HCO_3^-]$ in arterial blood and CSF at sea level after bilateral carotid sinus nerve section; there was also a corresponding decrease in P_{aO_2}. This mild hypoxia, in the absence of carotid chemoreflexes, did not induce increases in ventilation as seen in normal animals exposed to an equivalent altitude, and is not consistent with the suggestion that ventilatory acclimatization may not involve the participation of arterial chemoreceptors. Also, central sensitization does not seem to occur in the absence of peripheral chemoreceptors. Peripheral chemoreceptor-dependent mesencephalic reticular activation as described by Hugelin et al. (31) may be a part of the mechanism.

Residual Ventilation

Acute removal of hypoxia after acclimatization is not followed by a prompt return of ventilation to the preacclimatization level (32). The stimulus for this residual ventilation is generally thought to reside in CSF and arterial $[H^+]$ (33). It has, however, been found that the residual ventilation in unanesthetized goats is not accounted for by CSF and arterial $[H^+]$, because the pH of the fluids was not more acidic than normal. Thus, the phenomenon seems to be related to central sensitization. Both central sensitization and residual ventilation seem to depend upon intact peripheral chemoreceptors.

Role of Carbon Dioxide

The fact that alveolar or arterial P_{CO_2} is low in the altitude-acclimatized sojourners and residents has been well documented (34, 35) since Fitzgerald (36), and it has repeatedly been found that the P_{CO_2} threshold for ventilatory stimulation is correspondingly decreased (22, 37, 38). It is generally found that ventilatory sensitivity to CO_2 (slope of the line relating ventilation ($\dot{V}_I$) to arterial P_{CO_2} (P_{aCO_2})) is increased in man acclimatized to high altitude (22, 37), although the reason for this is not clear. A decrease in CSF and blood $[HCO_3^-]$ as it occurs in acclimatized subjects would increase $[H^+]$ for a given P_{CO_2} increase and, therefore, stimulate ventilation, but an increase in P_{CO_2} is unlikely to have caused any greater increase in $[H^+]$ than at sea level. This reasoning is substantiated by the observation that the $\dot{V}_I$–P_{aCO_2} slope is not changed by mild metabolic acidosis in man (39), whereas $[HCO_3^-]$ is decreased to a similar degree. It is, however, important to note that CSF pH is better regulated in metabolic than in respiratory derangement. The two situations are, therefore, not strictly comparable. It seems that the increase in CO_2 sensitivity in acclimatized man is associated with hypoxia for some unknown reason. Such a notion is also supported by the observation that chronic hypoxia in man produces an extra ventilatory drive not accounted for by acid-base adjustment which is known to follow hypoxic hyperventilation (33, 40, 41).

On the other hand, altitude polycythemia increases the buffering power of blood and may, therefore, by itself decrease respiratory sensitivity to CO_2. But the slow time course of hemoglobin increase does not correspond to a rapid increase in CO_2 sensitivity in a few days of hypoxic exposure.

Acid-Base Homeostasis

Role of Cerebrospinal Fluid As alluded to previously, regulation of CSF pH during chronic hypoxic hypocapnia and its role in the regulation of ventilation are not well understood. Because of its physical properties, bicarbonate containing CSF is not as well buffered against carbonic acid as it is to metabolic acid-base changes. This difference between respiratory and metabolic disturbances as documented by various investigators has been reviewed recently (27). However, respiratory alkalosis increases metabolic acid accumulation in CSF because of the effect of pH on the glycolytic enzyme system, which in turn affords another line of defense against CSF alkalosis. Concomitant hypoxia perhaps enhances this effect indirectly (28).

According to the Henderson-Hasselbach equation, regulation of CSF pH is dependent upon a proportionate change in $[HCO_3^-]$ and P_{CO_2}, so that the $[HCO_3^-]$ to $[H_2CO_3]$ ratio remains near constant. Also, a better regulation of CSF pH compared to arterial blood means a change in the ratio of CSF and plasma $[HCO_3^-]$. To what extent this ratio changes during altitude acclimatization is not clear, although current data show that the change is considerably less than predicted from the metabolic acid-base derangement (27).

In recent years, considerable attention has been given to CSF pH regulation (27, 42–46). Current theories attribute the regulation to 1) active transport of HCO_3^-/H^+ (because of the electrochemical disequilibrium of these ions, CSF being 3–4 mV positive to plasma, HCO_3^- is either pumped out of or H^+ into the CSF); 2) passive distribution according to a continuous outflow of H^+ from brain tissue cells; and 3) the metabolic property of brain tissue, such as pH-dependent production of ammonia.

Role of Cerebral Blood Flow and Metabolism Acute hypoxia increases cerebral blood flow, but the concomitant respiratory alkalosis makes this increase self-limiting because acute hypocapnic alkalosis decreases the flow. Thus, normocapnic acute hypoxia results in a greater increase in blood flow than does hypocapnic acute hypoxia. Fencl et al. (47) investigated ventilation and cerebral blood flow (indirectly by arteriovenous O_2 and CO_2 differences) during acute arterial acid-base changes and concluded that both are a unique function of CSF pH. But the similarity of responses between ventilation and cerebral blood flow ends there. During acclimatization to hypoxia in man, cerebral blood flow diminishes (48, 49), whereas ventilation remains increased (32, 50). The decrease in cerebral blood flow, in spite of a less alkaline CSF and blood pH, is inconsistent with the notion of a synergistic control of cerebral blood flow and ventilation. The initial transient increase in cardiac output (51) during acclimatization may contribute to the initial increase in cerebral blood flow (47), but it is not supposed to do so in the face of autoregulation. It is also unlikely that a small increase in P_{aO_2} through hyperventilation should have caused this gradual but profound change in cerebral blood flow.

It is not known for certain how cerebral blood flow reacts to prolonged or lifelong hypoxia. The study of Milledge and Sørensen (49) on cerebral arterio-venous O_2 difference implied that it was low in high altitude natives (4,300 m) and was reduced further when they breathed 100% O_2. Roy (52), however, found a relative increase in cerebral blood flow in high altitude natives. Lahiri and Milledge (53) reported that the CSF pH in native residents of the Himalayas was closer to that of the lowlanders at sea level than to lowlanders acclimatized at high altitude. Sørensen and Milledge (54) reported a more acidic CSF pH in the Andean high altitude natives. More recently, Blayo et al. (55) measured acid-base of both cisternal and lumbar CSF and total cerebral blood flow in high altitude natives (3,500–4,800 m). They found that lumbar CSF pH was 7.330 compared to 7.357 of cisternal CSF and confirmed the findings of Lahiri and Milledge (53). They also found that the cerebral blood flow was low, whereas the cerebral oxygen uptake was similar to that at sea level. It has consistently been found that CSF lactate concentration is higher at high altitude, suggesting a metabolic change which seems to be associated with low CSF P_{O_2}. The greater acidity in CSF and arterial blood in the highlanders than in sojourners would constitute a greater component of their ventilatory drive if their respiratory sensitivity to CO_2–H^+ were the same as in the sea level natives.

Hypoxic Sensitivity

In the previous volume (8), it was recorded that the balance of evidence indicated that no important change in ventilatory hypoxic sensitivity occurred during early acclimatization to chronic hypoxia. The notion has been reiterated more recently (56). It has now been repeatedly confirmed that hypoxic ventilatory sensitivity in the adult natives of high altitude is blunted as was documented earlier.

Euoxic Ventilatory Drive in the High Altitude Natives

Although little or no ventilatory response to hypoxia is found in the high altitude natives (57–59) during conventional steady state testing, they do respond to the transient O_2 test during rest and exercise. In transient tests, alveolar P_{O_2} is momentarily changed by inhaling a breath or two of pure O_2, and the transient change in ventilation is recorded breath by breath. This prompt transient ventilatory response reflects activity of peripheral chemoreflex uncomplicated by secondary changes which follow in the course of time. They also lower ventilation as a result of O_2 inhalation during exercise. Thus, the high altitude natives do have a hypoxic drive, although apparently small (S. Lahiri, unpublished observations).

Not only do high altitude natives have a greater hypoxic stimulus by virtue of their lower arterial P_{O_2} at a given high altitude, but also a greater central stimulus as a result of a more acidic CSF pH than in lowlanders acclimatized to the same altitude. They also show a lower cerebral blood flow, but a similar cerebral O_2 uptake.

Sørensen and Milledge (54) proposed that the low CSF pH was sufficient to account for the ventilatory acclimatization of the high altitude natives in the absence of a functioning hypoxic drive from peripheral chemoreceptors. Sørensen (28) substantiated this contention by showing a similar decrease in CSF $[HCO_3^-]$ in intact and bilaterally chemoreceptor-denervated rabbits after exposure to hypoxia (P_B = 470 mmHg) for 24 hr, and suggested that this cerebral acidosis, which may not be compensated, provides a persistent respiratory stimulus. This hypothesis contradicts the one put forward by Severinghaus et al. (23) proposing that it is the active regulation of CSF pH from alkaline to its normal acidity which promotes ventilatory acclimatization to chronic hypoxia. The weakness of the hypothesis which purports that intact peripheral chemoreceptors are not essential to the ventilatory acclimatization to hypoxia is exposed by the following argument. If a given ambient hypoxia causes the same ventilatory acclimatization in normal and chemoreceptor-denervated animals, it follows that they will have the same arterial P_{O_2}, which is expected to cause the same cerebral acidosis. It seems, however, that central excitation potentiates the peripheral chemoreflex effect of hypoxia on ventilation. Accordingly, intact animals will achieve a greater ventilation and consequently a greater arterial P_{O_2} in response to the same hypoxic environment. Thus, intact and peripheral chemoreceptor-denervated animals cannot achieve the same ventilatory acclimatization at the same ambient hypoxia. It seems that Sørensen's interpretation of reduction of CSF $[HCO_3^-]$, taken alone as a measure of ventilatory acclimatization in his two preparations, may not be entirely valid. Recently, Bouverot and his colleagues (60, 61), emphasized the important role of carotid chemoreceptors in hypoxic acclimatization and contradicted Sørensen. Our own observations (29) on the effects of CSF and arterial acid-base on ventilation during acclimatization to hypoxia in goats before and after carotid chemodenervation are consistent with the conclusion of Bouverot. Gilfilan et al. (62) had previously indicated a similar importance for peripheral chemoreceptors. The important point is that highlanders do not respond in the same way as did Sørensen's denervated rabbits; highlanders do hypoventilate relative to sojourners.

It is well documented that at high altitude, native subjects hyperventilate in steady states while breathing O_2 at sea level P_{IO_2} (63, 64). The mechanism of this phenomenon is unclear. However, Sørensen and Milledge (54) found that in high altitude natives breathing 100% O_2 an increased cerebral arteriovenous O_2 difference occurred, suggesting a decrease in cerebral blood flow. This might have increased central $P_{CO_2}-H^+$ stimulus which in balance stimulated ventilation, because the contribution of hypoxic drive was small to start with. An investigation of the medullary acid-base metabolism and blood flow during transition from hypoxia to hyperoxia would be of great interest.

Of interest are the observations that the P_{aO_2} at which ventilation is stimulated is lower in the highlanders than in the lowlanders and that the degree of hypoxic insensitivity is directly proportional to the level of chronic hypoxia.

This relationship seems to be reminiscent of the Weber-Fechner law, which states that the intensity of response is proportional to the logarithm of the intensity of the stimulus.

The difference between highlander and lowlander in hypoxic response may lie in the peripheral chemoreceptors as well as in the central nervous system. In order to localize the difference, an animal model would be useful.

Effect of Lifelong Hypoxia at Sea Level

Husson and Otis (65) previously compared the effect of right to left shunt hypoxia and altitude hypoxia in lowlanders and provided evidence that lifelong hypoxia caused an increased P_{aCO_2} . Inspection of these and other data on cyanotic subjects at sea level and comparison with those of high altitude natives have suggested (34, 66) that adult cyanotic subjects do hypoventilate for a comparable degree of hypoxemia as do the high altitude natives. Accordingly, Sørensen and Severinghaus (67) tested hypoxic sensitivity in patients with tetralogy of Fallot after surgical correction and found that normal sensitivity was not recovered even after 1–7 years of normoxia. These results agreed with the idea of an irreversibly blunted hypoxic response in high altitude natives. Edelman et al. (68) studied subjects with cyanotic congenital heart disease (10–37 years old) with the use of transient N_2 tests and found that the subjects were similar to high altitude natives. But, several months after surgical correction, their hypoxic sensitivity had returned to that of the normal sea level natives. These observations have been confirmed more recently (69), showing that blunted hypoxic sensitivity has developed as early as 8 years of age in subjects with congenital cyanotic heart disease. The diminished response was partially reversed after corrective surgery. These observations show that, while lifelong hypoxia causes blunting of respiratory response to hypoxia, it need not be irreversible. On the other hand, they do not necessarily mean that the phenomenon was the same as that found in high altitude natives.

Environmental Versus Genetic Effect

Another line of work consists of studying sea level natives who have been exposed to hypoxia for various durations at various ages. Compiling and analyzing the blood gas and alveolar gas data of adult human subjects, Lahiri (66) found that P_{aO_2} and P_{aCO_2} were similar in short-term and long-term residents at high altitudes, whereas P_{aO_2} was lower and P_{aCO_2} higher in high altitude natives. Only Chiodi's (70) data appear to suggest that this hypoventilation might also have developed in the migrant long-term residents, but uncertainty as to the ancestry of the subjects made it difficult to arrive at a definite conclusion. Weil et al. (71) suggested that the decrease in ventilatory sensitivity was a gradual process and was independent of the age of exposure.

Byrne-Quinn et al. (72) reported that the hypoxic sensitivity of children aged 9–10 years born in Leadville, Colorado, U. S. A. (3,100 m), was similar to that of a control group living in Denver, Colorado (1,600 m). They concluded that the loss of hypoxic sensitivity at high altitude occurs after a prolonged hypoxia. However, the children living at 3,100 m had a P_{aCO_2} level 2–3 mmHg higher than those at 1,600 m; in some, the P_{aCO_2} was 35–38 mmHg, which is in the range normally found at sea level. These data alone suggest that Leadville children hypoventilated in spite of hypoxia. However, the authors determined hypoxic response by the method of "isocapnic hypoxia" at a single P_{aCO_2} – higher (mean 34.3) for the Leadville group and lower (mean 32.8) for the Denver group. The use of a higher P_{aCO_2} for the Leadville group might have raised their response unduly for a valid comparison. In addition, the difference in altitude between 1,600 and 3,100 m was not large enough for a clear documentation of the phenomenon. Thus, further work is needed to come to a definitive conclusion.

Loss of Hypoxic Drive

In a recent expedition to Peru, the problem has been investigated at high altitude (Puno, 3,850 m; barometric pressure (P_B) = 484 torr), as well as at the sea coast (Tacna, 800 m; P_B = 715 torr) (73). Most young subjects (5–10 years) showed normal ventilatory response to hypoxia. Adolescent subjects and young adults showed a gradual decrease of response with age. All adults showed blunted response. These results were independent of the ancestry of the subjects. Offspring of high altitude natives born at the coastal region showed a normal ventilatory response from childhood to adulthood. Thus, it may be concluded that blunted hypoxic ventilatory drive in high altitude natives is an acquired phenomenon. The length of hypoxic exposure necessary to develop the phenomenon seems to depend upon the intensity of hypoxia. The Leadville (3,100 m) studies indicated that it developed after the age of 12–15 years. There are indications that at a higher altitude or with more severe hypoxemia it develops at an earlier age with a greater intensity (69). This information is summarized in Figure 1. Ventilatory sensitivity to hypoxia is shown as a function of duration of hypoxia at two different strengths (altitude). These strength-duration curves show that at higher altitude a) the process of losing hypoxic drive begins earlier in age, b) much of the loss also occurs at an early age, and c) the extent of the loss is greater.

How the loss of hypoxic drive relates to the age at which exposure occurs is an often-asked question. It has been observed that adult subjects from sea level exposed to 3,000–4,000 m for 5–7 years did not develop blunted response. This is not surprising in view of the fact that children up to 10 years of age at 3,850 m did not develop a diminished response. Adults exposed to hypoxia for a longer period seem to develop it, but the intensity of blunting seems to be less than for the high altitude natives. Thus, one is forced to conclude that blunting of hypoxic ventilatory drive can occur at any age, but younger subjects show it more fully, suggesting a developmental aspect of the phenomenon.

Reversibility of Blunted Hypoxic Drive

Lahiri and colleagues (74, 75) reported that the blunted hypoxic response in highlanders was not reversed by 6 weeks of residence at sea level. Nor was it

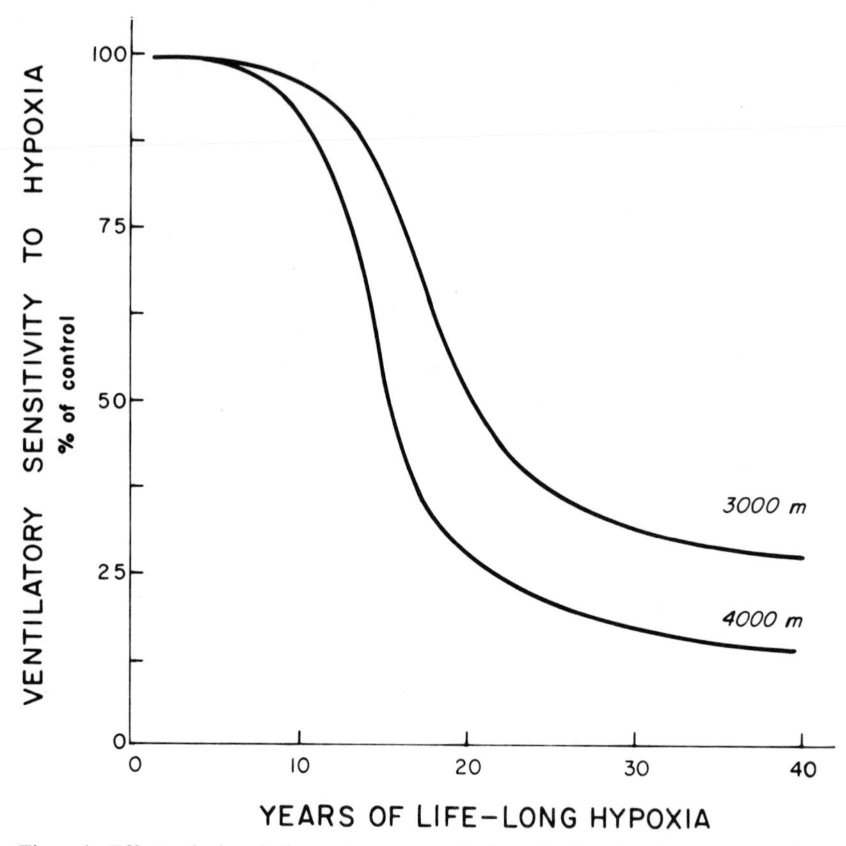

Figure 1. Effect of chronic hypoxia on control of ventilation. Ventilatory sensitivity to hypoxia is plotted as a function of lifelong hypoxia at 3,000 and 4,000 m with P_{aO2} = 65 and 50 torr, respectively. In early childhood, the sensitivity is normal at both altitudes, but it begins to diminish thereafter, reaching stable values in adulthood. At higher altitude, the onset of diminution occurs earlier in life, and the total loss of ventilatory drive is greater.

regained in three natives of the Peruvian Andes in 10–12 months at sea level. Sørensen and Severinghaus (76) examined nine high altitude natives who had migrated to sea level and lived there for 2–16 years. They concluded that the blunted hypoxic response was irreversible. However, variability of results and paucity of subjects kept the question open. Recently, 59 migrants from high altitude (above 3,500 m) to sea coast (800 m) in Peru were examined. The more recent adult migrants showed blunted response, confirming our previous findings. However, those adult migrants who had spent 10–25 years at the coastal area showed a better response than the recent migrants and were not distinguishable from the local population. These results clearly indicate that the diminished ventilatory response to hypoxia in adult, high altitude natives is reversible.

Animal Model

The need to investigate in depth the blunting of the human ventilatory response to hypoxia led to a search for an animal model. High altitude animals, llamas and yaks, studied at high altitude (3,880–4,300 m) showed normal hypoxic responses (77, 78), as did the goats and sheep born at high altitude. Lefrançois et al. (79) examined anesthetized dogs at 4,800 m and Hornbein and Sørensen (80) decerebrated cats at 4,300 m and found normal responses comparable to sea level values. Tenney et al. (81), however, found that blunted response to acute hypoxia developed in intact, adult, sea level cats when exposed to 6,000 m for several weeks. Midcollicular decerebration, however, practically returned the hypoxic response to normal. Tenney's results suggest, therefore, that even if the cats of Hornbein and Sørensen were actually insensitive to the hypoxic stimulus decerebration abolished the blunting. However, the question has been re-examined in intact anesthetized cats at 3,850 m (S. Lahiri, unpublished observations). Their ventilatory response to hypoxia was normal, and bilateral section of carotid sinus nerves caused profound ventilatory depression which was relieved by high O_2. Thus, the peripheral chemoreflex seemed to be normally active in high altitude cats, as was the activity of carotid chemoreceptors (17). However, Tenney's cat preparation provides an interesting model in which central phenomena were crucial. On the other hand, more recent studies seem to implicate peripheral chemoreceptors. Barer et al. (82) reported a diminished ventilatory response and enlarged carotid body in chronically hypoxic rats. Studies on carotid chemoreceptors in these models are highly desirable.

Grover et al. (83) made the interesting observation that some cattle species, when moved to high altitude, decreased their P_{aCO_2} only transiently. The authors interpreted this reversal of P_{aCO_2} to normal sea level values as a loss of ventilatory response to chronic hypoxia. This type of P_{CO_2} change is peculiar to ruminants, and its interpretation is complicated because of rumen acid-base transport (S. Lahiri, unpublished observations). The species of cattle studies by Grover et al. (83) also developed relentless pulmonary hypertension at high altitude. This study was extended by Bisgard et al. (84, 85), who reported that ventilation of the calves at 4,000 m responded normally to acute changes in P_{aO_2}. Furthermore, carotid body denervation in these species increased P_{aCO_2} remarkably, i.e., about 20 mmHg, suggesting a strong hypoxic drive to ventilation at 4,000 m. This presence of normal hypoxic ventilatory drive in the calves is in conformity with the observation of Lahiri (78) in 1–3-year-old cows and yaks.

Role of Neonatal Hypoxia

It is well known that the P_{aO_2} of carotid arterial blood in human fetuses at sea level is in the range of 40 mmHg when maternal P_{aO_2} is around 100 mmHg. At a maternal P_{aO_2} of 50 mmHg at high altitude, fetal carotid arterial P_{O_2} is expected to be 30–35 mmHg (86). In the normal neonates at sea level, P_{aO_2} increases to 60–70 mmHg immediately after birth and gradually increases to the order of 80–90 mmHg in about 2 weeks as the ductus arteriosus closes. At high altitude, right heart and pulmonary arterial pressure remain elevated after birth for a longer period of time (87), presumably because of a lack of sufficient oxygen-induced contraction of the ductus (88, 89). This results in a delayed closure of the ductus and a prolonged arterio venous shunt; at 4,500 m, neonatal P_{aO_2} would thus be expected to be lower than 30–35 mmHg. This low P_{aO_2} is expected to depress ventilation of the newborn according to the effect of acute hypoxia on ventilation in the human neonates at sea level (90).

Nothing was known, however, regarding ventilatory response to hypoxia in the human newborn at high altitude until recently (91). It was found that full-term neonates (8 hr to 5 days old) at 3,850 m show only a feeble transient response to changes in alveolar P_{O_2}. An increase in P_{O_2} showed a small transient decrease in ventilation, followed by an increase; a decrease in P_{O_2} showed a small transient stimulation of ventilation, followed by a depression which was relieved by high P_{O_2}. Infants (2–5 months old), however, showed a sustained decrease in ventilation with hyperoxia and a sustained increase with hypoxia. Thus, hypoxic reflex ventillatory response was not fully developed in the newborn; it developed after birth. These results were similar to those seen in the newborn near the sea level region in Peru. The ancestry of the offspring of the subjects had no influence. What maintained ventilation in the newborn at high altitude is not clear. Central sensitization due to chronic hypoxia may be postulated, but it is merely speculative.

In fetal life, hypoxic depression presumably keeps respiratory movement low. Following birth, intense hypoxia causes intense discharge of chemoreceptors, offsetting the depressant effect and producing the first breath. Increased oxygenation of the brain and the consequent removal of the depressant effect of hypoxia allow respiratory neurons to respond to peripheral chemoreceptor activity normally. Dawes et al. (92) reported increasing respiratory movement in fetal lambs with advancing gestation. These movements were suppressed by mild hypoxia. It has also been reported that hypoxia at the level of carotid chemoreceptors stimulates ventilation in the fetal lamb (93, 94). Direct recordings from the carotid sinus nerves do show that the carotid chemoreceptors are active in fetal lamb (95) and fetal kittens (S. Lahiri, unpublished observations), but a quantitative assessment suggests a relative insensitivity and slow response to arterial stimuli.

In the awake neonates of cats, goats, and sheep, a depressant effect of severe hypoxia on ventilation can be demonstrated. A similar depressant effect can also be seen in anesthetized adult cats in which the carotid sinus nerves were either cut (97) or central hypoxia was created independently of carotid chemoreceptors (98). In the cut chemoreceptor nerves, the activity increased as ventilation was depressed due to hypoxia. In the intact animals, these impulses were transformed into a ventilatory effect counteracting the central depressant effect of hypoxia.

It is interesting to note that at birth cardiovascular, respiratory, and hematological parameters are similar in the newborn at sea level and at high altitudes. This implies that during intrauterine life fetal tissues at high altitude are not a great deal more hypoxic than at sea level. This must mean that a more effective transfer of O_2 from maternal to fetal blood occurs through the placenta. There are reports that placental weight and the ratio of weight of placenta to fetus is considerably greater at high altitude (99, 100).

Effect of Erythrocythemia

The contribution of erythrocythemia to blunted ventilatory response seen in the high altitude natives is likely to be small, because the subjects show the same blunted response at sea level and at high altitude, independent of their red cell concentrations. There is, however, intrinsic interest in this subject because the increased O_2 content of the blood during erythrocytosis would raise P_vO_2 at a given P_{aO_2}, other conditions remaining constant (101).

If the peripheral chemoreceptors are sensitive to such changes in P_{vO_2} as a result of erythrocythemia, chemoreceptor and consequently ventilatory response to arterial hypoxia may decrease. The question of carotid body tissue P_{O_2} remains to be solved. The role of oxygen delivery to the central component of the reflex is also an important consideration.

Carotid Body and Chronic Hypoxia

Evidence has accumulated which shows more often than not an enlargement of the carotid body in chronic hypoxia in man and animals (82, 102, 103). This enlargement consists of dilation and engorgement of blood vessels and both hypertrophy and hyperplasia of carotid body cells. Electron microscopic studies of rabbit carotid body showed (103) an increase in the number of dense-cored vesicles and mitochondria in the glomus cells. Apparently no changes were observed in the nerve endings. The functional significance of these changes is not known. Barer et al. (82) reported both enlargement of carotid body and a diminished ventilatory response to hypoxia in rats exposed to 390 torr. In this animal model, a direct measurement of carotid chemoreceptor activity would help to locate the blunted response. Olson et al. (104), however, did not confirm a blunted hypoxic response in chronically hypoxic rats.

The function of the glomus cells of the carotid body is not known. It has been suggested that they are dopaminergic (105). Because vascularly administered dopamine inhibits carotid chemoreceptor activity, it has been speculated that the amine provides a means for modulating the activity (106, 107). In addition, the change in dopamine content of the carotid body during acute hypoxia seems to depend upon an intact carotid sinus nerve (108). Thus, one line of argument is that the modulation occurs through efferent fibers to the glomus cells. Efferent activity is reported to be increased by an increased activity of chemoreceptor afferents, as well as by alkaline CSF pH (109, 110). There is, therefore, a potential mechanism for peripheral chemoreceptor inhibition during hypoxia, but what role it may play in the ventilatory acclimatization and adaptation to chronic hypoxia is uncertain.

Hypoxic Chemoreflex Versus Central Depression

Although the transient N_2/O_2 and CO_2 tests did show decreased ventilatory responses in the high altitude natives, the physiological tests did not identify the site(s) responsible for the blunted respiratory sensitivity. These tests do not distinguish between desensitization at the peripheral receptor sites and inhibition (or block) in the reflex pathways central to the receptors. However, these tests do demonstrate that the blunted response involves reflex pathways and are not the result of an excessive cerebral medullary depression as may happen during steady state hypoxia. Nonetheless, a high arterial P_{O_2} consistently increases ventilation in high altitude natives. A similar phenomenon is seen in newborns. These observations seem to indicate that a hypoxic brain plays an important role in the control of ventilation. It is interesting to note that carboxyhemoglobinemia can depress ventilation in anesthetized cats, even when carotid chemoreceptors are stimulated (S. Lahiri, unpublished observations). It is a matter of common experience that a progressive hypoxia during rebreathing of air with a CO_2 scrubber can cause unconsciousness in man.

Summary

With respect to the control of ventilation at rest, it can be stated that 1) at sea level, hypoxic ventilatory drive and response play a small part and CO_2-H^+ drive predominates; 2) during acute hypoxia, hypoxic drive and response predominate, whereas CO_2 drive is negligible; 3) in the course of acclimatization to hypoxia, CO_2 drive increases and the response eventually exceeds the sea level value, whereas hypoxic response is maintained, the combination of the two stimuli producing a multiplicative effect; 4) in adult, high altitude natives or man with lifelong hypoxia at sea level, the hypoxic response is diminished, whereas the CO_2 response alone appears to be normal; 5) the blunted hypoxic response develops after birth during a prolonged exposure to chronic hypoxia following a strength-duration relationship.

HYPOXIC EXERCISE

It is generally known that exercise increases the stimulating effect of acute hypoxia on ventilation at sea level (111). Although the mechanism is not understood, the enhanced effect clearly involves the chemoreflex mechanism, because it disappears promptly with a breath of O_2 (19). A similar conclusion was reached by studying the effect of chronic hypoxia on the sojourners to high altitude (112). This enhancement of the hypoxic effect on ventilation during exercise occurred even though hyperventilation caused respiratory alkalosis (113). If the alkalosis were prevented by raising P_{aCO_2}, ventilation increased further. High altitude natives, however, ventilate relatively less than sojourners and maintain their arterial P_{CO_2} and pH during moderate exercise (114). It seems that hypoxic sensitivity determines the ventilatory augmentation during hypoxic exercise, and arterial pH follows this response without visibly influencing ventilation. Exercise per se does not increase carotid chemoreceptor activity (115, 116), nor does it increase ventilatory hypoxic sensitivity (113, 117).

It has been indicated previously that CSF is more acidic in the high altitude natives at high altitude (53). If increased CSF acidity were responsible for the augmented ventilation in hypoxic exercise, there should have been a pronounced increase in ventilation in the high altitude natives during exercise. The fact that this is not so suggests that factors other than a chemical agent may be involved.

TEMPERATURE REGULATION: INTERACTION BETWEEN HYPOXIA AND COLD

Acute exposure of a newborn infant to hypoxia at sea level promptly decreases his ventilation and lowers body temperature and O_2 uptake, particularly if the ambient temperature is below 34–35°C (thermoneutral environment) (118). The fall in body temperature is presumably due to the inability of the newborn to increase its metabolism in response to low ambient temperature. At high altitude, particularly in the Himalayas, the newborn is immediately exposed to a very low temperature because of the lack of adequate heating facilities. This combination of cold temperature and hypoxia of high altitude presumably lowers the body temperature of the newborn. This initial response of the newborn to altitude environment, hypoxia and cold, may alter its temperature regulation in adult life.

ACUTE MOUNTAIN SICKNESS

An insufficient tolerance to the hypoxia of high altitude is often shown by newcomers (119). The symptoms, which usually consist of headache, nausea, lassitude, insomnia, gastrointestinal disorders, etc., appear within a few hours of arrival and disappear in a few days, depending upon the intensity of hypoxia. For example, the symptoms may never disappear at 6,000 m.

Susceptibility of individuals varies enormously, although it is thought that lack of physical fitness is an important determinant. Needless to say, hypoxemia is the initial factor, but there is no understanding of what physiological response or lack of response makes the symptoms appear or disappear. Hence, there is no real treatment. Inhalation of oxygen is no doubt a cure, but it is also an escape from high altitude. Of the agents used for symptomatic treatment, acetazolamide stands out prominently and is used most frequently. Acetazolamide interferes with CO_2 transport primarily by inhibiting carbonic anhydrase (120). Tissue P_{CO_2} is probably increased, and a decrease occurs in arterial and CSF bicarbonate and an increase in arterial acidity (121). There is an increase in resting ventilation, which increases alveolar and arterial P_{O_2}, thereby lowering the effective altitude. The net result is beneficial under resting condition.

In the context of this beneficial effect against hypoxia, it is perhaps worth pointing out that acetazolamide diminishes the activity of carotid chemoreceptors (122, 123). Figure 2 shows the effect of acetazolamide on carotid chemoreceptor activity and ventilation in an anesthetized cat. Administration of acetazolamide was promptly followed by a decrease in chemoreceptor activity as well as in ventilation. However, in the course of several minutes, ventilation increased above normal while chemoreceptor activity remained depressed. This

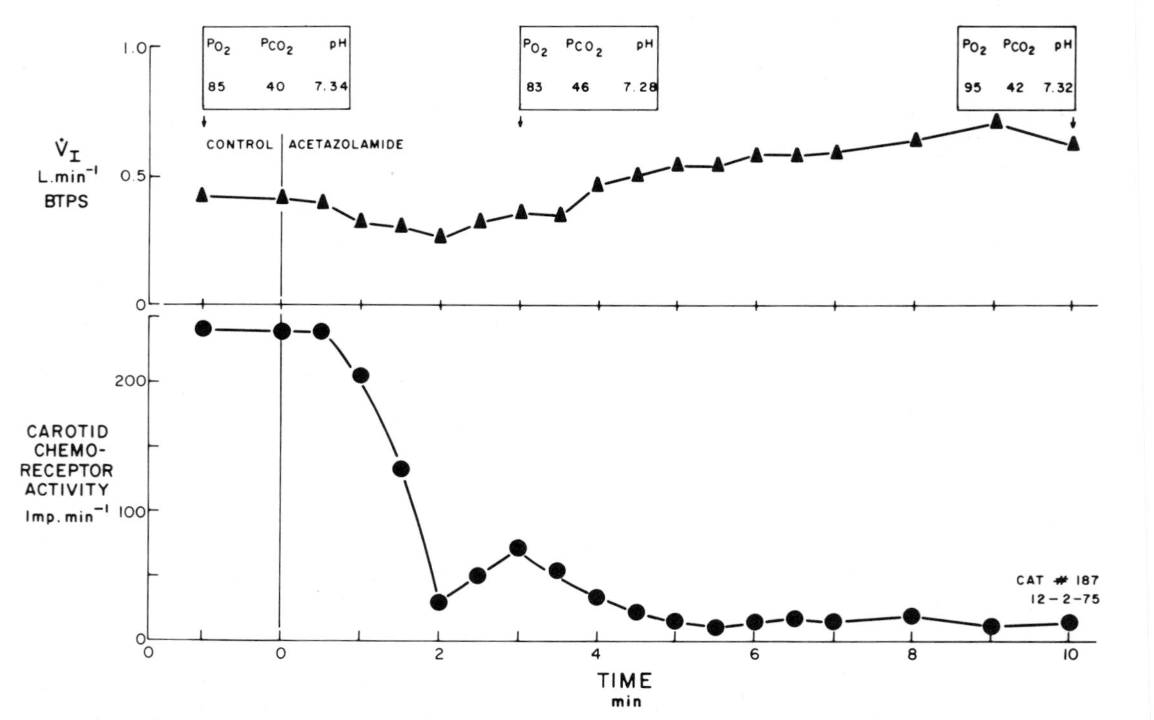

Figure 2. Effects of acetazolamide on carotid chemoreceptor activity and ventilation in cat anesthetized with chloralose. Intravenous administration of acetazolamide (50 mg kg^{-1}) was followed, within a minute, by a decrease in chemoreceptor activity and ventilation. Thereafter, ventilation continued to increase above the control value, although chemoreceptor activity remained low. A part of the initial decrease in the activity of chemoreceptors and ventilation was presumably due to the alkalinity of acetazolamide solution. Later in the time course, as carbonic anhydrase was inhibited more completely, CO_2 transport between plasma and red cells in the pulmonary capillary blood was slowed down, although the gas pressures between the capillary blood and alveolar air were in equilibrium as before the administration of acetazolamide. After leaving the capillary, plasma P_{CO_2} and $[H^+]$ continued to increase. Thus, carotid chemoreceptors "saw" a higher P_{CO_2} and $[H^+]$ than were present in the pulmonary capillary blood, and the arterial P_{CO_2} and $[H^+]$, as measured several minutes after the collection, were presumably greater than those present in the carotid arterial blood. However, by using alveolar P_{CO_2} and the equivalent in vivo plasma $[H^+]$, it can be shown that the carotid chemoreceptors were inhibited by acetazolamide, as was the hypoxic chemoreflex.

diverse effect at the two sites of respiratory receptors may not be directly related to arterial H^+. Thus, central stimulation was responsible for an increased ventilation, despite a decrease in the activity of the oxygen sensor. Therefore, an agent which blunts the chemosensing mechanism for oxygen need not afford a good protection against hypoxia.

MOUNTAIN DISEASES AND MOUNTAIN MEDICINE

Chronic mountain sickness is not a perpetual form of acute mountain sickness; it is another form of a disease which develops over a long period of chronic hypoxia (124). There is no known treatment except to move the patient to lower altitude. The outstanding diagnostic features are profound hypoventilation, arterial hypoxia, erythrocythemia, and pulmonary arterial hypertension. All patients examined so far have shown very little ventilatory response to hypoxia, and consequently they also retain carbon dioxide, unlike patients with acute mountain sickness. Not only the native altitude dwellers, but lowlanders as well develop this disorder. (Two such lowlander patients were discovered at 4,540 m in the Peruvian Andes, 1966–67, during preliminary screening procedures.) If chronic mountain sickness is a consequence of loss of hypoxic ventilatory sensitivity, it may be surmised that these lowlanders also decreased their sensitivity. However, it is possible that all these lowlanders were potentially poor responders but did not show it under sea level normoxic conditions. When patients with chronic mountain sickness are moved to sea level, they gradually lose their pulmonary hypertension and erythrocythemia. There are instances at sea level in which some lowlanders may develop a similar disease without any initial hypoxic exposure at all (125).

More than 25 million people live about 3,000 m, but there has been little organized effort to develop treatment for diseases at high altitude, although the physiological differences between sea level and high altitude dwellers are now well known. The physiological and biochemical norms on which much of the medical practice is based are clearly different; the treatment developed for the sea level population cannot be applied in toto to the population at high altitudes, just as medicine developed for adults is not applicable to life in the early developmental period. The subject deserves much needed attention.

SLEEP AND HYPOXIC RESPONSE

Although much of the neonatal life and about a third of adult life are spent sleeping and sleep has a profound effect on physiological processes, very little information is available regarding the effect of sleep on high altitude adaptation. Reed and Kellogg (126) showed that ventilation and ventilatory sensitivity to CO_2 is diminished by sleep at high altitude. Subsequently, Honda and Natsui (127) showed that hypoxic ventilatory drive is diminished as well. Thus, arterial hypoxia is bound to increase during sleep. It is also a common experience of sojourners at high altitude that during sleep a periodic breathing tends to set in

(128, 129) (S. Lahiri, unpublished observations). Its mechanism is not clear. However, it is known that ventilatory sensitivity to CO_2 is increased in the sojourners at high altitude. The respiratory drive from the peripheral chemoreceptors is obviously increased at high altitude due to hypoxia. The state of sleep is unlikely to influence the activity of peripheral chemoreceptors directly. However, a central initiation of abnormal respiratory rhythm with apneic spells results in a corresponding oscillation of arterial blood gases and the activity of arterial chemoreceptors, which in turn produce respiratory effects and accentuate and perpetuate the periodic breathing. Respiratory alkalosis due to hypoxic hyperventilation and a consequent reduction of central stimulus level below its threshold may be a contributory factor. In that case, correction of alkalosis should eliminate the phenomenon. It is interesting to note that high altitude natives who do not show alkalosis also seldom show periodic breathing.

It is noteworthy that in awake human subjects the phenomenon is not usually manifested, presumably because of man's voluntary influence on rhythmic breathing. Conscious influence on breathing is also evident in the lack of apnea in many human subjects, even after a passive hyperventilation (129, 130). This influence is absent in awake goats; passive hyperventilation is invariably followed by apnea in this species (S. Lahiri, unpublished observations). Thus, consciousness alone is not a factor. Experience obtained through normal breathing presumably prevents most men from going into apnea or periodic breathing even when chemical stimuli so dictate.

O_2 AND CO_2 TRANSPORT BY LUNGS AND BLOOD

This section deals with the structural and functional adaptations of the cardiovascular and pulmonary systems to hypoxia. Since the last review (8), little change in outlook has occurred.

It is expected that with the growth of the whole body from youth to adulthood its component parts, including the lungs, also grow in some approximate proportion. It is also reasonable that large animals will generally have lungs with a large surface area so that a large amount of O_2 can diffuse through, according to the demand of the metabolic rate. Tenney and Remmers (131) pointed out that, although these general expectations hold good, small animals which have a greater specific O_2 consumption (O_2 uptake per unit of mass) have a disproportionately large lung volume and surface area. As a corollary to the concept of optimal design for functional needs, Tenney and Remmers also questioned whether changes can be induced in the morphological dimensions of lung by environmental stress, such as oxygen want.

This speculation is substantiated by observations in chronically hypoxic animals (132–134). It seems that growth of the lungs was not retarded by chronic hypoxia, whereas body growth was. Increased environmental P_{O_2}, on the other hand, retarded lung growth and decreased its gas exchange surface area.

Hurtado (135) noted great thoracic development and large residual volume in high altitude natives of the Peruvian Andes, and Baker (136) reported a slow postnatal growth in this population. Working on boys of 11–20 years of age in Peru, Frisancho et al. (137) reported that chest circumference and forced vital capacity (FVC) are greater in the young adult native subjects at 4,300 m, in spite of their smaller body stature. Boyce et al. (138) confirmed and extended these studies. Their data, relating FVC to age or body stature, show that the curve for highlanders is parallel to that of lowlanders. The extrapolated intercept suggests that the difference may have occurred at birth. However, it was found recently that the ratios of crying vital capacity and body weight of newborns at 3,850 m and at sea level were similar (139). The difference occurred, therefore, during postnatal development. Sherpa subjects in the Himalayas do not often show a large lung volume (S. Lahiri, unpublished observations). This is presumably due to exposure to tobacco and other smoke throughout their lives (S. Lahiri, unpublished observations). Saldaña and Garcia-Oyola (140) found that the alveoli of adult, high altitutde natives (3,840 m) are large in diameter (286 versus 264 μm) and greater in number (335×10^6 versus 283×10^6) than those of sea level dwellers of the same size. This form of adaptation is unlike the adaptation in small animals which achieve a greater surface area by greater internal partitioning for a given lung volume.

Diffusing capacities of the lungs for CO of the native residents of the Andes, Himalayas, and Rockies have been found to be large compared to those of the sea level natives (141–144). Most investigators are inclined to believe that the diffusing capacity in adult sojourners after several months of exposure to altitude does not increase. However, years of hypoxic exposure in young and adult lowlanders increased their diffusing capacity for CO and total lung volumes. If exposed during childhood, sojourners achieved lung diffusing capacities similar to those of native residents of high altitude. Postnatal increase in alveolar number (over 10-fold) continues up to age 8–20 years and in diameter (1–2 times) for the next 10–15 years. Hypoxia may, therefore, influence both these determinants of diffusing capacity in high altitude natives and migrants during

Table 1. Respiratory adaptation to high altitude

Duration of hypoxia	Ventilatory response to acute hypoxia	Increase vital capacity from predicted sea level values
Birth to 1 week	±	0
2–6 months	+++	0
5–8 years	++++	+
8–12 years	+++	++
12–20 years	++	+++
20–25 years	+	+++

Reproduced from *Nature,* 1976, 261:133–135, by permission.

this period of development. However, it is difficult to understand how hypoxia can influence total lung volume beyond the growth period of lungs. The diffusing capacity measurements need not necessarily reflect lung volume and surface area.

The high diffusing capacity of the high altitude native was maintained after several weeks of deacclimatization at sea level. It is interesting to note that the growth of lungs and loss of hypoxic drive to ventilation in the natives of high altitude are reciprocally related during the developmental period. This relationship is depicted in Table 1 (73). As ventilatory drive due to hypoxia is diminished from the teenage years onward, a larger lung surface area seems to compensate for the loss. The result appears to be an economy in the work of breathing.

HIGH ALTITUDE ANIMALS

Lenfant et al. (145) were among the first to show a correlation between the increases in $P_{50_{7.4}}$ and 2,3-DPG (diphosphoglycerate) concentration in red blood cells in the sojourners. Weiskopf and Severinghaus (146), however, did not confirm the finding. In a review, Bullard (147) emphasized that most high altitude species showed lower $P_{50_{7.4}}$ for the same body weight, and that fetal and early neonatal $P_{50_{7.4}}$ values for most mammals are also lower. It is generally known that these species with a low P_{50} tolerate hypoxia better, and Bullard commented, "This negative correlation of P_{50} and hypoxic tolerance has been forgotten with the modern emphasis on DPG." This question has been discussed in depth elsewhere (101).

The shape and position of the O_2 equilibrium curves of blood provide a fine example of evolutionary adaptation in the transport of O_2. According to this evolutionary design, full O_2 saturation of arterial blood under the ambient conditions of normal residence is expected. Thus, the species which show full O_2 saturation at high altitude can be considered fully adapted to high altitude. According to this reciprocal relationship between high O_2 affinity and low arterial P_{O_2}, humans are newcomers to high altitude and have not had a long enough altitude history for adaptation to evolve. This evolutionary adaptation is expected to lie in the hemoglobin molecule rather than in the red cell constituents which modify oxygen-hemoglobin affinity. The latter changes seem to represent a short-term adjustment.

It is interesting to note that the animals which have a high $P_{50_{7.4}}$ and which tend to increase it at high altitude also respond by increasing total mass of circulating red blood cells, unlike the high altitude animals with low $P_{50_{7.4}}$. This empirical relationship is not, however, well documented; nor is the integrative relationship between O_2 affinity and production of hemoglobin in response to chronic hypoxia known.

Distribution of mammalian population at high altitude depends upon their ability to survive. Among other factors, physiological control is certainly an important determinant. For example, it has been presumed that small animals which show a higher oxygen uptake per unit of body weight and a somewhat higher $P_{50_{7.4}}$ than large animals are less suitable for high altitude (D. Robertshaw, personal communication). A comprehensive knowledge of this distribution of mammals and their physiology should illuminate the pathways nature preferred to follow. However, by and large, this information is lacking (148).

PHYSICAL PERFORMANCE CAPACITY

In the past (8), numerous investigators have studied the effect of high altitude acclimatization on maximal working capacity of subjects with different physiological and ethnic backgrounds. Most of the objective measurements involved cardiorespiratory parameters during short bursts of activity. Most workers agree that maximal O_2 uptake decreases with increasing altitude, and P_{IO_2} equivalent to sea level value promptly increases maximal $\dot{V}_{O_2}$ but does not quite fully reverse it.

Maximal oxygen uptake is most affected by acute hypoxia. With acclimatization, it is improved and stabilized in a few days. With prolonged acclimatization, a further improvement in $\dot{V}_{O_2}$ may not be noticeable, but exercise tolerance has been reported to increase.

There are observations showing that the maximal $\dot{V}_{O_2}$ of the native residents of high altitude is comparable to that of sea level natives at sea level, although it is not generally accepted (149, 150). Enormous variability of maximal $\dot{V}_{O_2}$ uptake from 40–80 ml kg^{-1} min^{-1} in man makes such comparisons difficult unless a large number of subjects are studied and several determinants of maximal $\dot{V}_{O_2}$ are taken into account (8).

However, one is struck by the fact that Sherpas in the Himalayas are capable of doing a great deal more strenuous work over a longer period of time than the visiting climbers. Even those Sherpas, including Sherpanis (Sherpa women), who may show a relatively low maximal oxygen uptake have been found to do sustained physical work beyond the capacity of athletic sojourners at great altitudes. This fact seems to have been well demonstrated by the superior performance of the altitude dwellers at Mexico City (2,250 m) during the 1968 Olympics. Presumably, the 1975 Davis Cup Tennis Tournament at Mexico City tells the same story. Thus, endurance tests rather than maximal $\dot{V}_{O_2}$ is likely to establish the difference clearly.

It is a fortunate coincidence that ambient temperature and humidity decrease as altitude increases. This natural phenomenon allows adapted man to perform better at high altitude than if hypoxia, heat, and humidity were combined (149).

As documented previously (8), the cardiovascular system in children at high altitude retains its fetal character unlike that at sea level. This is likely to produce deficient O_2 transport and consequently an impaired physical performance among the young children native to high altitude. However, there is no documentation of this phenomenon.

Hypoxia during the developmental period most certainly influences growth of lungs, ventilation, and other physiological parameters concerned with physical performance. Thus, a difference in physical performance between the adult subjects who grew up at high altitude and the sea level subjects who migrated to the same altitude at the adult age is not unexpected. Measurements at Leadville (3,100 m) did not bring out this difference clearly (8). However, more recent measurements in Cuzco City (3,400 m) (151) showed that maximal oxygen uptake was greater in subjects who were native or migrated in youth to Cuzco. This aerobic capacity was well correlated with the duration of residence between 3 and 21 years, indicating that the adaptation process is rather slow. The extent of adaptation was less in those who migrated at an older age. These results are consistent with our observation on the developmental effect of hypoxia on the control of ventilation and lung growth (73).

CELLULAR AND SUBCELLULAR ADAPTATION

The ultimate objective of the blood gas transport processes is to help bring the molecular O_2 to the mitochondrial surface, where it is reduced by electron transport from the substrate through the final common pathway of respiratory chain to produce H_2O. This respiratory function is coupled to, and controlled by, the phosphorylating system. Because in vitro the P_{O_2} at which cytochrome oxidase reacts at half its maximal rate (K_m) has been found to be less than 1 mmHg (152), it has been assumed that the normal oxidative rate is impaired only around such a low P_{O_2}. However, in vivo relationships may be different (153). It has been reported more recently that after acute hypoxia of 30–60 min respiratory activities in state 3 (glutamate + malate + ADP) per unit of cytochrome oxidase ($a + a_3$) increased in mitochondrial preparations from liver, heart, and brain of rats. Because this change is not observed with mitochondria from normoxic tissues which have been exposed to hypoxia after isolation, it is presumed that some in vivo cellular factor is necessary to bring about the observed increased mitochondrial respiratory activity in intact hypoxic rats. There was no change in the enzyme content during the acute state of hypoxia.

Chronic hypoxia of several days or longer duration is known to induce an increase in the number of mitochondria per unit of cell mass (154). The mitochondrial mass increase is also associated with the increases in succinic oxidase, succinic dehydrogenase, ADP-stimulated respiration (153, 155), and cellular synthesis of nucleic acid and protein (156). Cytochrome oxidase ($a + a_3$) and cytochrome *c* per mg of mitochondrial protein appeared to decrease, but the turnover number increased. Thus, arterial hypoxia (P_{aO_2} = 30–40 torr) which is far above the in vitro K_m value of mitochondria shows a clear regulatory effect on mitochondrial mass and its respiratory activity. Because oxygen uptake of intact animals under the given hypoxic state is not reduced, it is reasonable to assume that mitochondrial P_{O_2} was more than adequate for its normal function. Thus, the immediate cause for the mitochondrial response is not obvious. According to Dr. Leena Mela (personal communication), some other cellular mediator may be involved. The physiological meaning of the mitochondrial

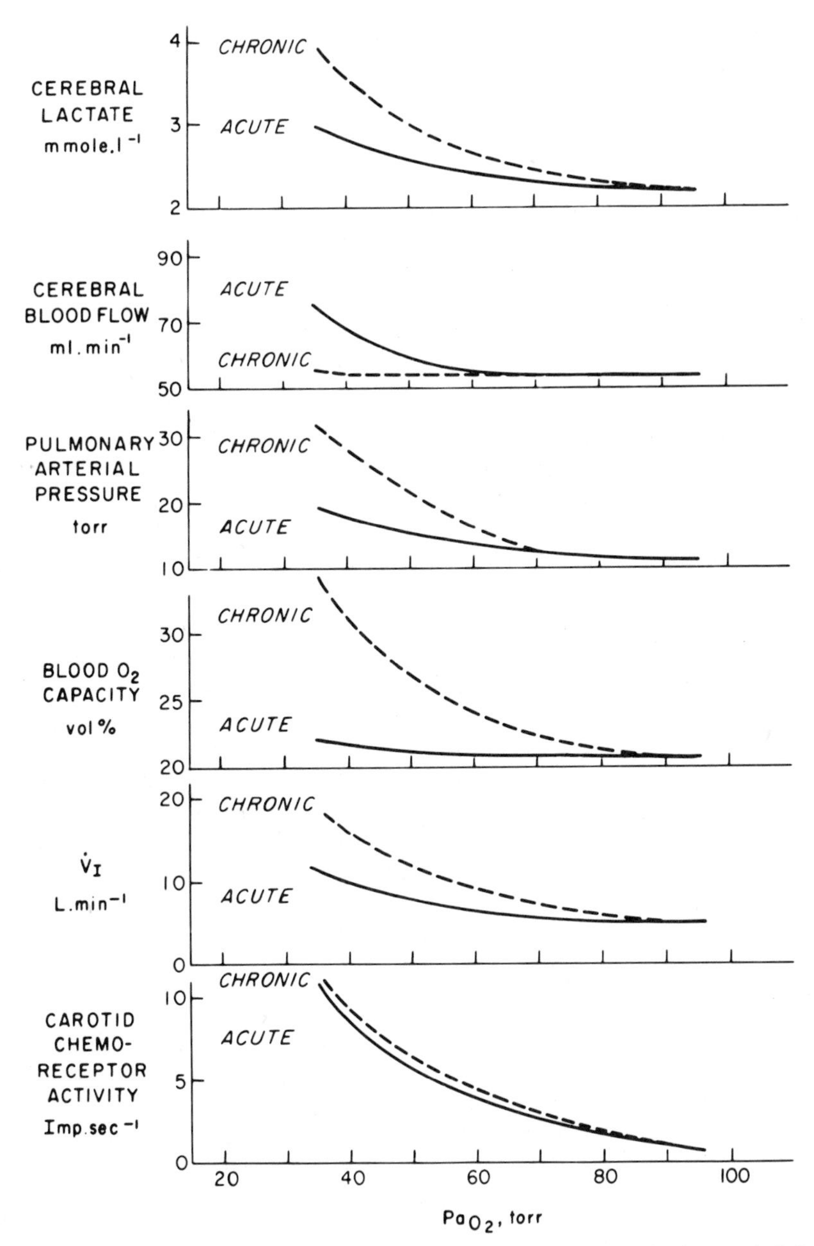

Figure 3. Physiological responses to acute and chronic hypoxia. The basic characteristic is a hyperbolic relationship between the stimulus and response. Between acute and chronic states, arterial chemoreceptor response does not seem to change; ventilation increases initially during chronic states, but eventually may become blunted (not shown here); blood O_2 capacity and pulmonary artery pressure are increased perpetually under chronic hypoxia; cerebral blood flow may diminish, whereas cerebral lactate concentration remains increased during chronic hypoxia.

response is also not clear. According to Dr. B. Reynafarje (personal communication), an increased apparent substrate affinity of the dehydrogenase system is helpful in utilizing the substrates at their lower concentrations.

GENERAL CONCLUSION

Problems of response and adaptation to oxygen want are encountered at many levels in the organism. The responses to low environmental P_{O_2} can be divided into two main categories: 1) the fast response, examples of which are reflex hyperventilation mediated by the peripheral chemoreceptors and the increase in total and regional blood flow, e.g., in brain, heart, and kidney and 2) slow and sustained responses in the form of hyperplasia and hypertrophy of tissues, cells, and subcellular materials. The fast responses, however, tend to subside in time as a complex function of duration and strength of the stimulus (see Figure 3). All slow responses show an approximately hyperbolic relationship with the intensity of hypoxic stimulation. The mitochondrial cellular contents which are directly concerned with O_2 transport and utilization also respond. We do not know the molecular basis of the initiation and maintenance of these responses.

Intuitively the role of the autonomic nervous system is bound to be important in the responses to hypoxia, but there is not much published material on the subject. However, there are instances which suggest that in the absence of normal autonomic function some of the physiological responses to hypoxia may not be sustained.

Where the response ends and adaptation begins is difficult to define. This problem is further complicated by the fact that there are processes which are linked in series as well as others which develop in parallel. Eventual adaptation depends upon the pattern of the multiphasic responses. Hypoxic adaptation not only leads to an increase in tolerance to hypoxia but increases tolerance to other stressful stimuli, thereby showing the phenomenon of cross-adaptation (156).

It is emerging more clearly now than before that the physiological differences between the native population at high altitude and the sojourners are chiefly acquired in time. The adaptation is prolonged over many years, and the balance of evidence suggests that the degree of this adaptative response depends upon exposure to chronic hypoxia during the developmental period. Much work, however, is needed to clarify the issues.

There are obviously limits to the adaptive process determined by the limits of homeostasis. An equivalent altitude (5,800 m) of half an atmospheric pressure seems to be the limit for human beings. Among other processes, rapidity of aging and loss of fertility which leads to the fear of extinction are the final determinants.

ACKNOWLEDGMENTS

This review owes much to many discussions with my colleagues and to Miss Marianne Yachinsky for her secretarial assistance.

NOTE ADDED IN PROOF:

The literature on the regulation of CSF and brain pH continues to grow. Arief et al. (157) among others followed pH of arterial blood, CSF, brain, and skeletal muscle in dog during 3-hr exposure to respiratory alkalosis and acidosis (P_{aCO_2} = 15 to 85 mm Hg). While arterial and muscle pH remained acid, the CSF and brain intracellular pH returned towards the control values in 3 hr. These data clearly show that $[H^+]$ of the brain environment is guarded more closely than skeletal muscle, for example, by some special mechanism(s). An appropriate respiratory response may follow the stimulus level of H^+ at the central chemoreceptors. The role of central chemoreceptors in respiratory control was updated at a symposium (158) recently.

Of great interest is a recent claim by Morpurgo et al. (159) that Sherpas living permanently at 4,000 m do not have increased hemoglobin parameters and have a higher affinity of blood for oxygen than Sherpas living at 1,200 m. Since these characteristics of blood of high altitude Sherpas are similar to those of Andean high altitude animals (see Ref. 8) the authors concluded that the Sherpas are genetically better adapted than the native Indians of the Andes. However, it is important to note that high altitude Sherpas do respond and show high hemoglobin concentrations when exposed to altitudes higher than their permanent residence (53, 160). On the question of oxygen affinity of blood, it is intriguing that $P_{50_{7.4}}$ for Sherpas increased from 22.6 mm Hg to 36.7 mm Hg as the altitude decreased from 4,000 m to 1,200 m. This is a large change in the reverse direction not even shown by native high altitude animals when moved to sea level (see Ref. 8). The observation of Morpurgo et al. (159) is so radical and exciting that it needs verification and extension.

REFERENCES

1. Cherniack, N. S., and Longobardo, G. S. (1970). Oxygen and carbon dioxide stores of the body. Physiol. Rev. 50:196.
2. Clark, J. M., and Lambertsen, C. J. (1971). Pulmonary oxygen toxicity: a review. Pharmacol. Rev. 23:39.
3. Haugaard, N. (1974). The effects of high and low oxygen tensions on metabolism. *In* O. Harashi (ed.), Molecular Oxygen in Biology: Topics in Molecular Oxygen Research, p. 163. North Holland Publishing Co., Amsterdam.
4. Berkner, L. V., and Marshall, L. C. (1964). The history of oxygenic concentration in the earth's atmosphere. Discuss. Faraday Soc. 37:122.
5. Gerscham, R. (1963). Biological effects of oxygen. *In* F. Dickens and E. Neil (eds.), Oxygen in the Animal Organism, p. 475. Pergamon Press, Oxford.
6. Shanklin, D. R. (1969). A general theory of oxygen toxicity in man. Perspect. Biol. Med. 13:80.
7. Gilbert, D. L. (1963). Atmosphere and evolution. *In* F. Dickens and E. Neil (eds.), Oxygen in the Animal Organism, p. 641. Pergamon Press, Oxford.
8. Lahiri, S. (1974). Physiological responses and adaptations to high altitude.

In D. Robertshaw (ed.), MTP International Review of Science, Physiology Series One, Vol. 7, Environmental Physiology, p. 271. Butterworths, London.

9. Petropoulos, E. A., and Timiras, P. S. (1974). Biological effects of high altitude as related to increased solar radiation, temperature fluctuations and reduced partial pressure of oxygen. *In* S. W. Tromp (ed.), Progress in Biometeorology, Vol. 1, p. 295. Swets and Zeitlinger, Amsterdam.
10. Frisancho, R. A. (1975). Functional adaptation to high altitude hypoxia. Science 187:313.
11. Badeer, H. S. (1973). Cardiomegaly at high altitudes. Aerosp. Med. 44:1173.
12. Hornbein, T. F., Griffo, Z. J., and Roos, A. (1961). Quantitation of chemoreceptor activity: interaction of hypoxia and hypercapnia. J. Neurophysiol. 24:561.
13. Eyzaguirre, C., and Koyano, H. (1965). Effects of hypoxia, hypercapnia and pH on the chemoreceptor activity of the carotid body *in vitro*. J. Physiol. (Lond.) 178:385.
14. Biscoe, T. J., and Sampson, S. R. (1970). The frequency of nerve impulses in single carotid body chemoreceptor afferent fibers recorded *in vivo* with intact circulation. J. Physiol. (Lond.) 208:121.
15. Lahiri, S., and DeLaney, R. G. (1975). Stimulus interaction in the responses of carotid body chemoreceptor single afferent fibers. Respir. Physiol. 24:249.
16. Åstrand, P.-O. (1954). A study of chemoreceptor activity in animals exposed to prolonged hypoxia. Acta Physiol. Scand. 30:335.
17. Hornbein, T. F., and Severinghaus, J. W. (1969). Carotid chemoreceptor response to hypoxia and acidosis in cats living at high altitude. J. Appl. Physiol. 27:837.
18. Dejours, P. (1975). Principles of Comparative Respiratory Physiology, p. 202. North Holland Publishing Co., Amsterdam.
19. Cunningham, D. J. C., Spurr, D., and Lloyd, B. B. (1966). The drive to ventilation from arterial chemoreceptors in hypoxic exercise. *In* R. W. Torrance (ed.), Proceedings of the Wates Foundation Symposium on Arterial Chemoreceptors, p. 301. Blackwell, Oxford.
20. Nielsen, M., and Smith, H. (1952). Studies on the regulation of respiration in acute hypoxia. Acta Physiol. Scand. 24:293.
21. Loeschcke, H. H. (1974). Central nervous chemoreceptors. *In* J. G. Widdicombe (ed.), MTP International Review of Science, Physiology Series One, Vol. 2, Respiratory Physiology, p. 167. Butterworths, London.
22. Kellogg, R. H. (1963). Effect of altitude on respiratory regulation. Ann. N. Y. Acad. Sci. 109:815.
23. Severinghaus, J. W., Mitchell, R. A., Richardson, B. W., and Singer, M. M. (1963). Respiratory control at high altitude suggesting active transport regulation of CSF pH. J. Appl. Physiol. 18:1155.
24. Severinghaus, J. W. (1964). Electrochemical gradients for hydrogen and bicarbonate ions across the blood-CSF barrier in response to acid-base balance changes. *In* McC. C. Brook, F. F. Kao, and B. B. Lloyd (eds.), Cerebrospinal Fluid and the Regulation of Ventilation, p. 247. Blackwell, Oxford.
25. Dempsey, J. A., Forster, H. V., and DoPico, G. A. (1974). Ventilatory acclimatization to moderate hypoxemia in man. J. Clin. Invest. 53:1091.
26. Dempsey, J. A., Forster, H. V., Gledhill, N., and DoPico, G. A. (1975). Effects of moderate hypoxemia and hypocapnia on CSF $[H^+]$ and ventilation in man. J. Appl. Physiol. 38:665.

27. Siesjö, B. K. (1972). The regulation of cerebrospinal fluid pH. Kidney Int. 1:360.
28. Sφrensen, S. C. (1970). Ventilatory acclimatization to hypoxia in rabbits after denervation of peripheral chemoreceptors. J. Appl. Physiol. 28:836.
29. Lahiri, S., Cherniack, N. S., Edelman, N. H., and Fishman, A. P. (1975). Role of carotid body, arterial and cisternal fluid pH in the acclimatization to chronic hypoxia. Physiologist 18:284.
30. Nielsen, M. (1936). Untersuchungen über die Atemregulation beim Menschen. Scand. Arch. Physiol. 74 (Suppl. 10):83.
31. Hugelin, A., Bonvallet, M., and Dell, P. (1959). Activation réticulaire et corticale dórigine chémoceptive au cours de l'hypoxie. Electroencephalogr. Clin. Neurophysiol. 11:325.
32. Åstrand, P.-O. (1954). The respiratory activity in man exposed to prolonged hypoxia. Acta Physiol. Scand. 30:343.
33. Lahiri, S. (1972). Dynamic aspects of regulation of ventilation in man during acclimatization to high altitude. Respir. Physiol. 16:245.
34. Lahiri, S. (1968). Alveolar gas pressures in man with life-time hypoxia. Respir. Physiol. 4:373.
35. Kellogg, R. H. (1968). Altitude acclimatization, a historic introduction emphasizing the regulation of breathing. Physiologist 11:37.
36. Fitzgerald, M. P. (1913). The changes in the breathing and the blood at various high altitudes. Philos. Trans. R. Soc. Lond. (Biol. Sci.) 203:351.
37. Tenney, S. M., Remmers, J. M., and Mithoefer, J. C. (1963). Interaction of CO_2 and hypoxic stimuli on ventilation at high altitude. Q. J. Exp. Physiol. 48:192.
38. Michel, C. C., and Milledge, J. S. (1963). Respiratory regulation in man during acclimatization to high altitude. J. Physiol. (Lond.) 168:631.
39. Cunningham, D. J. C., Shaw, D. G., Lahiri, S., and Lloyd, B. B. (1961). The effect of maintained ammonium chloride acidosis on the relation between pulmonary ventilation and alveolar oxygen and carbon dioxide in man. Q. J. Exp. Physiol. 46:323.
40. Forster, H. V., Dempsey, J. A., Virduk, E., and DoPico, G. A. (1974). Evidence of altered regulation of ventilation during exposure to hypoxia. Respir. Physiol. 20:379.
41. Eger, E. I., Kellogg, R. H., Mines, A. H., Lima-Ostos, M., Merrill, C. G., and Kent, D. W. (1968). Influence of CO_2 on ventilatory acclimatization to altitude. J. Appl. Physiol. 24:607.
42. Leusen, I. (1972). Regulation of cerebrospinal fluid composition with reference to breathing. Physiol. Rev. 52:1.
43. Vogh, B. P., and Maren, T. H. (1975). Sodium, chloride and bicarbonate movement from plasma to cerebrospinal fluid in cats. Am. J. Physiol. 228:673.
44. Wichser, J., and Kazemi, H. (1975). CSF bicarbonate regulation in respiratory acidosis and alkalosis. J. Appl. Physiol. 38:504.
45. Hornbein, T. F., and Pavlin, E. G. (1975). Distribution of H^+ and HCO_3^- between CSF and blood during respiratory alkalosis in dogs. Am. J. Physiol. 228:1149.
46. Choma, L., and Kazemi, H. (1976). Importance of changes in plasma HCO_3^- on regulation of CSF HCO_3^- in respiratory alkalosis. Respir. Physiol. 26:265.
47. Fencl, V., Vale, J. R., and Broch, J. A. (1969). Respiration and cerebral blood flow in metabolic acidosis and alkalosis in humans. J. Appl. Physiol. 27:67.

48. Severinghaus, J. W., Chiodi, H., Eger, E. I., Branstater, B. B., and Hornbein, T. F. (1966). Cerebral blood flow at high altitude. Circ. Res. 19:274.
49. Milledge, J. S., and Sørensen, S. C. (1972). Cerebral arterio-venous oxygen difference in man native to high altitude. J. Appl. Physiol. 32:687.
50. Rahn, H., and Otis, B. (1949). Man's respiratory response during and after acclimatization to high altitude. Am. J. Physiol. 157:445.
51. Moret, P., Covarrubias, E., Coudert, J., and Duchosal, F. (1972). Cardiocirculatory adaptation to chronic hypoxia. Acta Cardiologica 27:283.
52. Roy, S. B. (1972). Circulatory and ventilatory effects of high altitude acclimatization and deacclimatization of Indian soldiers. General Printing Co., Delhi.
53. Lahiri, S., and Milledge, J. S. (1967). Acid-base in Sherpa altitude residents and lowlanders at 4880 m. Respir. Physiol. 2:323.
54. Sørensen, S. C., and Milledge, J. S. (1971). Cerebrospinal fluid acid-base composition at high altitudes. J. Appl. Physiol. 31:28.
55. Blayo, M. C., Marc-Vergnes, J. P., and Pocidalo, J. J. (1973). pH, P_{CO_2}, and P_{O_2} of cisternal fluid in high altitude natives. Respir. Physiol. 19:298.
56. Gabel, R. A., and Weiskopf, R. B. (1975). Ventilatory interaction between hypoxia and $[H^+]$ at chemoreceptors of man. J. Appl. Physiol. 39:292.
57. Severinghaus, J. W., Bainton, C. R., and Carcelen, A. (1966). Respiratory insensitivity to hypoxia in chronically hypoxic man. Respir. Physiol. 1:308.
58. Sørensen, S. C. (1971). The chemical control of ventilation. Acta Physiol. Scand. Suppl. 361:1.
59. Milledge, J. S., and Lahiri, S. (1967). Respiratory control in lowlanders and Sherpa highlanders at altitude. Respir. Physiol. 2:310.
60. Bouverot, P., Candas, V., and Libert, J. P. (1973). Role of the arterial chemoreceptors in ventilatory adaptation to hypoxia of awake dogs and rabbits. Respir. Physiol. 17:209.
61. Bureau, M., and Bouverot, P. (1975). Blood and CSF acid-base changes, and rate of ventilatory acclimatization of awake dogs to 3,550 m. Respir. Physiol. 24:203.
62. Gilfilan, R. S., Cuthbertson, E. M., Hansen, J. T., and Pace, N. (1966). Surgical excision of the canine carotid bodies and denervation of the aortic bodies. J. Surg. Res. 7:457.
63. Hurtado, A. (1964). Animals in high altitudes: resident man. *In* D. B. Dill (ed.), Adaptation to Environment, Handbook of Physiology, Section 4, p. 843. Am. Physiol. Soc., Washington, D.C.
64. Lahiri, S., and Edelman, N. H. (1969). Peripheral chemoreflexes in the regulation of breathing of high altitude natives. Respir. Physiol. 6:375.
65. Husson, G., and Otis, A. B. (1957). Adaptive value of respiratory adjustments to shunt hypoxia and to altitude hypoxia. J. Clin. Invest. 36:270.
66. Lahiri, S. (1971). Genetic aspects of the blunted chemoreflex ventilatory response to hypoxia in high altitude adaptation. *In* R. Porter and J. Knight (eds.), Ciba Found. Symp. High Altitude Physiology: Cardiac and Respiratory Aspects, p. 103. Churchill, Edinburgh.
67. Sørensen, S. C., and Severinghaus, J. W. (1968). Respiratory insensitivity to acute hypoxia persisting after correction of tetralogy of Fallot. J. Appl. Physiol. 25:221.
68. Edelman, N. H., Lahiri, S., Braudo, L., Cherniack, N. S., and Fishman, A. P. (1970). The blunted ventilatory response to hypoxia in cyanotic congenital heart disease. N. Engl. J. Med. 282:405.

69. Blesa, M., Lahiri, S., Rashkind, W., and Fishman, A. P. (1977). Normalization of the blunted ventilatory response to acute hypoxia in congenital cyanotic heart disease. New Engl. J. Med. 296:237.
70. Chiodi, H. (1957). Respiratory adaptations to chronic high altitude hypoxia. J. Appl. Physiol. 10:81.
71. Weil, J. V., Byrne-Quinn, E., Ingvar, E., Filly, G. F., and Grover, R. F. (1971). Acquired attenuation of chemoreceptor function in chronically hypoxic man at high altitude. J. Clin. Invest. 50:186.
72. Byrne-Quinn, E., Sodal, I. E., and Weil, J. V. (1972). Hypoxic and hypercapnic ventilatory drives in children native to high altitude. J. Appl. Physiol. 32:44.
73. Lahiri, S., DeLaney, R. G., Brody, J. S., Velasquez, T., Motoyama, E. K., Simpser, M., and Polgar, G. (1976). Relative role of environmental and genetic factors in respiratory adaptation to high altitude. Nature 261:133.
74. Milledge, J. S., and Lahiri, S. (1967). Respiratory control in lowlanders and Sherpa highlanders at altitude. Respir. Physiol. 2:310.
75. Lahiri, S., Kao, F. F., Velasquez, T., Martinez, C., and Pezzia, W. (1969). Irreversible blunted respiratory sensitivity to hypoxia in high altitude natives. Respir. Physiol. 6:360.
76. Sφrensen, S. C., and Severinghaus, J. W. (1968). Irreversible respiratory insensitivity to acute hypoxia in man born at high altitude. J. Appl. Physiol. 25:217.
77. Brooks, J. G., Jr., and Tenney, S. M. (1968). Ventilatory response to llama to hypoxia at sea level and high altitude. Respir. Physiol. 5:269.
78. Lahiri, S. (1972). Unattenuated ventilatory hypoxic drive in ovine and bovine species native to high altitude. J. Appl. Physiol. 32:95.
79. Lefrançois, R., Gautier, H., and Pasquis, P. (1968). Ventilatory oxygen drive in acute and chronic hypoxia. Respir. Physiol. 4:217.
80. Hornbein, T. F., and Sφrensen, S. C. (1969). Ventilatory response to hypoxia and hypercapnia in cats living at high altitude. J. Appl. Physiol. 27:833.
81. Tenney, S. M., Scotto, P., Ou, L. C., Bartlett, D., Jr., and Remmers, J. E. (1971). Suprapontine influences on hypoxic ventilatory control. *In* R. Porter and J. Knight (eds.), Proc. Ciba Found. Symp. High Altitude Physiology: Cardiac and Respiratory Aspects, p. 89. Churchill, Edinburgh.
82. Barer, G. R., Edwards, C., and Jolby, A. I. (1972). Changes in ventilatory response to hypoxia and in carotid body size in chronically hypoxic rats. J. Physiol. (Lond.) 221:27P.
83. Grover, R. F., Reeves, J. T., Will, D. H., and Blount, S. G., Jr. (1963). Pulmonary vasoconstriction in steers at high altitude. J. Appl. Physiol. 18:567.
84. Bisgard, G. E., and Vogel, J. H. K. (1971). Hypoventilation and pulmonary hypertension in calves after carotid body excision. J. Appl. Physiol. 31:431.
85. Bisgard, G. E., Ruiz, A. V., Grover, R. F., and Will, J. A. (1974). Ventilatory acclimatization to 3400 meters altitude in the Hereford calf. Respir. Physiol. 21:271.
86. Sobrevilla, L. A., Cassinelli, M. T., Carcelen, A., and Malaga, J. (1971). Tensión de oxigens y equilibrio ácido-base de madre y teto durante el parto en la altura. Estudios Sobre la Gestation Yel Recien Nacido en la Altura. Separata de Ginecologia y Obstetricia 17:45.

87. Peñaloza, D., Arias-Stella, J., Scine, F., Recavarren, S., and Marticorena, E. (1964). The heart and pulmonary circulation in children at high altitudes. Pediatrics 34:568.
88. Kennedy, J. A., and Clark, S. L. (1942). Observations on the physiological reactions of the ductus arteriosus. Am. J. Physiol. 136:140.
89. Heymann, M. A., and Rudolph, A. M. (1975). Control of the ductus arteriosus. Physiol. Rev. 55:62.
90. Brady, J. P., and Ceruti, E. (1966). Chemoreceptor reflexes in the newborn infant: effects of varying degrees of hypoxia on heart rate and ventilation in a warm environment. J. Physiol. (Lond.) 184:631.
91. Lahiri, S., Brody, J. S., Simpser, M., Motoyama, E. K., and Velasquez, T. (1976). Hypoxic ventilatory drive in neonates at 3850 m. Respir. Dis. 113:211.
92. Dawes, G. S., Fox, H. E., Leduce, B. M., Liggins, G. C., and Richards, R. T. (1971). Respiratory movements and rapid eye movement sleep in the fetal lamb. J. Physiol. (Lond.) 220:119.
93. Pagtakahn, R. D., Faridy, E. E., and Chernick, V. (1971). Interaction between arterial P_{O_2} and P_{CO_2} in the initiation of respiration in sheep. J. Appl. Physiol. 30:382.
94. Woodrum, D. E., Parer, J. T., Wennberg, R. P., and Hodson, W. A. (1972). Chemoreceptor response in initiation of breathing in the fetal lamb. J. Appl. Physiol. 33:120.
95. Biscoe, T. J., Purves, M. J., and Sampson, S. R. (1969). Types of nervous activity which may be recorded from the carotid sinus nerve in the sheep fetus. J. Physiol. (Lond.) 202:1.
96. DeLaney, R. G., and Lahiri, S. (1972). Neonatal hypoxia and the development of attenuated ventilatory sensitivity to hypoxia. Fed. Proc. 31:390 (Abstr.).
97. Lahiri, S., and DeLaney, R. G. (1974). Hypoxic depression of ventilation. Physiologist 17:269.
98. Lee, L. Y., and Milhorn, H. T., Jr. (1975). Central ventilatory responses to O_2 and CO_2 at three levels of carotid chemoreceptor stimulation. Respir. Physiol. 25:319.
99. McClung, J. P. (1969). Effects of High Altitude on Human Birth: Observations on Mothers, Placentas and the Newborn in Two Peruvian Populations. Harvard University Press, Cambridge, Massachusetts.
100. Krüger, H., and Arias-Stella, J. (1970). The placenta and the newborn infant at high altitudes. Am. J. Obstet. Gynecol. 106:586.
101. Lahiri, S. (1975). Blood oxygen affinity and alveolar ventilation in relation to body weight in mammals. Am. J. Physiol. 229:529.
102. Edwards, C., Heath, D., Harris, P., Castillo, Y., Krüger, H., and Arias-Stella, J. (1971). The carotid body in animals at high altitude. J. Pathol. Bacteriol. 104:231.
103. Møller, M., Møllgard, K., and Sørensen, S. C. (1974). The ultrastructure of the carotid body in chronically hypoxic rabbits. J. Physiol. (Lond.) 238:447.
104. Olson, E. B., Jr., Wanner, M. A. B., and Dempsey, J. A. (1976). Human-like ventilatory adaptation to chronic hypoxia in the rat. Physiologist 19:316.
105. Fillenz, M. (1975). The function of the Type I cell of the carotid body. *In* M. J. Purves (ed.), The Peripheral Arterial Chemoreceptors, p. 133. Cambridge University Press, Cambridge.

106. Mills, E., Slotkin, T. A., and Sampson, S. R. (1975). Carotid body chemoreceptors. Nature 258:268.
107. Osborne, M. P., and Butler, P. J. (1975). New theory for receptor mechanism of carotid body chemoreceptors. Nature 254:701.
108. Yates, R. D., Chen, I.-L., and Duncan, D. (1970). Effects of sinus nerve stimulation on the carotid body glomus cell. . Cell Biol. 46:544.
109. Willshaw, P. (1975). Sinus nerve efferents as a link between central and peripheral chemoreceptors. *In* M. J. Purves (ed.), The Peripheral Arterial Chemoreceptors, p. 253. Cambridge University Press, Cambridge.
110. O'Regan, R. G. (1975). The influences exerted by the centrifugal innervation of the carotid sinus nerve. *In* M. J. Purves (ed.), The Peripheral Arterial Chemoreceptors, p. 221. Cambridge University Press, Cambridge.
111. Asmussen, E., and Nielsen, M. (1958). Pulmonary ventilation and effect of oxygen breathing in heavy exericse. Acta Physiol. Scand. 43:365.
112. Lahiri, S., Milledge, J. S., and Sørensen, S. C. (1972). Ventilation in man during exercise at high altitude. J. Appl. Physiol. 32:766.
113. Masson, R. G., and Lahiri, S. (1974). Chemical control of ventilation during hypoxic exercise. Respir. Physiol. 22:241.
114. Lahiri, S., Kao, F. F., Velasquez, T., Martinez, C., and Pezzia, W. (1969). Respiration of man during exercise at high altitude: highlander vs. lowlander. Respir. Physiol. 8:361.
115. Davies, R. O., and Lahiri, S. (1973). Absence of carotid chemoreceptor response during hypoxic exercise in the cat. Respir. Physiol. 18:92.
116. Aggarwal, D., Milhorn, H. T., Jr., and Lee, L. Y. (1976). Role of the carotid chemoreceptors in the hyperpnea of exercise in the cat. Respir. Physiol. 26:147.
117. Cunningham, D. J. C. (1974). The control system regulating breathing in man. Q. Rev. Biophys. 6:433.
118. Cross, K. W. (1961). Respiration in the newborn baby. Br. Med. Bull. 17:160.
119. Acosta, I. (1590). Hisotria Natural y Moral de las Indias, p. 535. Iuan de Leon, Seville.
120. Tomashefski, J. R., Chinn, H. I., and Clark, R. T., Jr. (1954). Effect of carbonic anhydrase inhibition on respiration. Am. J. Physiol. 177:451.
121. Kronenberg, R. S., and Cain, S. M. (1968). Effects of acetazolamide and hypoxia on cerebrospinal fluid bicarbonate. J. Appl. Physiol. 24:17.
122. Hayes, M. W., and Torrance, R. W. (1975). An effect of acetazolamide on the steady state responses of arterial chemoreceptors. J. Physiol. (Lond.) 244:56P.
123. Lahiri, S., DeLaney, R. G., and Fishman, A. P. (1976). Peripheral and central effects of acetazolamide in the control of ventilation. Physiologist 19:261.
124. Monge, C. M., and Monge, C. C. (1966). High Altitude Diseases, p. 32. Thomas, Springfield, Illinois.
125. Pare, P., and Lowenstein, L. (1956). Polycythemia associated with disturbed function of the respiratory center. Blood 11:1077.
126. Reed, D. J., and Kellogg, R. H. (1960). Effect of sleep on hypoxic stimulation of breathing at sea level and altitude. J. Appl. Physiol. 15:1130.
127. Honda, Y., and Natsui, T. (1967). Effect of sleep on ventilatory response to CO_2 in severe hypoxia. Respir. Physiol. 3:220.

128. Barcroft, J. (1925). The Respiratory Function of the Blood. Part I, Lessons from High Altitude, pp. 1–168. The University Press, Cambridge.
129. Haldane, J. S., and Priestley, J. G. (1935). Respiration, pp. 189, 191, 297. The Clarendon Press, Oxford.
130. Tawadrous, F. D., and Eldridge, F. (1974). Posthyperventilation breathing patterns after active hyperventilation in man. J. Appl. Physiol. 37:353.
131. Tenney, S. M., and Remmers, J. E. (1963). Quantitative comparative morphology of mammalian lungs: diffusing areas. Nature 197:54.
132. Bartlett, J., Jr., and Remmers, J. E. (1971). Effects of high altitude exposure on the lungs of young rats. Respir. Physiol. 13:116.
133. Burri, P. H., and Weibel, E. R. (1971). Morphometric estimation of pulmonary diffusing capacity. II. Effect of P_{O_2} on the growing lung; adaptation of the growing rat lung to hypoxia and hyperoxia. Respir. Physiol. 11:247.
134. Cunningham, E. L., Brody, J. S., and Jain, B. P. (1974). Lung growth induced by hypoxia. J. Appl. Physiol. 37:362.
135. Hurtado, A. (1932). Respiratory adaptation in the Indian natives of the Peruvian Andes: studies at high altitude. Am. J. Phys. Anthropol. 17:137.
136. Baker, P. T. (1969). Human adaptation to high altitude. Science 163:1149.
137. Frisancho, R. A., Velasquez, T., and Sanchez, J. (1973). Influence of developmental adaptation of lung function at high altitude. Hum. Biol. 45:583.
138. Boyce, A. M., Haight, J. S. J., Rimmer, D. B., and Harrison, G. A. (1974). Respiratory function in Peruvian Quechua Indians. Ann. Hum. Biol. 1:137.
139. Brody, J. S., Lahiri, S., Simpser, M., Velasquez, T., and Motoyama, E. K. (1976). Lung growth at high altitude. Respir. Dis. 113:48.
140. Saldaña, M., and Garcia-Oyola, E. (1970). Morphometry of the high altitude lung. Lab. Invest. 22:509.
141. Velásquez, T., and Florentini, E. (1966). Maxima capacidad de defusión pulmón en nativos de la altura. Arch. Inst. Biol. Andina 1:197.
142. Guleria, J. S., Pande, J. N., Sethi, P. K., and Roy, S. B. (1971). Pulmonary diffusing capacity at high altitude. J. Appl. Physiol. 31:536.
143. DeGraff, A. C., Grover, R. F., Johnson, R. L., Hammond, J. W., and Miller, J. M. (1970). Diffusing capacity of the lung in Caucasians native to 3100 m. J. Appl. Physiol. 29:71.
144. Dempsey, J. A., Reddan, W. G., DoPico, G. A., Cerny, F. C., and Foster, H. V. (1975). Determinants of acquired changes in pulmonary gas exchange in man via chronic hypoxic exposure. *In* H. Herzog (ed.), Progress in Respiration Research, Vol. 9, p. 180. S. Karger, Basel.
145. Lenfant, C., Torrance, J., English, E., Finch, C. A., Reynafarje, C., Ramos, J., and Faura, J. (1968). Effect of altitude on oxygen binding by hemoglobin and on organic phosphate levels. J. Clin. Invest. 47:2652.
146. Weiskopf, R. B., and Severinghaus, J. W. (1972). Lack of effect of high altitude on hemoglobin oxygen affinity. J. Appl. Physiol. 33:276.
147. Bullard, R. W. (1972). Vertebrates at altitudes. *In* M. K. Yousef, S. M. Horvath, and R. W. Bullard (eds.), Physiological Adaptations: Desert and Mountain, p. 209. Academic Press, New York.
148. Folk, G. E. (1974). Introduction to Environmental Physiology, p. 213. Lea and Febiger, Philadelphia.
149. Lahiri, S., Weitz, C. A., Milledge, J. S., and Fishman, M. C. (1976). Effects

of hypoxia, heat, and humidity on physical performance. J. Appl. Physiol. 40:206.
150. Balke, B. (1972). Physiology of respiration at altitude. *In* M. K. Yousef, S. M. Horvath, and R. W. Bullard (eds.), Physiological Adaptations: Desert and Mountain, p. 195. Academic Press, New York.
151. Frisancho, R. A., Martinez, C., Velasquez, T., Sanchez, J., and Montoye, H. (1973). Influence of developmental adaptation on aerobic capacity at high altitude. J. Appl. Physiol. 34:176.
152. Chance, B. (1965). Reaction of oxygen with the respiratory chain in cells and tissues. J. Gen. Physiol. 49:163.
153. Mela, L., Goodwin, C. W., and Miller, L. D. (1976). In vivo control of mitochondrial enzyme concentrations and activity by oxygen. Am. J. Physiol. 231:1811.
154. Ou, L. C., and Tenney, S. M. (1970). Properties of mitochondria from hearts of cattle acclimatized to high altitude. Respir. Physiol. 8:151.
155. Tappan, D. V., Reynafarje, B., Potter, V. R., and Hurtado, A. (1957). Alterations in enzymes and metabolites resulting from adaptation to low oxygen tensions. Am. J. Physiol. 190:93.
156. Meerson, F. Z. (1975). Role of synthesis of nucleic acids and protein in adaptation to the external environment. Physiol. Rev. 55:79.
157. Arief, A. I., Kerian, A., Massry, S. G., and DeLima, J. (1976). Intracellular pH of brain: alterations in acute respiratory acidosis and alkalosis. Am. J. Physiol. 230:804.
158. Acid Base Homeostasis of the Brain Extracellular Fluid and the Respiratory Control System. (H. H. Loeschcke, ed.), Thieme-Edition Publishing Sciences Group, Inc., Stuttgart.
159. Morpurgo, G., Arese, P., Bosia, A., Pesearmona, G. P., Luzzana, M., Modiano, G., and Krishna Ranjit, S. (1976). Sherpas living permanently at high altitude: a new pattern of adaptation. Proc. Natl. Acad. Sci. 73:747.
160. Cerretelli, P. (1976). Limiting factors to oxygen transport on Mount Everest. J. Appl. Physiol. 40:658.

Index

Abdomen
  cold receptors in, 92
  temperature sensors in, 81
Acclimatization
  to cold, 194, 195, 197, 198, 222
  to heat, 4, 107, 108, 109
Acetazolamide, 233
Acetylcholine, 36, 208–209
Acidosis, 221, 243
Adaptations to high altitude, 219–242
Adenosine diphosphate, 48
Adenosine monophosphate, 48
Adenosine triphosphate, 45
Adenylate cyclase, 48
Adipose tissue, triglycerides in, 38. *See also* Brown adipose tissue; White adipose tissue; Fat
Adrenal cortex, 192, 193
Adrenal gland, nerve supply to, 193
Adrenal medulla, 36, 38
  animals without functional, 196
  method of studying, 192–194
  pharmacological inactivation of, 194
  role in thermoregulation, 189–212
  secretion of catecholamines from, 40
  secretions of, and heat conservation, 195
Adrenergic receptors, 207, 209, 211. *See also* Adrenal medulla
Aerobic metabolism, 7
Alkalosis, respiratory, 129, 134, 135, 142, 205, 220, 221, 223, 236, 243
Altitude, 6
  high, adaptions to, 219–242
  high, diseases at, 233–235
Aluminum foil, 8–9
Aluminumized fabrics, 9
Alveoli, effect of high altitude on, 237
Amino acid
  imbalance, 51
  in muscle, free, 46
  in muscle, labeled, 47
  as substrate for gluconeogenesis, 55
Animals. *See also* Aquatic animals; Birds; Desert animals; Fish; Insects; Mammals; Primates; *names of specific animals*
  body weight of, 122, 123
  at high altitudes, 238–239, 243
  running, 122, 134, 135, 136
  size and heat loads, 138
  size of, in relation to heat production, 123
  speed of, 122
  surface area of, 123
Antelope, 133
Antidiuretic hormone, 57–58
Apocrine glands. *See* Sweat glands
Aquatic animals, 11. *See also* Fish
Armadillo, sleep state of, 152, 155
Atrichial glands. *See* Sweat glands
Atropine, 209
Automatic nervous system, 189–192, 200
Autonomic blocking drugs, 200

Baboons. *See* Primates
Baroreceptor, activity and shivering, 201
Bats, during hibernation, 168, 199
Basal metabolic rate. *See* Metabolic rate
Birds
  heat loss of, 127
  ravens, 10
Blood
  arterial, 102
  bicarbonate buffer system in, 129
  -brain barrier, 211, 212
  fatty acids in, 50
  flow during hypoxia, 223–224
  flow in primates, 85–92
  free fatty acids in, 201
  glucose in, 201
  pH, 205, 219, 220, 223, 224

Blood—*continued*
plasma, 58
temperature of, 152, 153
transporting carbon dioxide in, 236–238
transporting oxygen in, 236–238
Blood circulation, 17–18, 36, 56–57
in adipose tissue, 38
changes of, during sleep, 152
hypothalamic, 201
during hypoxia, 223–224, 225
to nasal mucosa, 130
during nonshivering thermogenesis, 200
to skin, 43, 85, 142
temperature changes and, 86
temperature-produced changes in, 91
in thermoregulation, 85–92
in tongue of dog, 133
Blood flow. *See* Blood; Blood circulation; Cutaneous blood flow
Blood pressure, 35, 37, 56, 80, 206, 211
mountain sickness and, 235
Blood vessels. *See also* Cutaneous blood vessels
in adipose tissue, 37
in gut, 38
locally cooled, 36
during shivering, 56
of skin, 38, 194
veins, 81
Body rhythms, 148–149
Body temperature. *See also* Core temperature; Hypothalamic temperature; Thermal neutral zones; Temperature
of animals, 2, 3, 42, 45
of cats, 154, 159–160
daily cycle of, 148–149
during heat exposure, 204
during hibernation, 168–179
and heat loss, 124, 127
hypoxia and, 233
of mammals, 71
of rats, 196
rectal, 211
and sleep, 149, 150, 155, 156
thermoregulation of, 16–20
Bradykinkin, 90
Brain, 16, 30, 33, 54. *See also* Hypothalamus
adrenal medulla, and relationship to thermoregulation in, 191
blood flow in, 223–224
blood-brain barrier, 211, 212
cold center in, 32
cooling of, 201
frontal areas of, and cold reception, 33
medullary chemosensitive areas in, 220
midbrain and shivering, 34
midbrain and thermoregulation, 78
somatosensory cortex, 31
temperature of, in sleep, 150, 151–152, 154, 155, 165
thermal-sensitive areas in, 75
thermocouples in, 152, 153
thermodes in, 103
Breathing. *See* Respiration; Exhaled air; Inhaled air
Brown adipose tissue, 33. *See also* Fat; White adipose tissue
during cold exposure, 37–40, 53, 56, 84, 92, 173–174, 198, 199–200
heat production in, 47–49
Buccal cavities, 133, 134, 135, 136, 139
Buffalo, 138. *See also* Water buffalo

Calf, 9, 97, 199. *See also* Cattle
Calorimeters, 3, 4, 5, 98
Calorimetry
direct, 6, 15, 171
indirect, 171
Carbohydrate metabolism, 7, 54–55
Carbon dioxide, role in regulating ventilation, 222–223, 236
Carboxyhemoglobinemia, 232
Cardiovascular changes. *See also* Heart; Vasoconstriction
induced by warming, 80
responses to maintain thermal balance, 84–92
Carotid arteries, 152, 203, 229
Carotid bodies, 219, 229, 231
Carotid chemoreceptors, 220, 225, 231
Carotid retes, 152, 153
Carotid sinus, 35, 220
Cat, 75, 76, 91, 110
carotid sinus in, 220

evaporative cooling in, 133
kittens and hypoxia, 230
as model for hypoxic responses, 229
panting in, 155
skin temperature of immobilized, 159–160
sleep state of, 150, 151, 152, 153, 154, 155, 156, 158, 159, 164
sympathectomized, 200
ventilation in anesthetized, 232, 233
Catecholamines, 40, 46, 47–48, 50, 201, 208, 211. *See also* Epinephrine; Norepinephrine
circulating, 192, 200, 202, 203, 204, 205, 207, 210, 212
excretion of, 195, 204
free, 192
in ox, 204
in shivering, 200–201
in sweating, 209
thermogenic action of, 211
Cattle, 9, 14, 43, 44, 45, 54, 56. *See also* Ruminants
evaporative cooling in, 138
Cellular and subcellular adaptation, 240–241
Central nervous system, 74, 109, 171, 190, 221, 223. *See also* Brain; Spinal cord
hypoxic response in, 226
temperature sensors in, 75–77
Cerebrospinal fluid, 219, 221
in homeostasis, role of, 223
in natives of high altitudes, 224–226, 232
Chemoreceptors, 231
carotid, 220, 225, 231
in panting, 206
peripheral, in hypoxia, 226
during pulmonary ventilation, 219–220
in shivering, 201
Cheetah, 122
Chemoreflexes, 219–220
Chinchilla laniger, 166
Chipmunks, 10
during hibernation, 170
Chlorisondamine, 201
Choline, 53
Cholinergic nerve fibers, 90
Citellus beechyi, 178
Citellus beldingi, 166, 171
Citellus lateralias, 165–166, 171, 179
Citellus tridecemlineatus, 166, 171
Clothing, 15–16
Coats, animal, 2, 8, 37
color of, 9, 10
heat transfer through, 13, 14
as insulation, 43–45
of primates, 96
shedding of, 45
winter, 42, 44
Cold
acclimation to, 194, 195, 197, 198, 222
interaction between hypoxia and, 233
Cold exposure
and brown adipose tissue, 47–49
brown fat of animals during, 199
changes in efferent hormones during, 40–43
and circulation, 56–57
coat types and heat loss during, 44
effect on hair, 37
effector organs during, 43, 44, 45
effects on adrenal cortex, 42
effects on adrenal medulla, 40, 195
effects on thyroid, 41–42
epinephrine excretion in urine during, 193, 194, 195
epinephrine secretion during, 200
food intake during, 41
gut, response to, 51–52
the heart during, 56–57
intolerance of animals to, 196
the kidney during, 57–58
muscle response to, 45
nerve centers as cold centers during, 32–33
piloerectin during, 203
physiological effects of, 29–58
protein metabolism in, 55–56
reflex pathways stimulated by, 33
role of afferent nerves in, 30–32
role of efferent nerves in, 33–35
secretion of plasma cortisol during, 42
short term, and metabolic rate, 202
splanchnic nerves during, 38–40
sympathetic nerves during, 35–37
and white adipose tissue, 50–51
Cold receptors, 31, 32, 33, 74, 75, 92, 98

Cold-reflex activity, 32, 36
Cold sweat, 191, 212
Cold vasodilation, 36
Conductance, 122, 123
Conduction, 2, 122
Conductive heat transfer, 8
  through animal coats, 14, 44
Control theory of temperature
    regulation, 72
Convective heat transfer, 8
  in air, 10–11
  in water, 11
  within animal coats, 14
Convection, 2, 122
Cooling. *See* Evaporative cooling;
    Hypothalamic cooling; Nonevaporative heat loss; Heat loss.
Copper manikin technique, 15
Core temperature, 11, 71, 76, 90, 92,
    102, 103, 126
Cori cycle, 54
Cortex. *See* Brain; Adrenal cortex
Countercurrent heat exchange, 17–18
Cutaneous blood flow, 18, 85–87, 90,
    91, 203. *See also* Blood; Blood
    circulation; Skin
Cutaneous blood vessels, 35, 87, 90.
    See also Skin
Cutaneous heat exchange, 18, 43, 138.
    See also Heat exchange; Heat
Cutaneous temperature sensors,
    74–75. *See also* Blood vessels; Skin
Cytochrome *c*, 47

Deep body temperature sensors,
    77–81
Dehydration, 57, 125
Desert. *See* Environments
Desert animals, 9, 124
Diet. *See* Food intake
Dog, 16, 43, 47, 51, 54, 75, 78, 91,
    122, 176
  blood flow in tongue, 133
  in desert, 124
  evaporative cooling in, 136, 139
  exercising (running), 134, 135, 136
  heat exposure and, 204
  model of hypoxia responses, 229
  nasal heat exchanger in, 130
  panting in, 131, 133, 134, 139
  respiratory alkalosis and acidosis in,
    243
  saliva production of, 135
  sleep state of, 151, 152
  Steno's gland in, 132–133
  sweating in, 139, 207
Donkey, in desert, 124
Dopamine, 231
Drugs. *See also specific drug*
  autonomic-blocking, 200–201
  acetylcholine, 208–209
  cholimimetic, 208
  ganglionic-blocking, 200–201

Eccrine gland. *See* Sweat glands
Efferent hormones, 40–43
Eland, 138
Elephant, 122
Electroencephalogram (EEG), 150,
    151, 155, 156, 165, 166
Electromyography (EMG), 93, 155,
    156, 165, 178
Endocrine, 47. *See also specific glands*
  gland, 48
Energy budgets, 3–6, 7
Energy exchange, 3–16
Environmental heat loads, 124–142
Environmental time cues, 149
Environments
  cold, 156
  cool, 150, 153, 204
  desert, 6, 71, 121, 124, 125,
    137–138
  heat exchange with, 2
  hot, 107, 125, 139, 142, 204
  inhabited by mammals, 71
  micro-, 106
  thermoneutral, 233
  warm, 111, 154
Enzyme systems, 17
Epinephrine, 36, 46, 190, 191, 192,
    193, 194, 195, 196, 197, 199,
    200, 201, 202, 203, 204, 205,
    206, 207, 208, 209, 210, 211
Epitrichial glands. *See* Sweat glands
Eptesicus fuscus, 199
Erythrocythemia, 235
  effects on ventilation, 231
Esophageal temperature. *See* Core
    temperature
Evaporation, 2, 6, 9, 11–13, 121
  calculating rates of, 12
  cutaneous, 138
  in man and animals, 124–125

resistance to, 15
from respiratory tract, 129–130, 131, 133
Evaporative coefficient, 12
Evaporative cooling mechanisms, 138–142
Evaporative heat loss, 81, 82, 84, 96–111, 120, 122, 123, 124, 125, 128–142, 154, 204, 205, 209. *See also* Heat loss
Exchange. *See* Nasal countercurrent exchange
Exercise. *See also* Flying; Running; Swimming; Walking
energy budget for, 7
and environmental heat loads, 120–142
at high altitude, 224
hypoxic, 232–233
and radiant heating, 101
sweat glands during, 209–210
thermal equilibrium in, 4
Exhaled air, 126–128, 130. *See also* Respiration

Fat, 7, 37, 38, 51. *See also* Brown adipose tissue; White adipose tissue
Fatty acids, 50, 53. *See also* Free fatty acids
unsaturated, 51
volatile, 51
Feedback
anticipatory loop, 101
negative, 72
signals during sleep, 155
First Law of Thermodynamics, 2–3
Fish
propoise fluke, 17
respiratory heat exchange in, 11
swimming muscles, 18
Flying, 8
Food intake
animal diet lacking choline, 53
in animals, 41, 42, 45, 51
during cold exposure, 54–55
fasting, 52
high fat-cholesterol diet in rats, 54
influence on body temperature, 148
Free fatty acids, 38, 45, 47, 49, 50, 51, 52, 201
Fur. *See* Coats

Ganglionic-blocking drugs, 200–201
Gastrointestinal tract, 52. *See also* Gut; Intestine
Gazelle. *See* Thompson's gazelle
Gluconeogenesis, 54, 55, 200
Glucose, 40, 45, 49, 53, 201, 202
Glycogen, 45, 54, 97
Glycogenolysis, 201
Goat, 75
evaporative cooling in, 138, 139, 140
high altitude, response to, 221, 225
hyperventilating, 236
panting in, 131, 133, 136, 139
young and hypoxia, 230
Ground squirrel
during hibernation, 165–167, 170, 171, 172, 174, 176, 177, 199
entering hibernation, 178, 179
Guinea pig, 33, 49, 75, 177
Gut, 41
blood flow in, 38, 39
glucose from, 54
tissue and eating, 51–52
Glomus cells, 231

Hair. *See* Coats; Human
Haldane transformation, 6
Hamsters, 49, 93, 174
Heart, 41
affected by cold stimulation, 36–37
cardiovascular changes and breathing, 80
diseases of, 226
during hyperthermia, 206
output in cold environment, 56–57
rate during hibernation, 170
rate during sleep, 151
thermal balance and, 84–92
Heat. *See also* Heat Production; Heat transfer
acclimatization to, 4, 107, 108, 109
basal heat production, 92
conservation and adrenal medulla, 195, 202, 203
conservation in body, 38
dissipation and epinephrine, 205
exchange between core and peripheral tissue, 82
extra, from eating, 51–52
generating mechanism in cold-acclimated animals, 47

Heat—*continued*
  local, of skin, 105
  loss exceeds production, 196
  origin of, 121–122
  production, 7, 55
    while hibernating, 171–172
    while resting, 123
    while running, 122
    while sleeping, 155
  radiative, 44
  redistribution of, 11
Heat exchange. *See also* Heat transfer
  countercurrent, 117–18
  cutaneous, 18
  nasal mechanism for, 126
  radiant, 9
  rate of evaporative, 11–13
Heat exposure, effects of, 204–212
Heat flow in animal coats, 2
Heat loads. *See also* Exercise
  exercise and environmental, 120–142
Heat loss
  of adrenal-demedullated rats, 202–203
  and body posture, 44
  from buccal surfaces and tongue, 133–137
  cardiovascular adjustments to, 206
  catecholamines, levels of during, 204
  evaporative, 81
  hypothermia and, 196
  in man and animals, 122–125
  mechanisms, 191
  nonevaporative, 122
  through nose, 126–128
  rates of, 5, 43
  during sleep, 155, 158
  of sweating animals, 137–138
Heat production, 7, 55
  of adrenal-demedullated rats, 202–203
  while hibernating, 171–172
  hypothermia and, 196
  during nonshivering thermogenesis, 200
  while resting, 123
  while running, 122
  while sleeping, 155
Heat storage within the body, 2, 3
Heat stress. *See* Stress, thermal
Heat transfer
  through coats, 13–14
  by convection in air, 10–11
  by convection in water, 11
  by evaporation, 11–13
  by radiation, 8–10
  sensible, 8–11
Hedgehog, during hibernation, 166
Hematocrit, 58
Hemoglobin, 223
Hepatic sinus, 18
Hibernation
  adrenal medulla, role of in, 200
  ambient temperatures during, 168–170
  brown fat in animals during, 199
  nonshivering thermogenesis and, 196
  as extension of sleep, 165–168
  temperature regulation during, 173–179
  thermoregulation during, 168–171
High altitude. *See also* Hypoxia
  adoption to, 219–242
  diseases at, 233–235
  physical performance at, 239–240
Homeostasis, 219
  acid-base, 223
  cardiovascular responses to thermal balance, 84–92
Homeotherms, 2, 71–72, 73, 81, 92, 111
  hypothetical, 82
Homeothermy in higher primates, 82–111
Hormones. *See also names of specific hormones*
  antidiuretic, 57–58
  efferent, 40–43
Horse, 122
  respiratory evaporation in, 138
  sweat glands of, 137, 204, 207
  sweating of, 208
"Huddling" during cold exposure, 44
Human
  antidiuretic hormones in, 42
  as aquatic animal, 11
  clothing of, 15–16
  energy budgets of, 3–6
  fetal hypoxia, 229
  hair covering, 207
  hair and heat loss, 44
  heart, in cold environment, 56–57
  infants, 3
  "lower critical temperature" in, 94
  metabolic rate of, 7

nasal mucosa in, 139
neonates, 93, 229
rate of convective heat transfer of, 10
rate of heart storage, 3
rate of oxygen consumption in, 6
resting, 121
skin, 8, 18
sleep state of, 155
sweat glands in, 96, 137
thermoregulation of, 18–19, 20
thermoregulatory responses in, 83
thermoneutral zones in, 82
vasomotor control in, 91
ventilation in newborn, 230
water lost by evaporation, 124
Hummingbird, 6, 148, 149, 171
Hygrometry, resistance, 103, 108
Hypertension. *See* Blood pressure
Hyperthermia, 204, 205, 206, 207, 208, 209
Hypoglycemia, 201–202, 209, 212
Hypothalamic changes during sleep, 155–159
Hypothalamic control of salivation, 125
Hypothalamic cooling, 156, 164, 172
Hypothalamic heating, 204–205
Hypothalamic temperature, 17, 72, 77, 78, 81, 91, 103, 110, 134, 135, 150, 155, 156, 158, 201
effect on sleep, 159–165
Hypothalamus, 16, 31, 32, 33, 36, 37, 38, 42, 75, 91, 92, 134, 162, 164, 173–174, 177, 191, 204, 207, 211
localized cooling of, 194, 201
neural information from, 93
thermodes in, 103
Hypothermia. *See* Cold; Cold exposure; Body temperature; Heat loss
Hypothermic states, 71
Hypoxia, 218–219, 220
acclimatization to, 221, 223, 225
adults and children exposed to, 227
carotid bodies and, 231
chronic, 221–222
depressed ventilation during, 232
during development, 239
effects on growth of body, 236–238
effects on size of lungs and chest, 239–240
environmental versus genetic effects in, 226–227
hypocapnic alkalosis, 223
intolerance to, 233–235
neonatal, 229–231, 233
at sea level, 226–227
severe, 221
sensitivity, 224
ventilation of natives of high altitudes, 224–226
Hypoxic animal models, 228–229
Hypoxic central depression, 231–232
Hypoxic chemoreflex, 231–232
Hypoxic drive
blunted, reversed, 227–228
loss of, 227
Hypoxic exercise, 232–233

Infrared radiation, 8–9
Inhaled air
through the mouth, 133–135
through the nose, 126–128, 130
Insects
beetle flight muscle enzyme, 17
Insulation
animal coat types, 43–45
peripheral, 82
skin as insulator, 85
Insulin, 39, 40, 50, 54, 201, 209, 212
Insulin hypoglycemia. *See* Hypoglycemia.
Interneurons, 32
Intestine, 91. *See also* Gut; Gastrointestinal tract
Iodide, urinary excretion of, 41
Isoprenaline, 209

Kangaroo, 125, 139
Ketosis, 53
Kidney, 39, 41, 55, 56, 199
and water balance, 57–58

Lasiurus borealis, 168
Lactate production, 54–55
Lambs. *See* Sheep
Lipase, 49, 50, 53, 201
Lipid accumulation, 54
Lipid metabolism, 52–54
Lipoproteins, 53
Liver, 39–40, 41, 48, 52–56

Llamas, 228
"Lower critical temperature," 82, 84, 94, 95. *See also* Temperature
Lungs, 128, 129
 growth of, 237–238
 in natives of high altitudes, 237
 transport carbon dioxide in hypoxia, 236–237
 transport oxygen in hypoxia, 236–237

Macaca cyclopis, 80
Macaca fuscata, 84
Macaca mulatta. *See* Monkey, rhesus
Macaca speciosa, 193, 202
Mammals
 carotid and aortic bodies in, 219–220
 evaporative heat loss in, 206
 hibernation of, 165–179
 at high altitudes, 238
 homeothermic, 219
 neonates of, 93
 oxygen pressure and, 218
 panting of, 131
 placental, 109–110
 sleep state of, 155
 small, heat loss of, 127
 thermal neutral zones of, 82–83
 thermoregulation in, 78
Marmota flaviventris, 165
Marmots, during hibernation, 165, 166–167, 177
Marsupials, salivation in, 125
Medulla. *See* Adrenal medulla; Medulla oblongata
Medulla oblongata, 32, 34, 75. *See also* Brain
Mesocricetus auratus, 166
Metabolic defenses of hibernators, 171
Metabolic energy
 and physical work, 7
 turnover, 2
 storage of, 7
Metabolic heat gain, 5
Metabolic heat production, 2, 7, 84, 111, 125, 155, 168, 171, 173, 174, 177, 178
Metabolic rate, 6–7, 83, 94, 96, 111, 121, 131, 148, 201
 basal (BMR), 92, 93, 155, 196, 202, 203
 summit, 196, 197–198, 200, 202
 thermal neutral, 92, 94, 98
Metabolic responses to environmental temperatures, 81–111
 heat production in, 92–96
 of rhesus monkey below "lower critical temperature," 95
Mice, 51, 55
 hispid pocket mouse during hibernation, 171
 running, 122
Mitochondria, 48, 49, 50, 240, 242
Models of thermoregulation, 19–20, 73
Mole, sleep state of, 152
Monkey
 epinephrine excretion of, 193
 Formosan, 80
 Japanese, 84
 stump-tailed macaque, 193, 202, 209
 rhesus, 74, 75, 84, 94, 95, 96, 97, 103, 110, 111, 150, 199, 200, 202
 sleep state of, 151, 152
 squirrel, 84, 85, 105, 110
Mountain sickness
 acute, 233–235
 chronic, 235
Muscle, 81
 activity, 92
 brown adipose tissue in, 48, 50, 200
 cardiac, 53
 and heat in animals, 121
 lactate in, 54
 protein depleted, 55
 shivering role in, 45–47
 skeletal, 56, 92, 151, 156, 159, 196, 200, 243
Mus musculus, 166
Myotis lucifugus, 168
Man. *See* Human

Naphthalene diffusion model, 15
Naphthalene sublimation technique, 10
Nasal countercurrent exchanger, 128
 in dogs, 130
 in small rodents, 130
Nasal glands, 120
Nasal mucosa, 126, 127, 128–131, 132, 133, 139
Natural covers. *See* Coats

Neuroglandular transmitter substance, 105–106
Neurotransmitter, 36
Nerve fibers, 90
 adrenergic, 203
 sympathetic, 203
Nerves
 afferent, 30–32, 81
 blocking, 88–90
 efferent, 33–37, 38
 sensory in adipose tissue, 38
 somantic, 33–34
 splanchnic, 38, 39, 40, 81, 193, 204
 sudo-motor, 209
 sympathetic, 35, 36, 37, 155, 194
 vasoconstrictor, 87
Newton's Law of Cooling, 11
Nonevaporative heat loss, 122, 123, 133, 138, 139
Nonshivering thermogenesis, 93, 195, 196, 197–200, 203
Norepinephrine, 32, 36, 37, 38, 47, 48, 49, 51, 190, 192, 194, 195, 199, 200, 201, 202, 204, 206, 207, 211

Opossum, sleep state of, 152
Osmolality, 58
Ox, 78, 91
 panting in, 130, 131
 catecholamines in, 204
 sweat glands in, 207, 208
Oxidation
 of fatty acids, 50, 53
 of glucose, 54
Oxidative metabolism, 49
Oxygen
 amount absorbed in lungs, 129
 calorific value of, 7
 use of, by brown fat, 199
 consumption
  during cold exposure, 45, 46–47
  at high altitude, 224
  in hypoxia, 233, 236
  in liver, 52
  in man, 6, 7
 equilibrium, 238
 erythrocythemia, role in, 231
 fetal level of, 229
 neonate level, 229–230
 responses to low environmental, 242
 stores of, in human body, 218

Pachynoda sinuata, 17
Panting, 16, 76, 80, 81, 82, 83, 121, 142, 156, 191, 204, 205
 in cats, 154, 155
 as cooling mechanism, 139
 in exercise, 120, 133–135, 139
 of heat-exposed animals, 206–207
 shallow thermal, 120, 126–128, 130–131, 132–133, 135, 139, 206
 during sleep, 154, 155, 158
Penguins, 130
Perognathus hispidus, 171
Peromyscus, 171
Perspiration. *See* Sweating
Pheochromocytomas, 209, 210
Pig, 44, 51, 78, 91
 evaporative cooling in, 133
Pilocarpine, 106
Piloerection of hair, 37, 44, 195, 203
Placenta, 18, 109
 weight of, 230
Plasma cortisol, 42
Poikilotherms, 2
Polar bears, 10
Polycythemia, 223
Preoptic anterior hypothalamus. *See also* Hypothalamus
 cold sensitive units in, 76–77
 electrical activity of neurons, 76
 neurons, 75
Primates, 75, 78, 81, 82–95, 139, 151. *See also* Monkey; Human
 apes, 133
 baboons, 85, 205
 higher, sweating in, 96–111
Prolactin, 40, 42
Pronghorn deer, 122
Prostaglandin hyperthermia, 110
Protective covers. *See* Coats
Protein metabolism, 55–56
Phospholipids, 53, 54
Physical work rate, 2, 7–8
 of Sherpas in Himalayas, 239

Rabbits, 14, 16, 46, 76, 78, 110, 152, 158, 199, 231
Radiant heat
 and exercise, 101
 transfer, 8–10
Radiation, 2, 8–10, 84
Radiation reflectors, 9

Rats
activity cycle of, 149
adrenal medulla of, 196–197, 200–202
brown fat in, 47–48, 198–199
in cold environment, 37, 46, 50, 51, 55
cortisol in, 42
curarized, 197
hibernation of, 166
hyperthermic effects in, 110
kangaroo, 128, 156, 162
lipogenesis in adipose tissue of, 49
liver, heat production by, 52
liver, lipid accumulation in, 54
metabolic rate of, 7
in neutral environment, 56
salivation in, 125
shivering in, 125
sleep state of, 151
summit metabolism in, 202
thermodes in, 177
thermogenesis in, 197–201
thermoregulation in, 78
as thermoregulation model, 20
thirst of, 58
Rattus norvegicus, 166
Reflex depression in sweating, 101
Renshaw cells, 34, 35
Reptiles, heat loss of, 127
Resistance hygrometry, 99
Respiration. *See also* Alkalosis; Exhaled air; Inhaled air
core body temperature during, 126
dead space ventilation, 129
evaporative cooling during, 136
heat loss during, 111, 128
panting and, 120–142, 206
responses to hypoxia, 226
rate during sleep, 151, 153, 156, 236
rate in mammals, 80, 134
role of brown adipose tissue in, 48
role of white adipose tissue in, 50
as ventilation, 120
Respiratory tract, 84
evaporation from, 129–130, 136, 139
Rete mirabile, 18
Reticuloendothelial cells, 48
Rhea, 16
Rodents, 166. *See also* Rats
Ruminants, 54, 229. *See also specific animals*
Running, 122, 134, 135, 136
Saimiri sciurens, 84
Saliva, 125, 133–134, 135
as cooling mechanism, 139
Salivary glands, 120, 133
parasympathetic innervation of, 125
Salivation, 120
heat dissipation by, 125–126
Salts, 58
Sea lion, 20
Seals, 3
Sheep, 9, 14, 42, 45, 51, 52, 54, 56, 81, 110
evaporative cooling in, 133, 138
lambs and hypoxia, 230
lambs and shivering, 197
model for hypoxic responses, 229
panting in, 131
sleep state in, 152
Sherpas
hemoglobin, 243
lungs of, 237
work habits of, 239
Shivering, 11, 32, 33, 37, 78, 93, 150, 195, 197, 203
and animal's muscle, 45–47, 50
and blood pressure, 35
blood vessels during, 56
depressed, 81
and heat loss, 44
during hibernation, 178
metabolic rate and, 83
neural basis for, 33–34
role of adrenal medulla in, 200–202
and sleep, 153–154, 155, 156, 158, 159
thermogenesis by, 200–202
"Silver swaddler," 8–9
Skin. *See also* Blood flow; Coats, animal; Cutaneous blood flow
blood flow in, 85–92
denervation of, 208
as evaporative surface, 138, 142
hair covered areas of, 207
heat transfer from, 83
temperature of, 78, 100, 101, 102, 103–111, 142, 149, 150, 152
temperature during hibernation, 170, 178
temperature of immobilized cats, 159
temperature sensors in, 74–75
as thermal insulator, 85
Sleep
brain temperature during, 151

effects of ambient temperature on, 159–165
effects of body temperature on, 148–149, 151
effects of hypothalamic temperature on, 159–165
hypoxic response during, 235–236
similarity to hibernation, 165–168
states of, 151
thermoregulation during, 148–179
thermoregulatory changes related to, 149–151
Smoke, 237
Solar radiation, 6, 8, 9
Spine. *See* Spinal cord
Spinal cooling, 38
Spinal cord, 31, 32, 33, 34, 36, 38, 78, 80–81, 91, 92, 152
Spinal fluid. *See* Cerebrospinal fluid
Spleen, 39
Starvation, 50
Stefan-Boltzmann Law, 8
Steno's gland, 132–133
Stress, thermal, 96, 108, 133
Sudomotor activity, 97, 107, 209, 211. *See also* Sweating; Sweat glands
Summit metabolism. *See* Metabolic rate
Sweat glands, 120, 137
in animals, 139
of horse, 137, 204
in mammals and other animals, 137
of primates, 90, 96, 104, 105, 106, 108, 109
types of, 207–210
Sweating, 12, 18, 80, 81, 82, 83, 84, 90, 120, 121, 139, 142, 191, 204
acetylcholine-induced, 209
"cold," 191
in disease states, 210–212
and exercise heat loss, 137–138, 139–140
during hyperthermia, 207–210
in primates and man, 96–111
Swimming, 8, 11
Sympathectomy, 87, 201
Sympathetic nerve block, 88–90
Sympathetic nervous system, 200

Tamias striatus, 170
Temperature. *See also* Body temperature; Hypothalamic temperature
ambient, 3, 4, 83, 84, 91, 94, 95, 97, 99, 123, 142, 151, 152, 154, 155, 159–165, 168–170, 219, 233
average body, 3
body and heat loss, 124, 127
central nervous system sensors of, 75–77
deep body, 77–81
during hibernation, 165–179
of leg skin, 100–101
lethal, 125
"lower critical," 82
mean body, 3
neutral environmental, 56
rectal, 211
regulation in primates, 71–110
sensors, 73–74
of skin, 87, 90, 100–111
skin sensors of, 74–75
stress, 96
thermal neutral zones, 82–83, 84, 98
"upper critical," 83
Tetralogy of Fallot, 226
Thalamus, 31
Thermal equilibrium 2, 5
and exercise, 4
Thermal neutral zones, 82–83, 84, 98
Thermal radiation, 10
Thermal receptors, 98, 104, 152. *See also* Cold receptors; Warm receptors
Thermodes
in brain, 103
in dog, 134
in ground squirrel, 177
in hibernating animals, 173
in marmot, 177
in rats, 177
Thermodynamics, First Law of, 2–3
Thermogenesis, 92, 93, 94, 195
shivering, 200–202
Thermoregulation
adrenal medulla, role of in, 189–212
anticipatory control of, 20
in blood circulation, 17–18
cardiovascular responses during, 84–92
control mechanisms associated with, 72
digital computer simulation of, 20
of dogs and other mammals, 78
effector responses, 81
feedback, 20
during hibernation, 168–179

Thermoregulation—*continued*
  hypoxia and, 233
  mechanism for heat production, 93
  models of, 19–20, 73
  panting and, 120–142
  physical basis of, 1–20
  in primates, 82–111
  responses in vasomotor system, 38
  servomechanism, 20
  of skin in primates and man, 96–111
  skin sensors in, 74–75
  sleep, relation to, 149–165
  sweat glands, function of during, 137, 207–210
  and sweating, 211
  temperatures controlled in, 16–17
  through ventilation, respiration, and panting, 120–142
Thirst, 58
Thompson's gazelle, 122, 131, 136, 139, 140
Thyroid
  changes in during cold exposure, 40
  distribution of hormones from, 41–42
  hormones, 202
Thyrotropic hormone, 41
Tongue, 139
  of dog, 133, 134, 135, 136
Triglycerides, 45, 49, 50, 51, 53, 201
Triiodothyronine, 41
Tumors, 211, 212
Tyrosine hydroxylase enzyme, 40

Ultraviolet, 10
Ungulates, 122
"Upper critical temperature," 83, 84. *See also* Temperature
Urine
  excretion of epinephrine in, 193, 194, 195
  excretion of norepinephrine in, 194
  excretion of iodide in, 41
  free catecholamines in, 192
  output of,in animals, 57–58

Vasoconstriction, 32, 36, 195
  in animals, 45, 78, 80
  cutaneous, 43, 194, 195, 203, 204, 211
  during sleep, 152
  heat loss mechanism of, 191
  in monkeys, 84
  sympathetic, 36, 90
Vasomotor system, 38
  control responses, 91
  control zones, 82–83, 84
  cutaneous control in, 88
  innervation and thermal responses, 85
  in primates, 82–83
  responses of, 90
  and sleep, 152, 153
Ventilation, pulmonary, 219–220
  alveolar, 134, 135
  in anesthetized cats, 232, 233
  dead space, 129, 130, 134
  at high altitude, 219, 224–226, 231, 232, 236
  during severe hypoxia, 221
  hypoxic drive and, 227
  hypoxic threshold for, 220
  of human newborn, 230
  rate, 7
  residual, 222
  response to hypoxia, 228, 229, 233
  total, 134

Wakefulness, 154
  daily rhythm of, 149
  during hibernation, 166–167
  during sleep, 164, 165
  sleep similar to, 151
Walking, 7
Warm receptors, 31, 33, 74, 75, 98
Water. *See* Convective heat transfer; Evaporation; Saliva; Perspiration
Water buffalo, 9
White adipose tissue and cold exposure, 37–40, 50, 51, 53, 56, 201. *See also* Brown adipose tissue; Fat
Wind, 6, 10, 12, 44
Wool, 14, 44

Yak, model for hypoxic responses, 229